WATER USE EFFICIENCY
for Irrigated Turf and Landscape

GEOFF CONNELLAN

PUBLISHING

National Library of Australia Cataloguing-in-Publication entry

Connellan, G. (Geoffrey), 1946–

Water use efficiency for irrigated turf and landscape/by Geoff Connellan.

9780643094291 (hbk.)
9780643106888 (epdf)
9780643106895 (epub)

Includes bibliographical references and index.

Irrigation.
Turf management.
Landscapes.

635.9642

Published by
CSIRO PUBLISHING
36 Gardiner Road, Clayton VIC 3168
Private Bag 10, Clayton South VIC 3169
Australia

Telephone: [+613] 9545 8555
Local call: 1300 788 000 (Australia only)
Fax: +61 3 9662 7555
Email: csiropublishing@csiro.au
Web site: www.publishing.csiro.au

Front cover: Omni Tucson National Resort, Tucson, Arizona (Photo: Geoff Connellan)

Set in Adobe Minion Pro 10.5/13 and Stone Sans
Cover and text design by James Kelly
Typeset by Desktop Concepts Pty Ltd, Melbourne
Printed by Ingram Lightning Source

Feb26_RP_ILS

Contents

Acknowledgements

The information presented in this publication is the result of the efforts of many people involved in water management and irrigation. The expertise and experiences of horticulturists, industry specialists, natural resource scientists, engineers, landscape architects, arborists, educators and students have all contributed to advancing our knowledge to better manage water in the urban environment.

The preparation of this publication has been strongly supported through the involvement of the Royal Botanic Gardens Melbourne. The contributions of Peter Symes, Senior Curator, Horticulture, in numerous aspects of landscape water management and research has been extremely valuable.

The techniques and practices presented in this publication are based on the scientific principles advanced by professionals over the years. The understanding and analysis of the performance of sprinkler irrigation systems by J. E. Christiansen, California, in the 1940s has provided today's irrigation practitioners with the tools to use this technique with high efficiency. The Center for Irrigation Technology (CIT), Fresno, under the leadership of Dr David Zoldoske, has provided irrigators with a better understanding of irrigation technology and the tools to aid in smart water management decision making.

The early work carried out by Dr Fergus Black, at Knoxfield Research station, Department of Agriculture Victoria, in the 1960s and 1970s on microirrigation (trickle/drip), provided the basis for the major advance in irrigation efficiency through this new approach. The Australian irrigation sector has been very well supported through the publications of Kevin Handreck, formerly of CSIRO, which provide a sound scientific base to manage soil water systems.

In the many years spent at Burnley Campus, University of Melbourne, colleagues have assisted through the ready sharing of their excellent knowledge of all things horticultural.

Particular thanks to Dr Peter May and Dr Greg Moore for their advice and wisdom in understanding of the role of plants, soil, water and climate in urban horticulture.

Burnley colleagues, including Liz Denman, Ross Payne, Clare Scott, Jamie Pearson and David Aldous have also provided strong support for this publication.

The CRC for Irrigation Futures, which operated from 2003 to 2010, has advanced irrigation science and knowledge transfer in Australia. The cooperation and contribution of Associate Professor Basant Maheshwari, University of Western Sydney, CRCIF Urban, is acknowledged.

The Irrigation Australia Ltd (IAL), through its staff and members, provides all participants in urban water management numerous training, professional development programs and resources to continue to improve water use efficiency.

Staff of Irrigation Association (USA), including Brent Mecham, have provided technical information and resources to support the publication.

The line drawings in this publication have been expertly prepared by Stephanie Thompson of Stephanie Thompson Graphic Design.

Roxene Carroll, Rainlink Australia, provided several irrigation design plans.

The support in the form of technical editing and preparation of graphs and tables, provided by Dr Liz Denman, is very much appreciated.

Photographs, plans and other images are acknowledged at point of use.

The support and understanding of my wife Mary and children, Duncan and Jacqui, in this project is greatly appreciated.

Thanks to Ted Hamilton, CSIRO Publishing.

Preface

Currently there are major changes occurring in the urban landscapes of our cities and towns. Water availability, climate change and population growth are the main drivers for these changes.

There is a need to manage the process of change and plan for a future in which a healthy urban landscape is a core contributor to the health and wellbeing of urban communities.

Currently, significant volumes of potable water are used to maintain turf and landscape areas. This water is becoming less readily available and the pressure to ensure it is used wisely is increasing.

The achievement of sustainable turf and landscapes that can contribute to the wellbeing of the community requires decision making based on sound principles.

This publication brings together all of the key aspects that need to be considered when seeking to achieve high water use efficiency and irrigated landscape site that is sustainable. It provides the reader with a single reference that will inform them about the most appropriate techniques and approaches to be used in sound water management for urban green space.

The approach adopted is first to understand the water cycle of the site and then to determine the services to be provided by the site. A water-efficient landscape planting design is then prepared and the water needs are assessed from this base.

The technical areas covered in the book include assessment of water supplies – including water quality issues, plant water demand, irrigation technologies, best practice, system evaluation and benchmarking – and water management planning.

Key features of the publication

This book:

- covers all aspects of water management in irrigated urban open space areas and shows how to achieve high water use efficiency
- reviews water quality issues within the context of achieving sustainability
- shows how to determine water requirements and evaluate the performance of irrigated urban vegetation
- describes techniques for evaluating the performance of irrigated turf and landscape areas
- shows how to determine the water requirements of urban vegetation
- provides solutions to achieving high water use efficiency
- includes clear, authoritative treatment of key aspects of urban irrigation
- includes case studies that highlight high best practice and high efficiency
- provides guidance on preparing water management plans.

In summary, this publication provides a logical, scientifically sound approach to the responsible management of water in urban areas in Australia.

Chapter 1

Sustainable water use and efficiency

Water efficient approach

Planning and sustainability

Providing water for towns and cities has been a challenge for thousands of years. The residents of Rome and Pompeii were able to enjoy plentiful supplies of good quality water as a result of sound planning and the engineering skills of the Romans (Plate 1.1).

The challenge that is now faced is providing water to meet the increasing needs of communities at a time when demand is increasing and supplies are not only reducing, but there is also a lack of security, due to drought and climate change.

To meet this challenge requires sound planning and an understanding of the water cycle and the water needs of the community.

Water is needed to support human life. This requires good quality water for consumption, and also water to provide hygienic living conditions: for washing, cleaning and ablutions. Water is also needed to support the production of food, goods and services. Electricity generation from thermal power stations, for example, requires a significant amount of water. The standard of living we enjoy is strongly dependent on availability of a water source of high quality and from a secure supply.

The focus of this publication is on urban green space. In recent years, owing to drought, water to support urban environments in Australia has not been available to provide the standards of landscapes that have might have been expected in the past. The changing landscape environment is strongly influenced by the availability of water (Plate 1.2).

Green space in urban areas consists of vegetation, such as residential gardens, parklands, urban forests, sporting grounds, streetscapes, commercial landscapes, public gardens, botanic gardens, golf courses and racetracks.

Planning for sustainable use of irrigation water in urban areas to support green space is essential.

The aim of the approach outlined in this publication is to provide a knowledge base and understanding of all of the key factors and processes that are involved in achieving the efficient use of water in urban landscapes.

Efficiency applies to all processes involved in the management of the landscape. The maintenance of the urban landscape should be carried out efficiently. The use of fertilisers, labour, capital and the application of water all need to be managed astutely and resources used efficiently.

Sustainability – what does it mean?

The aim for all horticultural sites should be that they be sustainable in economic, social and environmental terms: they should be triple-bottom-line positive.

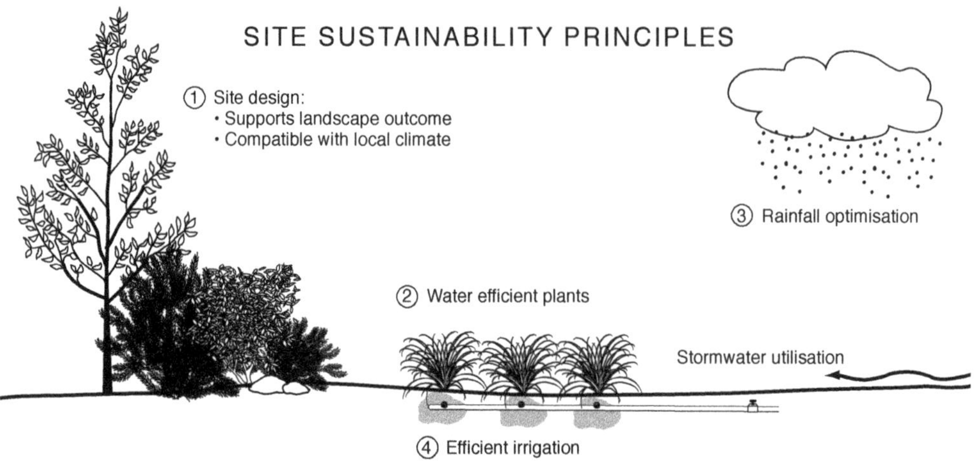

Figure 1.1. Sustainable water management principles

The term sustainable use is now liberally used in water management. A truly sustainable landscape would be one that involves no human-related inputs, including energy, water and chemicals. The purpose or role of most urban landscapes is to provide some form of service or outcome to the community. The space may be used in a passive way, such as a parkland, or it may be an active-use area, such as a sports ground.

A landscape that has no requirement for additional inputs and is in harmony with the environment could be considered truly sustainable. However, the purpose of most urban landscapes is to provide some type and level of service and so some inputs will be required. A landscape that provides a high level of services, such as intense sports activity, would be expected to have a high level of inputs and may not be considered sustainable, in the strict sense of the term.

There needs to be a balance between social benefits and environmental sustainability.

A key characteristic of a sustainable landscape is that it continues to perform and deliver the intended services or outcomes for the medium to long term: 20 to 50 years or more. It is not a short-term objective. A landscape in which water use is managed in a sustainable manner, but the site is degraded or fails, is not a sustainable landscape. An informed and integrated approach ensures that all factors that have an impact on sustainability are taken into account.

Sustainable water use is understood in this publication to mean the use of minimum amount of supplementary water to achieve the desired outcomes and that the use of water and other inputs are carried out in such a way that the environment is protected.

The efficient use of water to maintain a landscape is a critical part of achieving sustainability (Figure 1.1). It ensures that little or no water is wasted and that the impact on the environment is minimised. These core requirements are fundamental to sustainable turf and landscape areas.

Characteristics of sustainable irrigated landscapes

The sustainability of an irrigated area requires consideration of the health of the water source, the impact of the extraction of water on the environment and the health and viability of the irrigated turf and landscape areas.

Some features of sustainable irrigated landscapes include:

1. The extraction of water for irrigation does not have a negative impact on the water source and the environment.

2. Water security is achieved.
3. The irrigated area (landscape) is suited to its intended purpose.
4. The inputs (water, chemicals, energy and labour) used are minimised, used efficiently and managed so that environmental health is protected and the maintenance demands are minimised.
5. The capacity of the organisation is appropriate to manage the landscape, both in terms of resources used and outcomes or services, to be provided.
6. The business and functions of the site are viable.

Core elements of sustainable irrigated landscapes

The following are the core elements of an approach consistent with the achievement of sustainable use of water in urban landscapes.

1. The site is designed, including selection of plants, to minimise the demand for supplementary irrigation water.
2. The hydrology of the site is assessed and managed to optimise the use of rainfall.
3. The amount of supplementary water provided is matched to the site vegetation, soils and weather.
4. Irrigation is carried out efficiently.
5. An annual irrigation water requirement and water budgets are prepared for the site and used for ongoing monitoring and evaluation.
6. A secure non-potable water supply is a planning objective.
7. Resource management practices associated with the irrigated site, such as fertiliser and chemical, have no negative impact on the environment (thus protecting soil health and water bodies).
8. Water management is carried out to best management practices standards, within a framework, that encourages continual improvement.

The adoption of these core elements provides a sound basis for the achievement of sustainable landscapes.

The first stage in the development of an urban landscape, including turf areas, should be an objective analysis of the purpose and use of the landscape and the assessment of the need for supplementary water. This process should identify not only whether or not the site should be irrigated, but on what scale and to what level.

Landscape outcome

Approach

The starting point in the consideration of the development or ongoing maintenance of any urban landscape area that is proposed to be irrigated should be the identification of the outcome to be provided by the landscape or space (Plate 1.3).

The word 'outcome' is used, in this document, to describe the outputs or services of the landscape. These outputs may be aesthetic, functional (e.g. provide shade), active use (e.g. sporting activities), environmental modification, preservation of cultural and heritage values and conservation of botanical collections. High-quality landscapes can often be achieved without irrigation. This requires consideration of the plant species characteristics, local climate, the site conditions and landscape design.

The landscape outcome will vary greatly depending on the nature of the area. It may be a surface for active sports, it may be trees to provide shade or it may be an ornamental display. Clear identification of the landscape outcome, including detailing of the standards, qualities and properties to be achieved, should be the first step. This process requires close consultation

Figure 1.2. Parklands provide environmental and social outcomes

with all stakeholders in the site, including urban planners, landscape architects, horticulturists, service managers, asset managers, users, maintenance personnel, and sport and recreation management professionals.

Examples of some landscape outcomes are:

- the provision of a safe turf playing surface for contact sport
- a prominent display planting of high aesthetic quality
- street trees that provide high aesthetic quality and shade
- a border planting that is of moderate aesthetic quality and provides visual barrier
- treed parkland, with grass of medium aesthetic quality, suited to passive recreation use (Figure 1.2)
- a war cemetery garden that provides a space for reflection, a sense of peace and high overall amenity (Figure 1.3).

Having determined the landscape outcome of a site, then the site landscape design, including the plant selection, is undertaken to deliver the required outcome. In this approach, the plants and site requirements and their management are identified first and then the amount of water necessary to maintain the plants to the required performance level is determined. The driver of the process is the required performance of the site.

In the plant selection process, attention is paid to the underlying requirement that water use efficiency is a high priority. In some cases, this will mean that no long-term irrigation will be required. Some water may initially be required for establishment, say for several months or a year or two, and then the plant (e.g. a tree) will survive in this locality, subject to the climate of the locality and site conditions.

In the case of sports fields that are used intensively, some supplementary water to rainfall will be required in most parts of Australia. The nature of the surface (grass), and the need for strong growth and rapid repair, dictate that the plant selected (grass) will usually need irrigation to achieve the required surface standards. The performance level (landscape outcome) to be delivered will influence how much water is required.

A situation requiring vigorous grass growth will require more water than a situation where a moderate grass growth is required. The key, in terms of efficiency, is the selection of a grass

Figure 1.3. Providing a landscape space for reflection is an important use of urban water (War Cemetery, Perth)

species that is water use efficient. The adoption of warm season grasses in recent years is an example of the selection of plant species that deliver the required outcomes and are water efficient.

In the case of a turf area, the landscape outcome will be influenced by the type of sport and the level or grade of the sport. As an example, an area of turf that is to be used for a state-level sporting competition may have an annual requirement to deliver 40 games of AFL football, 25 games of cricket, and provide training services. This site has a higher performance requirement than an area used for local sports on an infrequent basis.

The delivery of the landscape outcome has specific implications for the site. The following is an example of the requirements of the site:

> 'The turf surface is required to be even, high quality with continuous grass cover, readily repair from wear and be safe to use'.

This information, together with knowledge of the water use characteristics of the particular grass species in use, allows the amount of water, appropriate to this site, to be determined.

Sports ground surface performance – hardness

The safe use of a surface is important in determining the outcomes or services to be provided by the irrigated recreational area. A key consideration is ground hardness. The Clegg Hammer is the instrument commonly used to measure ground hardness.

On the city website, Glen Eira, Melbourne provides updated readings of ground hardness, so that the users of the areas are aware of the condition of the grounds and changes in condition.

Table 1.1. Clegg Hammer reading for sports ground – Bailey Reserve 1, Glen Eira Council, Vic.

Year	Jan	Feb	Mar	Apr	May	Jun	Jul	Aug	Sep	Oct	Nov	Dec
2008	14.64	14.09	13.36	8.09	8.64	8.18	8.00	7.00	7.50	8.73	9.91	10.45
2009	12.91	15.18	14.91	11.00	9.27	7.73	7.00	7.27	–	9.55	9.82	9.91
2010	9.64	10.55	Renovations	9.64	8.18	9.43	6.57	7.43	Renovations			

Note: If Clegg Hammer reading is greater than 13 (CIV units), ground is rated as very hard
Source: Glen Eira website: http://www.gleneira.vic.gov.au

Table 1.1 presents readings over 3 years showing the typical range in ground condition over time. As expected in this part of Australia, the grounds increase in hardness over the summer months, with several months indicating the ground is very hard. This may require closure of grounds or changes in the management and planned use of the ground.

Knowledge of the areas to be maintained, performance standards required and agronomic conditions provide an informed basis on which to make management decisions. This type of reporting, together with precision irrigation management, provides a sound basis on which the site services can be delivered and water used efficiently.

An integrated approach to the management of the site is required.

Landscape outcomes and soil moisture level

The appearance or condition of a landscape generally relates to the soil moisture available to the vegetation. If water is continuously readily available, the plants, providing they are healthy, will generally grow and develop at an optimum rate. The landscape may tend to be lush in appearance. On the other hand, under low-soil-moisture conditions, which may cause plant stress, the landscape may not appear lush, but the appearance of the vegetation may be acceptable. It should be noted that this only applies to some plants. It is possible to have lush appearance, without supplementary irrigation, in many localities, but the maintenance of high soil moisture levels may require high applications of water to supplement the local rainfall.

The dependence of the landscape outcome on water consumption is outlined in Figure 1.4. A landscape that comprises a highly managed surface, such as a lush grass, will have a high supplementary (irrigation) water consumption requirement. A landscape that is maintained with rainfall only can be considered truly sustainable in terms of water.

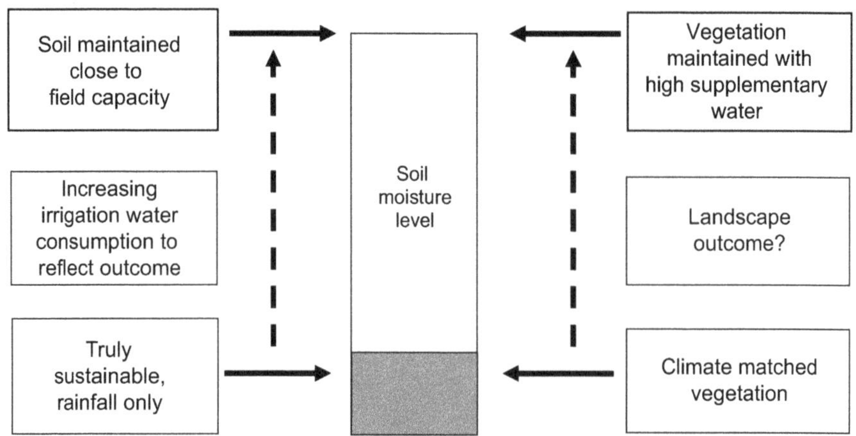

Figure 1.4. 'Landscape outcome' dependence on irrigation water use

Outline of efficiency approach

Fundamental to a sustainable approach is that water use is efficient. Minimising water wastage and loss is a core part of good water management.

The underlying principles involved in the approach adopted in this publication are:

- identifying the landscape outcome required for the site
- understanding the water requirements of the vegetation required to achieve the required outcome and preparing water budgets
- using efficient irrigation – high efficiency of application and high efficiency in the timing of irrigation
- measuring and evaluating water use
- adapting to changing climate and soil conditions.

Annual irrigation volume (AIV)

A key element of an approach to sustainable water use is the development of an annual irrigation volume (AIV). This is the estimated total amount of water that will be required to maintain the turf or landscape site, for the year or season. The AIV is determined, taking into account the landscape outcome of the site and calculated from the monthly water budgets.

An AIV sets a target value that can be used in the planning, management and evaluation of the water use at the site. Before undertaking the irrigation of an area, it is important to know how much irrigation water is required. The determination of a AIV is based on historical climate data. It is an amount that would be expected to be required under typical or average conditions. The actual conditions experienced will, of course, be different in a particular year. The role of the AIV is to provide a reference value for planning and water management.

The preparation of AIV for a site, involves evaluation of climate information, knowledge of the site area, the vegetation, the irrigation system performance and determining and specifying the landscape outcome.

Details of the calculation of an AIV and water budgets are presented in Chapter 5.

The AIV can be expressed as a total volume megalitre (ML) or kilolitre (kL). Water budgets can be also expressed as ML per hectare (ML/ha) and kL per hectare (kL/ha).

Reporting of water use without consideration or reference to a target volume is of limited value. Sound water management planning first involves establishing the amount of water that is expected to be required.

Plans and strategies

Planning for high efficiency

A sound planning framework provides the guidance and procedures required to achieve the above outcomes and progression towards sustainable turf and landscape sites.

There are numerous water-related management plans that are currently in use. They include: water management plans, irrigation management plans, drought management plans, drainage management plans and environmental management plans.

There are a range of codes and practice guidelines, such as best management practices, which provide guidance for each significant aspect of irrigation system design, operation and management. These are also important in achieving efficiency of water use.

The planning documents available will vary according to the nature and scope of water-related activities for which the organisation is responsible. A minimum requirement is a water management plan.

Water management plans

A water management plan brings together the key information that may have an impact on water decisions and the associated site management issues. It provides a framework for the ongoing water management of the site. It is also a powerful tool in communicating the strategies, actions and targets within the organisation and to external stakeholders.

The water management plan requires the organisation to initially identify the purpose and role of the site and then to carry out a thorough review and analysis of all water-related activities. A water management plan details the works and practices that will improve all water management, including irrigation, and water use efficiency for the site or enterprise. It identifies how water can be conserved and what strategies need to be put in place to ensure sustainability of water use in the future. The water management plan provides a reference framework within which the efficient use of water is a core element. It is a reference and recording document, in the overall achievement of efficiency.

The main benefits of a water management plan include: (a) recording the organisation's vision and water goals; (b) identifying the assets and information relevant to water at the site; (c) assessing the current water management practices; (d) identifying strategies for water conservation and efficiency improvements; and (e) assisting in securing future water availability, through sustainable practices.

Many local government organisations, such as natural resource management and environment departments, and water supply authorities, are required to prepare water management plans to meet the regulatory requirements of government agencies. Adoption of standardisation in reporting of water use and water use performance will assist in avoiding unnecessary duplication.

Details involved in the preparation of a water management plan are covered in Chapter 12.

Drought management plans

An outline of how the organisation would react to diminishing water availability is the basis of the drought management plan, sometimes called a drought response plan. It may be incorporated into the water management plan or be a stand alone document. It is basically a risk management plan.

The term drought generally refers to extended periods of below average rainfall. The Bureau of Meteorology defines drought as occurring when rainfall is below the lowest 10% of rainfall records for a period of 3 months or more (see http://www.bom.gov.au/climateglossary/drought.shtml). There is no single definition to cover all situations because any shortage of rainfall and water supply can be considered to be a drought. A plant growing in a container or pot may experience drought if it is not watered on a daily basis during summer.

The drought management plan includes responses to varying degrees of water scarcity or shortages. It outlines a progressive response to the stage, where potentially, no water, or only emergency water, is available. It identifies the risks associated with the drought conditions and how the site should be managed.

Some of the key aspects to be included are:

- identifying the climate or water supply conditions that constitute a 'drought'. For example, rainfall less than the 10 percentile for last 3 months during the irrigation season.
- prioritising areas or sites to be irrigated. This involves a hierarchy of areas to be progressively water reduced, or even be turned off, at the various stages of water restrictions or reduced water availability.
- identifying temporary water supplies, such as recycled, industrial waste water

- adjusting the irrigation control program and irrigation technology to comply with restrictions
- modifying the system control hardware, hydraulics and irrigation delivery
- developing techniques and procedures to monitor the condition of the landscape and sites, such as measuring the ground hardness
- employing communication strategies to keep all stakeholders informed.

In terms of managing risk, it is important to recognise that 'drought' conditions may occur even when the climate is normal. Failures in power, mechanical breakdown of pumping equipment, information technology (IT) meltdowns and supply water quality failure are all examples that may cause 'drought' for landscapes, that are strongly dependent on the use of irrigation water.

Environmental management plans

An environmental management plan is potentially very wide-ranging document. The subject area may cover not only land and water matters but also fauna protection, native vegetation, wetlands, fertiliser practices, pesticide storage and handling, and site water management.

Resources, such as *Improving the Environmental Management of NSW Golf Courses,* (Neylan 2003), describe the breadth of issues to be considered in managing green space to meet the required environmental standards and achieve a sustainable site.

Best management practices

There are many factors that can influence the efficiency of irrigation. The availability of guidelines, referred to as best management practice (BMP), which provide advice on how each aspect of irrigation should be carried out, is a major asset in achieving efficiency of irrigation water use.

Although BMPs provide guidance in good water management, they should not be seen as the end point in the process. Continued improvement in water use should be the underlying approach. With advancements in water management performance, there will be the need to prepare new goals and set new targets.

The BMPs cover the system design, performance standards, installation, maintenance and scheduling of irrigation and maintenance of systems. The relationship between the various best management practices that deal with water are shown in Figure 1.5. These guidelines are being increasingly adopted by government agencies, water authorities, the irrigation industry and irrigation managers. Key benefits of the adoption of BMP are that irrigation efficiency is increased and there is enhanced protection of the environment.

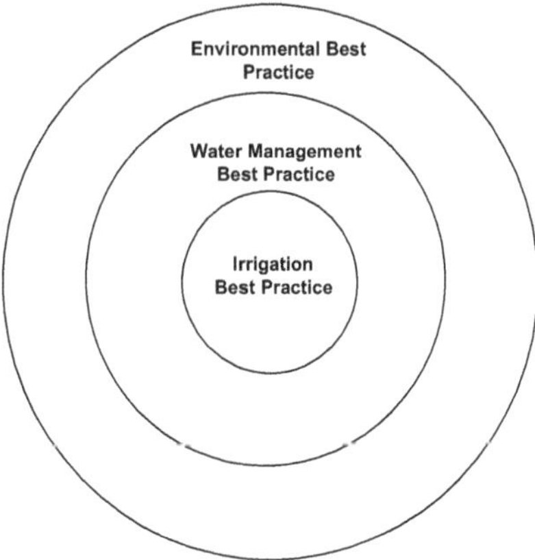

Figure 1.5. Relationship between irrigation, water management and environmental best practices

Achieving efficiency

Defining efficiency

There are two main measures of efficiency that describe how well water is used in the urban environment:

1. Water use efficiency – describes how well water, supplementary to rainfall, is used to maintain vegetation, including turf and landscape areas.
2. Irrigation efficiency – a measure of how well water is delivered, by the irrigation system, to the plants, over the irrigation season or year.

Water use efficiency (WUE)

The term water use efficiency (WUE) has its origins in the agricultural sector, where it is used, as a generic term, for any performance indicator to measure water use (Fairweather *et al.* 2002). It basically describes how well irrigation water has been used to provide yield or outputs.

In this situation, efficiency can be expressed in terms of yield (e.g. kg, tonne, $) per water volume used (e.g. kL, ML). The yield is a measurable quantity. In the case of ornamental horticulture, the 'yield' is not as readily quantified. It may be the aesthetic value or functional value (e.g. shade) of the irrigated landscape.

WUE is a measure of the level or value of landscape outcomes obtained, compared with the amount of supplementary (irrigation) water used. Improving water use efficiency will reduce the overall requirement for water. Selecting low water use plants is a strategy that increases WUE and reduces the amount of water required.

Managing the soil moisture level in a landscape so that there is some stress (using less supplementary water) will cause an increase in WUE, if the performance of the landscape is acceptable.

Although WUE incorporates the term 'efficiency', it is often not actually presented as a standard efficiency and expressed as a percentage. For example, the irrigation water used to maintain 2.0 hectares (ha) of parkland area may be 6 ML, whereas in the past it may have been 8 ML. In this case, WUE may be expressed as improving (reduction in water use) from 4 ML/ha to 3 ML/ha.

A further complication with the term WUE term, in the context of the urban environment, is that the replacement of potable water with an alternative water source, such as recycled water, is sometimes described as an increase in WUE. Because WUE generally describes the volume of water used, regardless of the source of the water, then changing the water source does not increase the WUE. If WUE is specifically defined to only relate to one water source, such as potable water, then the term could be used as a measure of increased efficiency.

The term WUE is useful in gauging the overall performance of supplementary water applied to the landscape. It is, however, not strictly a term for efficiency. In this publication, WUE will generally be used to describe the landscape outcomes delivered or produced per volume of irrigation water used.

Irrigation efficiency

Irrigation efficiency (IE) used in this publication is a measure of the amount of water effectively delivered into the plant root zone, compared with the amount applied.
Irrigation efficiency is defined in the following way:

$$IE = \frac{\text{water beneficially used by plants}}{\text{water applied to a field or area}} \times 100\,(\%)$$

Water applied to a field or area

The measurement of the water beneficially used by the plants is difficult to do, but it can be estimated. An acceptable simplification is to use the water effectively deposited in the plant root zone. It is assumed that it is all beneficially used. IE describes only how water is delivered from the irrigation system to the plant. Water savings achieved through reduced demand for water by plants, for example, are not strictly irrigation efficiency gains.

IE can be broken down into two parts. These are:

- the effectiveness of application of water to the plant root zone: this is referred to as the application efficiency (E_a)
- the timing of irrigation duration and event to meet plant needs; this is referred to as the water management efficiency (E_{wm}).

Each of these efficiency areas is prone to water losses. In the case of the application efficiency, water can be lost due to wind, evaporation, surface runoff and deep drainage (Figure 1.6). In the case of water management efficiency, water can be wasted through excessive application depths and incorrect timing of irrigation.

It is the combination of the two efficiencies that determines the overall efficiency of the irrigation water applied. IE describes the overall efficiency of the irrigation processes and is determined in the following way:

$$IE = E_a \times E_{wm}$$

The most common efficiency term used to describe and evaluate urban irrigation is the application efficiency (E_a). There are numerous factors that influence E_a and these are discussed later.

For irrigation to be highly efficient, both the E_a and the E_{wm} need to be high. Well-designed and well-managed systems can achieve high IE. To do this the irrigation system needs to be uniform in application and delivery, effective in application, operated at the right time and for the correct length of time.

Achievement of high irrigation efficiency is strongly recommended. It is important to recognise that both the application of water and the scheduling of the system operation need to be done well to achieve this goal. A poorly managed drip system, which potentially has a high E_a, will have a low E_{wm}, if it is operated for excessively long times and water is wasted.

The management of an irrigated site should be based on high efficiency and the setting of efficiency targets. The method of irrigation and the circumstances in which it is used will influence to some extent the efficiency that can be achieved. The following are suggested IE targets:

- sprinkler systems 80%
- drip irrigation systems 90%.

The achievement of these efficiency targets will require high application efficiency and astute water management.

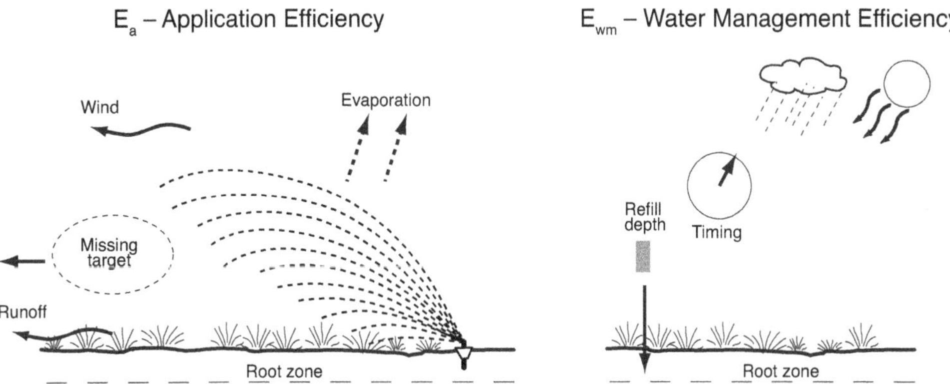

Figure 1.6. Factors influencing irrigation efficiency terms

Water savings and efficiency

There is now a strong emphasis on and promotion of water savings. These are often quoted as a measure of success of a particular strategy or technique. Saving water clearly results in less water being used. However, the savings achieved do not necessarily result in increased efficiency. For example, turning off the sprinklers watering a garden bed that subsequently dies is not an improvement in efficiency of watering. Turning off the water has resulted in a degradation of the existing landscape (landscape outcome diminished). However, it would be an increase in WUE if a replacement planting, which did not need as much irrigation, was installed and the site continued to deliver the required outcomes.

Water savings send a powerful message and they do encourage increased efficiency. In many situations, however, water savings involve reducing the use of potable water and converting to an alternative supply, such as recycled or stormwater. This assists in achieving sustainability of potable supplies for urban use. It is important that non-potable water supplies are also used efficiently.

Potential efficiency gains of an irrigated turf site

The opportunities for improvements in water use are demonstrated in the Table 1.2.

In this case, both improvements in WUE, through conversion to warm season grasses, and upgrading the irrigation system condition and method can reduce the water required very significantly. For the worst case scenario, the water use requirement is estimated to be in the vicinity of 10 ML/ha, compared with 4 ML/ha for a high irrigation efficiency system that is well managed and uses more water efficient grasses.

This example has been determined for a site that has an annual deficit between evaporation and rainfall of approximately 500 mm.

Precision irrigation application

The precision application of water is not only required to achieve water volume reductions, but also to achieve specific performance standards. High-quality turf surfaces require a very uniform application of water that is effectively delivered into the root zone. If application is uneven, the growth and surface will also likely be uneven.

Measurement of performance standards, such as hardness, compaction and infiltration rate, as well as meeting specified soil moisture levels, will require irrigation applications that are uniform, effective and timely. Monitoring of water application and soil moisture conditions should be a more common feature of irrigated sites in the future.

Table 1.2. Guide to potential water reductions through improved irrigation application efficiency (E_a)

Turfgrasses	Irrigation system and condition	Water application rate (ML/ha)
Cool season	Sprinkler – poor condition	10
Cool season	Sprinkler – good condition	8
Warm season	Sprinkler – poor condition	7
Warm season	Sprinkler – good condition	5
Warm season	Subsurface drip irrigation (SDI) – well designed system and suited to site soil properties	3.5 to 4.5

Note: These are generic values to provide a guide to potential relevant improvements. Site specific climate conditions will determine the actual value.

Evaluation of irrigation performance

Knowing how well the irrigation system is performing and the assessment of the water consumption are core components of efficiency and sustainability.

The evaluation of the performance of irrigation systems should be carried out regularly, rigorously and methodically. This procedure is essential in achieving high standards and also provides the information necessary to analyse and report on the use of water. The evaluation will assess the operating effectiveness of the system and how well the irrigation was managed over the irrigation season or year.

The main field technique used in evaluation is irrigation auditing. The process involves initially observing the site, assessing the condition of the irrigation system and then testing the uniformity of application or evenness of delivery. Weaknesses in the irrigation system that will reduce efficiency will be identified. Information required to determine the optimum irrigation watering schedule for the irrigation system and the site will also be obtained.

Implicit in a sustainable water use approach is that all water supplied or delivered to the irrigated site should be measured and evaluated and the results used in the overall water management planning process. Reporting of water use and benchmarking against industry best practice is a powerful instrument in improving water use efficiency, and communicating water performance to stakeholders and the community. Continuous monitoring and evaluation of irrigation performance is essential.

Benchmarking

Benchmarking is the process whereby the performance of water use is measured with reference to comparable situations and reported. This is valuable for the organisation or business, both internally and externally. Benchmarking requires the use of performance measures, known as key performance indicators (KPIs).

The following should be considered a minimum, in terms of evaluating and reporting the performance of irrigated areas:

- the total water consumption for the site (ML or kL)
- the water application rate or water consumption per unit area (ML/ha or kL/1000m^2)
- the annual irrigation volume or annual water budget for the site (ML or kL)
- the actual water consumption compared with the water budget (%)
- the irrigation index (Ii) or irrigation efficiency term, which is determined using the climate data for the irrigation season
- the uniformity of sprinkler application (DU) or uniformity of emitter delivery (EU).

In addition to being an aid to encourage overall efficiency, benchmarking is increasingly being used by water supply authorities, both for assessment of performance and for the allocation of water.

Sustainable irrigated turf and landscapes

Turf and landscape sustainability considerations

The adoption of the principles of sustainability by managers of irrigated turf and landscape areas requires each of the following aspects of the site water management to be considered and implemented:

- opportunities to increase water use efficiency through landscape site design, plant selection and site management

- replacement of potable supplies with alternative water sources
- minimal use and efficient use of chemicals – fertiliser, plant health chemicals, soil and water conditioners
- efficient application of irrigation water
- scheduling of irrigation to meet plant needs and climate conditions
- irrigation system design sympathetic to the site and environment and encompasses low maintenance requirements
- capacity and skills required to achieve the water management and landscape objectives
- monitoring of the site environmental conditions
- measurement and benchmarking of water use.

Examples of threats to sustainability

The following are examples of some of the water-related threats to the sustainability of urban landscapes:

- lack of water security – drought, water restrictions, water supply diminished or unavailable
- reduced water quality – recycled water that is not suited to sustainable use; other low water quality issues (e.g. salinity or toxicity)
- playing surface failure, due to overuse or excess hardness (carrying capacity not managed to protect the asset)
- degradation of aquatic ecosystems, through eutrophication (elevated nitrogen and phosphorus in particular)
- stormwater-based system failing, due to lack of rainfall
- the organisation's lack of capacity to manage to appropriate standards, such as key personnel depart or inadequate skills
- limited financial resources to upgrade technology and water infrastructure and resources to maintain the system at the required standard
- poor irrigation management practices, such as failure to water according to plant water needs and best practice.

Irrigation environmental risks

The establishment and ongoing maintenance of an irrigated site can potentially have an impact on the environment in a number of ways.

The modification of the natural landscape to implement irrigation may adversely affect:

- soil health
- plant biodiversity
- water source
- aquatic ecosystems
- groundwater.

Considering the breadth of these risks, it is most important that irrigation managers are aware of the potential risks and that the organisation has the policies and capacities to manage water in a way that protects the environment.

Irrigation should always be carried out within the context of a water management plan, which in turn, is supported by an environmental management plan.

Further details and specific examples of risks are outlined in Chapter 2.

Indicators for sustainability

The core aspects of the irrigated site that should be measured include the benefits generated or services provided, the water used and the assessment of the site conditions. Indicators can be used to assist in measuring the current status of the facility and provide a basis on which ongoing progress towards sustainability can be measured.

Table 1.3 is a summary of the broad sustainability strategy categories and associated examples that are available to organisations in managing irrigated sites and that contribute towards achieving sustainable irrigated sites.

Landscape design and potential sustainability strategies

The following principles should be adopted in planning a site and preparing a design so that high water efficiency and sustainability are achieved:

1. The site should be prepared and developed to optimise the natural water flows and make use of rainfall (Figure 1.7).
2. The existing native vegetation should be preserved wherever possible.
3. Irrigated areas should be matched to the site outcomes.
4. Low water use or climate matched species should be selected.

Table 1.3. Sustainability items and indicators for irrigated sites

Sustainability items	Sustainability category	Sustainability indicator
1. Outcomes – services	Social	Participants per site/venue, user/player hours per unit volume of irrigation water
	Environmental	Green space benefits – various
	Economic	Income per unit volume, employment per unit volume
2. Water use	Water source	Potable and non-potable, security of supply
	Water quality	Water quality parameter thresholds (physical, chemical, biological)
	Water used (volume)	Total volume, volume per unit area
	Efficiency of use	Water use efficiency, productivity, irrigation efficiency
	Effectiveness of delivery	Uniformity of application, uniformity of emission, precipitation rate
3. Site performance	Plant performance	Plant condition, aesthetic quality
	Surface (turf) performance	Surface hardness, turf coverage, surface evenness, carrying capacity
	Root system	Depth of roots
4. Soil health	Physical properties	Water storage, infiltration, water movement, aeration, compaction, soil bulk density
	Chemical properties	Soil acidity, soil salinity, nutrients, toxins
	Biological properties	Organic content
5. Environmental Impact	Soil	Soil salinity, soil toxins, waterlogging
	Groundwater	Water quality, yield, depth
	Waterways	Water quality (flora and fauna impact), quantity, flow rates, sediment load
	Energy	Greenhouse gas production
6. Organisation capacity	Policy	Appropriate water policies, water management plan
	Data	Accurate data (GIS), accessible data
	Skills	Training, qualifications
	Financial resources	Capital, operating expenses

5. Plants of similar water requirements should be grouped together.
6. Hydrozones should be identified for all areas that have similar water requirements
7. Irrigation should be carried out efficiently both in terms of application and management.
8. Best management practice should be adopted.
9. The site water use performance and site conditions should be monitored.

It is important that threats to sustainability, particularly during periods of low rainfall, are identified and strategies adopted to mitigate the risks and allow the site to continue, wherever possible, to deliver services. Examples of risk mitigation strategies are:

1. Management of site use
   - Modify the carrying capacity to suit the soil and climate conditions.
   - Adjust the landscape quality standards to reflect the climate conditions.
2. Protection of the water source
   - Ensure that water extraction for irrigation does not damage the health of the water source.
   - Secure non-potable supplies, if feasible.
   - Reduce dependence on potable supplies.
3. Protection of the environment
   - Minimise chemical use.
   - Use chemicals efficiently and appropriately to minimise contaminants in drainage solution (leachate).
   - Minimise water runoff and deep drainage.
   - Ensure that water application does not degrade the soil – erosion, sediment transport to waterways.
4. Use of 'fit for purpose' water efficiently to achieve outcomes
   - Match water quality to the requirement of the site and management practices ('fit for purpose').
   - Ensure high water use efficiency – landscape design and vegetation management, plant selection. Optimise rainfall usage and implement water sensitive urban design (WSUD) principles.
   - Use efficient irrigation (application and scheduling).
5. Skills and capacity to manage appropriately
   - Compile comprehensive database of sites, use and water infrastructure.
   - Develop capacity to manage irrigated sites and associated inputs.
   - Develop expertise in horticultural management (e.g. turf).

Sustainability case study – Salisbury, South Australia

The Paddocks Wetlands, in the northern suburbs of Adelaide, demonstrate how the needs of the community for areas such as parklands and sporting grounds can be satisfied and, at the same time, protect and enhance the natural landscape assets and water resources (Plate 1.4).

This area was developed in the 1970s through the initiative and leadership of Barry Ormsby, Principal Landscape Architect, City of Salisbury. Stormwater from nearby urban catchments is harvested and managed through a wetland system (Figure 1.8). Some water is injected into an underground aquifer during the winter months and then recovered over the warmer months for irrigation. This is known as aquifer storage and recovery (ASR). For further details see the Salisbury City Council website: http://cweb.salisbury.sa.gov.au.

Reporting sustainability progress – benefits

The benefits to an organisation, in reporting of sustainability progress, are outlined in Atkins *et al.* (2006). These benefits include:

Figure 1.7. Kerb inlet allows roadway stormwater to drain to rain garden

- demonstrating good corporate citizenship
- sharing information strengthens relationship with external stakeholders including suppliers, the market and the community
- improving internal operations through more efficient use of resources, including water savings
- building teams and internal relationships
- improving the capacity of the organisation.

Summary – planning for sustainable irrigated open space

Overall approach

The pathway to achieving sustainable irrigated areas involves defining the required landscape outcomes and then determining the horticulture required to support these outcomes. The management of the irrigation water should be based on scientific principles, the pursuit of effi-

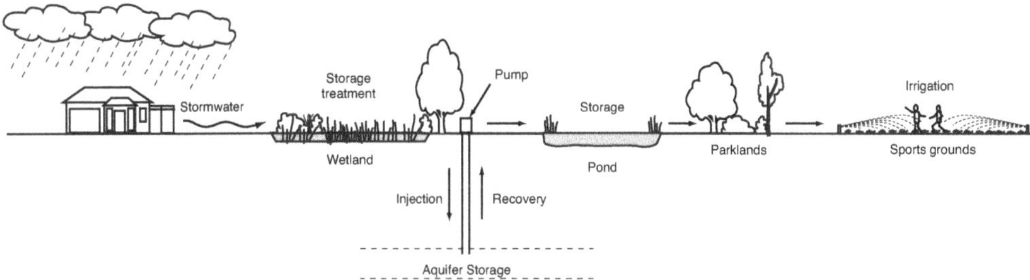

Figure 1.8. Integrated stormwater system for open space – City of Salisbury, South Australia

ciency of application and scheduling and the adoption of best management practices. The evaluation of performance, using KPIs appropriate to irrigated open space, is recommended.

The benchmarking of irrigated sports grounds, using water application rates (ML per hectare), efficiency (Irrigation Index) and services provided (user hours per ML of irrigation water) is an example of linking water use to the landscape outcomes and services provided.

Efficiency summary

The achievement of high efficiency of water use in the urban environment should be based on:

1. the adoption of landscapes and water management practices that deliver high outcomes per unit of irrigation water used
2. the implementation of irrigation strategies and practices that deliver the required amount of water to the plants without wastage
3. the setting and pursuit of targets.

Chapter 2

The urban water scene

Introduction

Water has become a focus for all sectors of the community. Recent drought, water restrictions, greater awareness of environmental issues and climate change have raised the awareness of water management to a level that would not have seemed possible a decade ago. The increased attention on water brings with it the need to consider how, where and why water is used.

Urban areas have become one of the hot spots in terms of where water use is continually questioned and discussed. There is now much debate about what should be watered, which raises the issue of the value of water used to maintain vegetation in the urban areas. The majority of urban water authorities have introduced measures to encourage water conservation and reduce overall water demand in an attempt to achieve sustainable supplies (Figure 2.1). Wherever restrictions have been introduced, the 'green' sector has been targeted to deliver significant water savings. As a consequence, the managers of urban landscapes, including parks and gardens and turf areas, have had to deal with a wide range of water management issues.

Figure 2.1. Many cities are dependent on rain-fed catchment storages, which dropped to very low reserves during drought conditions

The impact of water restrictions on urban horticulture has been great. The most visible example has been in the loss of the use of sports grounds and the death of urban trees (Plate 2.1). The severity of this loss is difficult to quantify. Many local government authorities have had to implement the watering of only a portion of the total sports grounds they manage. In some cases, a strategy of watering one ground in four has been applied. This approach obviously reduces the irrigated turf available to the community. One consequence has been the increased or more intensive use of fewer grounds. This may have long-term implications, both in terms of the cost of repair and maintenance and the overall sustainability of the site. Reduced carrying capacity of sports ground has been common in most towns and cities. In some locations, due to drought and water restrictions, no water has been able to be applied to the few grounds in the town.

Sporting grounds are not the only horticultural assets that have been affected. Parks and gardens have also suffered. Urban trees have been severely affected. In the 1982–83 drought period in Melbourne, many deaths of urban trees were reported. Local government authorities reported some tree deaths as a result of the recent extended dry periods and the introduction of water restrictions. Although tree stress, due to low soil moisture, is common, in many instances managers have been successful in preventing tree death by introducing a range of strategies, including manual emergency watering. In Adelaide, however, it was reported that the adoption of water saving practices may have contributed to the death of over 200 trees, including 50 'significant' trees (Hough 2005).

The trees in the City of Melbourne municipality have suffered owing to the dry conditions and water restrictions. In 2009, it was reported that 40% of the trees were exhibiting signs of significant stress. Nine hundred trees were lost which is three times the average loss rate. The City adopted a range of strategies to gain greater access to water to assist in preserving this valuable asset.

Many organisations responsible for urban landscapes have adopted policies that include significant reductions in water consumption. For example, in September 2004 the ACT Government outlined that it was taking steps to achieve a 40% reduction in water use. The maintenance of public open space in best possible condition was a requirement of this policy. In this case, the authority recognised that turning off the parkland irrigation would affect the health of some trees.

The nature of urban landscapes, including their tree populations, will, in the future, be influenced by the water supply and availability regimes that will exist in urban areas. The nature and character of the urban landscape is strongly influenced by the supplementary water that is applied to maintain these green spaces. An important question to be answered is 'What do we as a community expect from the urban green spaces?' Although it is expected that green space will change to reflect reduced water availability, there will be expectations that green space will continue to contribute positively to the wellbeing of the community and the quality of the urban environment.

The amount of water actually used for urban irrigation is not large in the context of national water consumption. It is, however, significant in terms of the potable water consumed in urban areas. According to the 'Strategies for managing sports surfaces in a drier climate' (GHD 2007) report prepared by the Municipal Association of Victoria, Victoria councils use the equivalent of 1.2% of all urban water usage to maintain turfed sports facilities.

One of the strong messages coming from governments is that potable water is precious and its use to maintain vegetation, both public and private (e.g. residential gardens), should be reduced (Figure 2.2). As a society, we cannot continue to expect to depend on potable water for the irrigation of home gardens and sports grounds. Some watering of urban vegetation, including sporting grounds, will continue with potable water, but the use of alternative supplies, such as recycled/reclaimed water and stormwater, will continue to be strongly encouraged in the

Figure 2.2. The use of potable water for turf and landscape irrigation may not be sustainable

future. It would be expected that the cost of potable water will continue to increase strongly. This will influence, to some extent, where and how water is used.

Some important water-related questions are currently being raised in regard to each industry sector, such as: 'How much water is used?', 'What is the source of the water?', 'What is the value or productivity of water used?', 'What is the efficiency of water use?' and 'How can water consumption be reduced?'. Each of these questions is relevant to each industry sector.

Australian water use

In terms of gaining an understanding of urban water use, it is valuable to have an appreciation of the total water scene in Australia.

As shown in Table 2.1, the volume of water potentially available on the Australian continent is large. However, much of this water is not readily accessible and is not in close proximity to the population centres of the eastern seaboard.

Maintaining a healthy environment has rightly become a high priority for Australian communities. It is important that there is a balance between water available for a healthy environment and the water extracted for community use and development, including irrigation.

Table 2.1. Water balance of Australian continent (Water Account 2004–05)

Water category	Australia 2004–05 (GL)	Proportion of total (%)
Rainfall on continent	2 789 424	100%
Evapotranspiration from surface of continent	2 510 482	90% (approx.) of rainfall
Runoff	242 779	8.7% of rainfall
Water extracted from environment	79 784	2.9% of rainfall
Water consumption – all purposes	18 767	0.7% of rainfall

Note: 1 GL = 1000 ML

Source: adapted from Australian Bureau of Statistics (2006)

Table 2.2. Water consumption by industry sector 2004–05

Industry sector	Water consumption 2004-05 (GL)	Water consumption 2004-05 (%)	Water consumption 2008-09 (GL)	Water consumption 2008-09 (%)
Agriculture	12 191	64.9	6996	49.6
Water supply	2083	11.2	2396	17.0
Household	2108	11.2	1768	12.5
Other industries	1059	5.6	1327	9.4
Manufacturing	589	3.1	677	4.8
Mining	413	2.2	508	3.6
Electricity and gas	271	1.4	328	2.3
Forestry and fishing	51	0.3	101	0.7
Total (GL)	**18 767**		14 101	

Source: adapted from Australian Bureau of Statistics (2010)

The majority of water that is used in Australia is used in agriculture and the greatest portion of this is used for irrigation. As shown in Table 2.2, agriculture, in the pre-drought period 2004–2005, represents 65% of all water consumed.

A diverse range of crops are produced using irrigation water. The total amount of water used depends on the irrigated area and the water application rates (Table 2.3). In terms of single crop water users (other than pasture), rice, cotton and sugar are the main consumers.

There is very limited detailed data available from the Australian Bureau of Statistics (ABS) on water used for recreational purposes. The ABS defines the relevant water category as 'cultural, recreational and personal services'. The category includes water use for parks, gardens, sports fields, golf courses and racecourses. The data for this category are included within a broader category 'other industries'. In the 2000–01 survey report, 395 049 ML (395 GL) were used for cultural, recreational and personal services. This represents approximately 2.5 % of total water used.

Turf industry and water use

The turf industry is a very important one in terms of economic value, numbers of people employed and the services provided to the community. According to the *Australian Turf*

Table 2.3. Crop water use and application rates in Australia (2009–10)

Crop	Irrigated area (ha)	Water applied (ML)	Application rate (ML/ha)
Rice	18 931	246 909	13.0
Horticulture	Not included		
– Vegetables	104 324	419 229	4.0
– Grapevines	162 602	515 484	3.2
– Fruit (excluding grapes)	134 221	654 663	4.9
Sugar	212 615	756 317	3.6
Cotton	153 189	851 950	5.6
Pasture	138 940 542 121	432 833 1 721 602	3.1 (hay) 3.2 (grazing)
Nurseries, cut flowers or cultivated turf	13 143	63 483	4.8

Source: Australian Bureau of Statistics (2011)

Industry Strategic Plan 2008–2011 (Moir 2004), the turf industry was valued at over $3 billion and employed approximately 80 000 people.

The main uses of turf include sports grounds, school playgrounds, golf courses, racetracks, bowling greens, tennis courts, parks and gardens. According to Aldous (2004), there were over 1500 golf courses, approximately 2000 lawn bowls clubs and approximately 750 local government authorities, who have responsibility for turf areas, including sports grounds.

A survey conducted by the Victorian Golf Association in 2005 (VGA 2005) provides comprehensive information on water issues and the nature of water use in the local golf industry. The lack of national statistics for golf was recognised and a national survey was conducted in 2007 by the Australian Golf Industry Council (AGIC 2007). This survey found that water consumption by golf in Australia was approximately 124 GL/year. Based on the estimated number of irrigated golf courses of 1000, this represents an average water consumption of 124 ML per 18-hole course. Further analysis shows that this is equivalent to a water application rate of 6.2 ML/ha/year, assuming there are 20 ha irrigated turf, on each course. The application rate will vary considerably and there are now many golf courses in southern Australia with application rates in the range of 3 to 4 ML/ha/year. These courses tend to be using more water efficient warm season grasses.

Climate change and urban water management

Predicted changes

The nature of rainfall is that it is highly variable over time and distance (Figure 2.3). Droughts and floods are events to be expected. The frequency and severity of these has a large influence on the nature of the landscape and the way in which the land is managed. The growing of plants and crops at a particular location is very strongly determined by the climate parameters of rainfall and evaporation and also, the extreme weather events that occur.

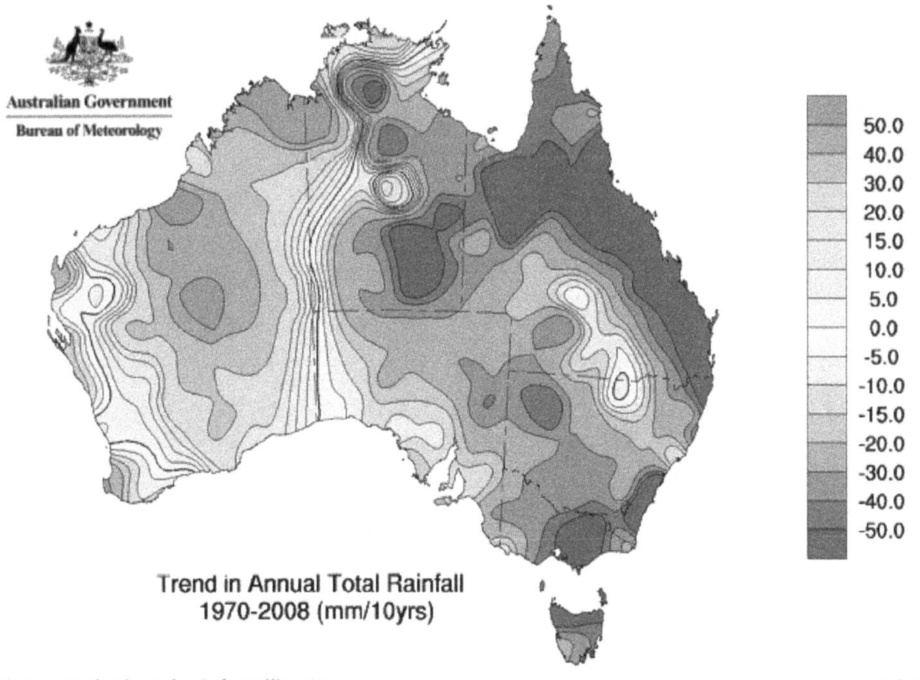

Figure 2.3. Rainfall trends in Australia 1970–2008 (Source: Bureau of Meteorology)

A change in the climate, even a small one, can have major consequences for the cultivation of plants and the health of ecosystems. Although the predicted changes due to 'climate change' are relatively small in magnitude, it can be the difference between some plants surviving or not surviving at some locations. The overall projections are that the future will be hotter and drier.

The average temperature in Australia has increased by 0.7°C in the past century. This rise is attributed to the increase in the amount of carbon dioxide and other greenhouse gases, which are causing more heat to be trapped within the Earth's atmosphere (the greenhouse effect). Projected temperature increases are 0.4°C to 2.0°C by 2030 and 1.0°C to 6.0°C by 2070. With the increases in temperatures come increased heatwaves, increases in evaporation and fewer frosts. The temperature increases are expected to be highest in inland areas.

It is predicted that droughts will be more extreme and more frequent. Some climate models predict that drought could be 20% more common in 2030 over much of Australia.

Rainfall has increased in the north-west of Australia in the past 50 years, but, importantly, has decreased in the south-west and south-east, where there is much development and population concentration and growth.

In south-eastern Australia, reductions in rainfall of up to 10% by 2030 and 35% by 2070 are predicted by some models. These are extreme values in terms of the predictions, but they do provide an insight into the potential severity of the issue.

Table 2.4. Estimated change by 2030 (relative to 1990) to annual average rainfall, average seasonal rainfall and average annual potential evaporation for regions in Australia

	Estimated change to feature					
Region	**Annual average rainfall**	**Summer average rainfall**	**Autumn average rainfall**	**Winter average rainfall**	**Spring average rainfall**	**Annual average potential evaporation**
North Western Australia	–1.5% to –3.5%	–1.5% to –3.5%	0%	NA	NA	+1.6% to +3.7%
South West Australia	–5 to –11%	–3 to –7.5%	–3 to –7.5%	–5 to –11%	–5 to –11%	+1.9 to +4.3%
Southern South Australia	–3 to –7.5%	–3 to –7.5%	–1.5 to –11%	–5 to –11%	–5 to –11%	+1.6 to +3.7%
Victoria	–1.5 to –3.5%	0%	–1.5 to –3.5%	–1.5% to –3.5%	–5 to –11%	+2.2 to 5%
Tasmania	+1.5 to +3.5%	–3 to –7.5%	0%	+1.5 to +3.5%	–1.5 to –3.5%	+1.9 to 4.4%
New South Wales	0%	+1.5 to +3.5%	+ 1.5% to + 3.5%	–3 to –7.5%	–3 to –7.5%	+2.4 to 5.6%
South-East Queensland	–1.5 to –3.5%	0%	–3 to –7.5%	–3 to–7.5%	–3 to -7.5%	+2.4 to 5.6%
North-East Queensland	–1.5 to –3.5%	+1.5to 3.5%	–3 to –7.5%	NA	0%	+1.6 to 3.7%
Central Australia	0%	0%	0%	NA	NA	+2.5% to 5.6%
Northern Territory (Top End)	0%	–1.5 to –3.5%	0%	NA	+1.5 to +3.5%	+1.6 to 3.7%

Source: CSIRO (2006)

The predicted changes in climate by 2030 for populated regions in Australia are presented in Table 2.4. Although these are predictions, and there are variations in the different climate models used, there are trends that should be recognised and taken into account in water management planning. Rainfall is predicted to decrease in all regions, except Tasmania. Evaporation is expected to increase in all regions, but by only relatively small amounts.

It should be noted that climate change predictions are continually being updated and local climate change websites should be consulted for most recent predictions (refer to the Australian Government Climate Change website: http://www.climatechange.gov.au).

The decrease in the runoff, which is used to replenish urban water storages, has major implications for all urban uses, not only urban horticulture.

The effect of reduced rainfall on a catchment is amplified in terms of reduced yield, because a base level of rainfall is required to maintain the vegetation, and to keep soil moisture at a level that will provide significant runoff. As a consequence, a 1% change in rainfall can cause a 2% to 3% change in runoff and catchment yield.

Climate change will add to the stresses currently being experienced by many landscapes. Natural ecosystems are vulnerable and often exist in a fine balance between natural resources available to support them and the pressures, both natural and human induced, exerted on them. It is expected that the incidence of pests, disease and weeds could be higher, and bushfire and forest fire risk will also increase, under climate change.

Urban water management

Climate change will have an impact on urban irrigation in two main ways. Firstly, lower rainfall will result in significantly reduced harvested water from catchments in some localities. This will mean less water for distribution to urban areas. Both Perth and Melbourne have experienced significant reductions in inflows into catchments. The Perth inflows have decreased from 173 GL for period 1975–2000 to 76 GL for 2006–2011 (Figure 2.4).

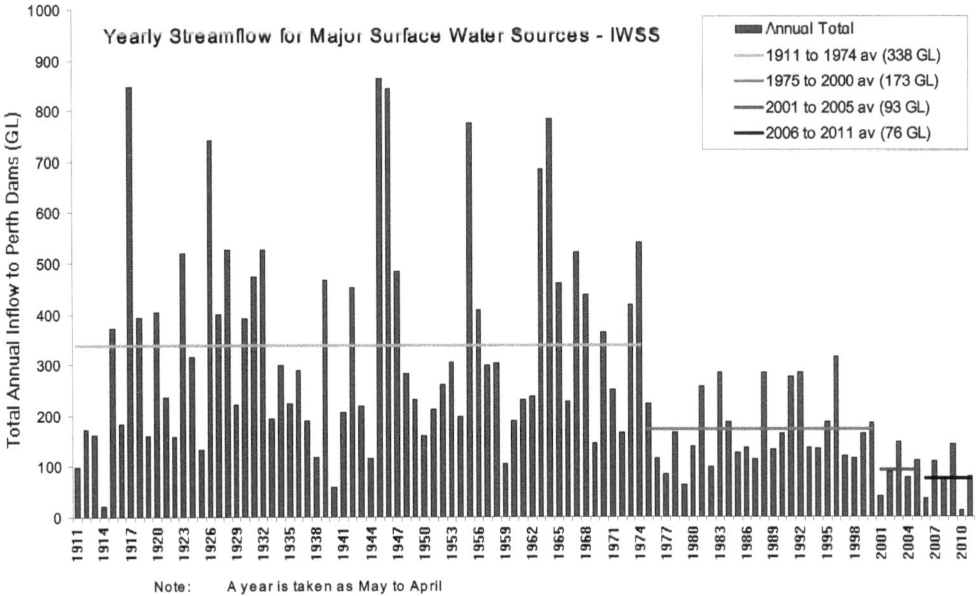

Figure 2.4. Storage inflows for Perth catchments (Source: Water Corporation of Western Australia)

Figure 2.5. Plantings of succulents or species that have silver or grey foliage tend to be more tolerant of drought and high temperatures

Climate change will require management of horticulture under different temperature, evaporation and rainfall regimes. In some cases, different plant species will be more suited to the local climate of higher temperatures and lower rainfall. The increased frequency of extreme high temperatures will impact negatively on some ornamental species (Figure 2.5).

Although native species such as eucalypts are expected to cope well, some of the exotic species used as street trees may not thrive in a more demanding environment. Many deciduous exotic street tree species, such as maples, birches and oaks, may not be as successful when exposed to higher average temperatures and drier conditions.

Another aspect of climate change is the increased deficit between evaporation and rainfall. Higher evaporation rates and lower rainfall will result in the need for greater amounts of supplementary water to grow some plant species. The re-assessment of the suitability of existing landscape plantings, in view of expected climate changes, should be part of the approach to achieving sustainable landscapes.

Landscape planning horizons need to be based on expected changes over the medium to long term, say over the next 20 to 50 years. Some plant species currently used in urban landscapes will not be suitable or sustainable in the future.

Although there is now significant investment in water supply infrastructure, such as desalination plants and stormwater recycled water systems, the urban water supply situation has changed for the long term. It is to be expected that some form of restrictions in the use of potable water will be in place for urban irrigation into the medium to long term.

Planning for climate change, both in terms of weather related change and the changes in government policies, need to be incorporated into the organisations water management plans. Planning for an environment in which there will be less potable supply water overall for irrigation is sensible.

Urban water supplies

Drought, during the 2000s, has highlighted the dependence of urban horticulture on reticulated mains supplies of potable water. The character and quality of urban horticulture has changed in response to less water available and, in many cases, there has been significant decline in the quality of urban landscapes.

There is much debate about water availability, lack of investment in water infrastructure, the setting of priorities, for access to water, and the pricing of water. The bottom line for the urban irrigator is that the conditions of supply have changed. The main conditions relate to time of access, duration of access and restricted use, such as the watering of lawns or the use of sprinklers. In some cases, the supply pressure has been decreased to reduce the risk of leakage from water mains and can have a negative impact on irrigation system performance.

The challenge faced by governments is to provide water supplies that can deliver the required level of service to the community in a period of increasing population demand, periodic drought and climate change.

Although planning and implementation of urban water infrastructure is highly technical, increasingly, it is the social aspects that will influence urban water supply and management. The community's expectations of how water is used and managed are important in determining the water availability for applications, such as irrigation.

The pressures that are now being experienced in urban water supply have been building for decades. The nature of many major urban areas is that the water supply is primarily sourced from dams, which are fed from catchments. Naturally, the yield from these catchments is directly dependent on rainfall and the storage capacity of the dams.

In very simplistic terms, the water available as inflow is decreasing, the storage capacity is fixed and there is increasing demand, due to population growth.

The gap between the demand and the supply is widening. The severity of the recent dry conditions experienced in many towns and cities has meant that the critical point, resulting in severe plant stress, has occurred earlier than anticipated. The solutions include increasing supply, sourcing alternatives, reuse and reducing demand or a combination of these. Increasing supply from rainfall-based systems is faced with the uncertainty of the climate and the prospect of constructing new dams. Building dams has not been popular with many governments in recent times.

In some discussions, groundwater is not considered to be rainfall dependent. Although the timeframe between the rainfall event and contribution to the groundwater at a particular location may be a long time – in some cases thousands of years – the groundwater is still dependent on rainfall, and the associated uncertainty, for replenishment.

The main potential supplies that are independent of rainfall are recycled water and desalination water.

A reduction in demand has been the primary focus of governments to achieve a balance between the supply and demand. The water yielded through water conservation generally has a lower cost than the cost of harvesting water, storing it in a dam, treating it and then distributing it.

A popular local catchment and storage strategy in recent times has been the installation of rainwater tanks (Figure 2.6). Although these are valuable at a local site level, the contribution in terms of the storage and supply at a city-wide scale is quite small. A rainwater tank may provide 5000 to 20 000 L of water annually to a residential property with a garden, compared with an annual consumption of 200 000 to 300 000 L, so the actual contribution to the total property water consumption is low: in this case, around 5%.

Figure 2.6. Rainwater tanks can provide supplementary water supplies for urban landscapes

It is a similar situation with sporting ovals. To some, a 100 000 L tank may sound like significant storage, but, in reality, it may not provide a full week of irrigation for a typical sports ground. Factors to be considered in modelling storage capacity and irrigation demand are presented in Chapter 3 under 'Rainwater harvesting'.

Characteristics of future urban water supply

In the future, the water supply in urban areas will reflect the need to achieve sustainable water use.

Some of the following could be expected to characterise future urban water supplies:

- Water supply catchment harvesting capacity will be limited, because many cities have adopted a 'no new dams' policy and climate change will have a negative impact on catchment water yield.
- There will be less potable mains water available for urban green space.
- The use of potable water for landscapes will be restricted.
- The cost of water will increase.
- There will be greater availability and use of recycled water (treated effluent).
- The technical sophistication of irrigation systems (e.g. communications and control systems) will increase.
- Lower quality water will be available for irrigation.
- The community will expect that water will be managed responsibly and in an environmentally friendly manner.
- The potential role of landscape sites for water harvesting, storage and treatment of stormwater will be recognised.
- There will be greater utilisation of stormwater.
- Long-term water solutions will require an integrated approach.
- Frequency of droughts or dry periods will increase.

Impact of restrictions on urban irrigation

The consequences of the recent limitations and constraints on the supply of water have been profound. The reduced availability of water has caused all water users to assess the context within which water is used. The character of the landscape and use of the facility are being examined to determine what is appropriate and what is sustainable.

The particular consequences of restrictions have been:

- reduced water used on irrigated sites
- some sites are not now irrigated
- a change of plant species (e.g. introducing warm season grasses)
- the active pursuit of alternative water sources
- changing irrigation technologies (e.g. increased use of drip irrigation)
- reprogramming and rescheduling of irrigation to fit water availability time windows.

These are major changes and, in many cases, they are costly. They represent a major advancement in the improved management of urban water resources.

The type and number of projects funded by the Victorian Government to assist country sporting clubs cope with drought show the range of strategies adopted (Table 2.5). In total, there have been over 500 projects and the two most popular have been water harvesting. including tanks (151 funded projects), and synthetic surfaces, (102 funded projects). This conversion from natural to synthetic surfaces is a very significant shift. The conversion from cool to warm season grasses has also been very popular.

One beneficial outcome from recent drought conditions has been closer working relationships between government, water authorities and urban irrigators. However, the water supply arrangements that have been put in place have not always been to the satisfaction and agreement of the green industry or irrigation sectors.

A challenge for the irrigation industry is to demonstrate efficiency of water use. The evaluation and reporting of irrigation performance should play a major part in an overall urban irrigation water conservation strategy. The technology is now available to deliver water effectively in a timely manner.

In addition to changes in technology, there is a need for irrigation systems to be more flexible (in delivery and control), so that they operate efficiently under water restriction regimes.

The design of the system to allow each hydrozone to be independently watered – for example, trees watered separate to lawn – and that programs are flexible in terms of timing and duration, are two aspects of irrigation systems that have become priorities under water restrictions.

Table 2.5. Drought assistance projects funded in Victoria 2007 to 2008

Type of project	Number of projects
Tanks/storage and water harvesting	151
Synthetic surfaces	102
Introduction of warm season grasses	97
Improved irrigation systems	69
Installation/connection of bores	52
Access to recycled water	45
Water carting	30
Identification and planning for local priorities	26
Pool blankets	16
Other water saving	16
Repair of leaks and cracks in swimming pools	14

Source: McKenzie (2009)

Figure 2.7. Stormwater often runs to waste, bypassing the tree root system

Another consequence of water restrictions has been in the design flow rate of irrigation systems. There has been a trend to maximise flow rates. High supply flow rates, which provide flexibility in irrigation applications, are beneficial in terms of being able to optimise the capture of rainfall and not be constrained by long watering times. In some cases, organisations have upgraded the size of the mains service water meters to access the higher flow rates and hence increased water volumes.

Climate implications for urban horticulture and landscape managers

The implications for urban horticulture managers range from regulatory through to technology issues. Some of the key implications are:

- There will have to be a greater focus on selecting species that are suited to the local conditions (climate, plants and soil).
- The principles of Water Sensitive Urban Design (WSUD) will need to be applied more broadly to achieve greater use of stormwater (Figures 2.7 and 2.8).
- The increased use of treated water will require consideration of water quality and its compatibility with urban horticulture. Sound water and soil management skills are required to manage this water effectively.
- Watering techniques will have to ensure effective application and compliance with water authority guidelines.
- There will need to be greater accountability for urban landscape water use.

Green space and benefits of urban irrigation

Urban green space

The impact of green space on the social and physical environment of towns and cities is being increasingly appreciated. The presence of vegetation, rather than hard surfaces such as concrete and pavements, provides numerous benefits. As pointed out by some experts, vegetation, and in particular trees, are the lungs of our cities.

Figure 2.8. Stormwater can be diverted for beneficial use, including reducing peak flows and improving water quality

Green space may include woodlands grasslands, forests, playing fields, public and private gardens, and streetscapes (Plate 2.2). The benefits of green space have been outlined in a report titled '*Irrigation of Urban Green Space: A Review of the Environmental, Social and Economic Benefits*' (Fam *et al.* 2008). Green space in an urban area usually consists of both irrigated and non-irrigated vegetation. The irrigated area is usually a smaller proportion of the total green space area. The areas that are irrigated tend (but not exclusively) to be those that are high value or high use, such as sporting grounds, and high-value landscapes, such as street trees, high-profile landscapes and heritage plantings.

The benefits of green space identified in the report include:

- microclimate modification – cooling by 2 to 8°C, with building energy consumption reduction by 7% to 47%
- improvements in urban hydrology (Figure 2.9)
- vegetation removes pollutants, such as phosphorus, nitrogen, lead and fine sediment
- provides 'hotspots' of biodiversity in cities
- positive in term of community health (physical and mental)
- strong social benefits through sports and recreational activities
- improvement of social cohesion and resilience.

The coverage of ground with vegetation reduces the likelihood of soil erosion and minimises the risk of sediment fouling water bodies and waterways (Figure 2.10).

Role and value of trees

The benefits of trees have been identified by McPherson (2005) and in some cases the benefits quantified. Benefits include improved air quality, energy conservation, reduced stormwater runoff, increase in property values, enhance business activity, reduce stress, increase healing and reduce crime.

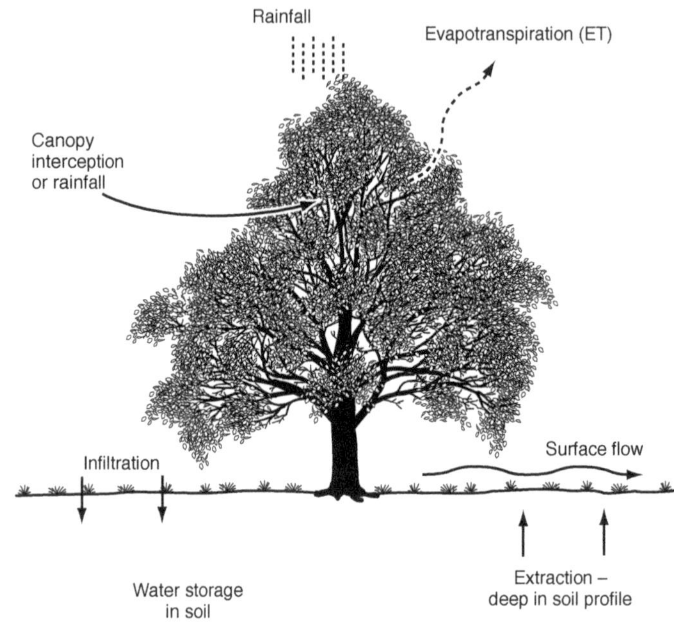

Figure 2.9. Modification of the hydrology of urban areas by trees

According to Dr Greg Moore, Honorary Fellow, School of Resource Management and Geography, University of Melbourne (Moore 2009), the role of trees is undervalued in the urban environment (Figure 2.11). The benefits provided by trees include: sequestration of carbon, saving in air-conditioning electricity costs, reduced carbon emissions, water saved

Figure 2.10. Maintaining vegetated cover assists in erosion control and protects waterway health

Figure 2.11. A single tree in a harsh urban environment is a symbol of the benefits of nature

from reduced electricity generation and prolonged life of pavements due to shading, which all provide an economic benefit to the community. According to Dr Moore, this value should be taken into account when making decisions about the allocation of resources for the maintenance of urban trees.

The value of the trees in Canberra's streets, parks and reserves has been studied by researchers from the Australian National University (Killy *et al.* 2008). These researchers found that the value of the ecosystem services, including carbon sequestration, provided by Canberra's 26 million m^2 of tree canopy area to be $23.5 million. This total is made up of $6 million saved annually in energy and air-conditioning costs, $12 million in pollution reduction and $5.5 million in stormwater mitigation and reduced infrastructure costs.

A comprehensive summary of the benefits provided by trees in Adelaide has been prepared by Killicoat *et al.* (2002). The study outlined a methodology for calculating the economic value of trees. It showed that for each $1 invested there was a return of $2.

The value of the energy savings resulting from the shading provided by a tree and the reduced air temperature can be readily determined.

The energy saving benefits of vegetation, particularly trees, are very significant (Figure 2.12). The shading of buildings and lower air temperatures in the area reduce the need for household air-conditioning. Temperature reductions of 2°C to 6°C are reported, as well as energy savings (due to reduced cooling loads) of up to 47% (Fam *et al.* 2008). The use of irrigation water to grow and maintain shade trees (preferably deciduous trees) is a positive contribution to the environment.

Benefits of green space and irrigated turf

Water supplied to maintain urban areas has a direct impact on the quality of those landscapes. A question of increasing importance is: 'What is the value of this water?' This is not a question

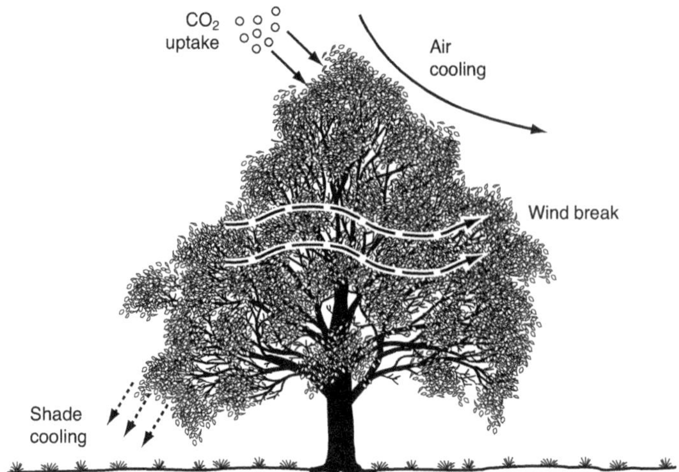

Figure 2.12. Multiple microclimate benefits of urban trees

that has been actively pursued in the past. Urban irrigation water has been taken for granted and has, until recent years, been treated as having a relatively low value. This has been reflected in the fact that, when restrictions are introduced in urban areas, the first water that is restricted is invariably water used for urban irrigation. It has often been considered as discretionary water.

Irrigated green space is a small part of the total open space in urban areas. Access to parks, public gardens and reserves is considered essential for the wellbeing of a community. The beneficial roles of open space are numerous, including enhancing community health, both physical and mental, environmental benefits, including flora and fauna, and modification of the urban climate. Each of these contributes to the economic benefits of urban open space (Plate 2.3).

The value of urban irrigation water used to maintain urban green space should be considered in terms of the environmental, social and economic impacts.

As cities become more urbanised there will tend to be less open space. This is reflected in open space report for Melbourne where the amount of open space per 1000 people is expected to decline from 5.8 ha per 1000 people in 2006 to 4.3 ha per 1000 people in 2026 (VEAC 2010).

Irrigated green space can play an important role in delivering open space services to the community.

Case study: energy benefits of turf – Parliament House, Canberra

A high-profile example of the use of irrigated green space to provide energy saving benefits is Parliament House in Canberra (Plate 2.4). The building has a number of grass ramps that form part of the roof of the building. These are important aesthetic features of the building. Temperature reductions during summer have been outlined by John Lloyd, Landscape Manager, Parliament House (Lloyd 2008 *pers. comm.*).

The ramps comprise two types of surfaces: grass and granite paving. Surface temperatures on the granite paving averaged 22°C higher than on the grass surfaces, over a 10-day period. This reduced surface temperature was reflected in lower temperatures beneath the superstructure: 23°C under the grass ramp compared with 31°C under the granite paving. The green roof of Parliament House provides a very significant benefit, in terms of reducing the demand for cooling energy, to the building. The main environmental benefits of a green roof are:

- temperature modification, both through direct shading and evaporative cooling
- a modified hydrological cycle, including reduced stormwater runoff
- sequestration of carbon

- an enhanced atmosphere, through absorption of CO_2, release of oxygen and absorption or entrapment of pollutants
- noise abatement
- dust suppression
- preservation of biodiversity
- soil stabilisation.

Determining the economic value of urban irrigation water

Determining an economic value of green space is a challenge. If the area or site is used for income-generating activities (e.g. entrance fees), then some form of assessment can be made. Putting an economic value on the less tangible benefits, such as reduced health-care costs as a result of people using or interacting with green space, is more difficult.

It is important that the strong link between the amount of water used and the value of the benefits gained is recognised. It is considered by many, in the green industries that, if a creditable value could be assigned, then decisions regarding the restriction of the use of urban potable water for the community and residential properties may be approached in a more balanced way.

There are a number of measures that can assist in allocating a value for this water. These include the return from sporting clubs, the employment associated with the existence of these facilities, the increased tourism, the value of the asset and the economic benefits flowing from the activity (such as the sale of mowers used to maintain irrigated turf areas). The large returns (from spectators, sponsorship, television, etc.) associated with major sporting events is another activity that highlights the value of water.

Methodologies and literature reviews appropriate to the development of the economic value of urban irrigation water are outlined in Fam *et al.* (2008).

The value of green space to the health of communities has been highlighted by the Victorian Department of Health (Dedman 2010). There are direct economic and social benefits through improved microclimates (Heat Island effect mitigation) and greater opportunity for activity, which leads to healthier lifestyles and reduced health costs.

Social outcomes and economic benefits

Sporting outcomes

The performance of irrigated agricultural enterprises is usually expressed in terms of productivity, as well as in terms of efficiency of water application. In the case of horticultural crops, this measure is relatively straightforward. For example, the water productivity for an orange crop may be 2500 kg/ML and for grapes 3700 kg/ML. This term can then be converted to the gross return per ML to provide a measure of the value of irrigation water. Gross returns per ML for grapes may be in the range of $2000 to $3000 per ML (Connellan 2007). The direct economic benefits of sports fields are not as readily determined.

In the case of ornamental plants and amenity outcomes it is more difficult to quantify the production benefits of the landscapes.

One output from a recreation area is the amount of active use provided by the area (Figure 2.13). A measure of use is the 'user (player) hour', which is the product of the number of participants and the hours of use by the participant. A sports field with an annual usage of 25 000 player hours and a total water consumption of 7.5 ML will provide 3333 user/player hours per ML (Connellan 2007). In this case, each ML of water is providing many hours of active use, which have direct benefits for the individuals and the community.

A large loss in recreational user hours on sports fields occurs, when urban green space cannot be maintained in a healthy condition.

The value of sports fields in South-East Queensland (SEQ) has been highlighted by Power (2007). There are approximately 2500 sports grounds using a total of 2.5 GL of water. The

Figure 2.13. Organised sports, such as soccer, are a key part of open space services

reported water application rate for these grounds was 5.6 ML/ha/year. These grounds provided venues for participation of approximately 1 000 000 users. The average ground provided 25 000 user hours per year to the community. The social return for the investment in sport infrastructure and water is high. For an average ground irrigated at 6 ML/ha/year this represents 4500 user hours per ML. The contribution to participants' health and overall community wellbeing is very significant. At $2.00 per kL ($2000 per ML), then the water cost to achieve the social benefit of 4500 user hours is $1500. Expressed at an individual person level, it represents 2.25 user hours per $1 of irrigation water. Alternatively, it represents a water cost per hour per person of $0.44. Although this is only one of the costs of maintaining turf areas, it shows the actual very low water cost involved in maintaining this type of facility.

The social benefits obtained through irrigation water cover all facilities that require irrigation as an essential input. For example, botanic gardens are, in most cases, irrigated to be able to grow and display plant collections that are to be preserved and also to provide green space for visitors to enjoy. In the case of the Royal Botanic Gardens Melbourne, this site attracts over 1.6 million visitors each year and has irrigation water consumption of around 130 ML/year (2008–2009 data). This represents 80 L of water per visitor per annum or 1 ML per 12 500 visitors.

Tourist precincts that incorporate irrigated landscapes to attract and retain visitors can be assessed in a similar manner.

Value of urban trees and maintenance water

Urban trees are widely acknowledged in terms of the aesthetics, function and overall amenity they provide, but their economic value is not as well appreciated.

The 50 000 or so street, boulevard and parkland tress managed by the City of Melbourne are considered to have a value in excess of $600 million (Shears, I 2009, *pers. comm.*). It is not uncommon to value a single urban tree in the range of $5000 to $20 000. Many of the tree stocks in our cities are in excess of 50 years old and are important in terms of the overall landscape structure and heritage.

The maintenance of these trees during drought periods may require an application of 30–50 kL/year. If potable water is used at $2.00 per kL, the cost is approximately $60–100/year; this is a very low cost compared with the value of the asset that may be put at risk as a result of not watering the tree.

The allocation of water for the protection of existing mature trees that provide valuable functional and environmental services should be embedded in water management strategies for urban areas.

Economic returns from golf

The contribution of irrigated green space to society can be gauged by examining the golf industry in Australia. According to a report prepared by Ernst & Young (2006), the value of the Australian golf economy was estimated at $2.71 billion: about 1.25 million people play golf and 23 000 people are directly employed within the golf industry. Water is a critical input to golf. The employment rate, a key social indicator, derived from golf is significant at approximately 200 employees per gigalitre (GL) of irrigation water used.

The value of irrigation in golf courses can be quantified, in water consumption terms, if the irrigation is deemed essential for revenue. As an example, a semi-private club with an income of $80 000 and a water consumption of 100 ML can be quantified as producing a gross return of $8000 per ML (Connellan 2007).

The benefits of increased biodiversity as result of the irrigated green space within areas, such as golf courses, are more difficult to quantify.

Future cities – essential green space

The importance of green space in urban areas is now widely acknowledged (Plate 2.5). The creation and maintenance of these areas will require significant amounts of water. The main question to be considered is 'how much green space is required, what type of function or purpose is it required to serve and how much water will be required?'

There will not be one answer, but there should be a minimum requirement that is essential for the wellbeing of the community. A minimum amount of active recreational space (sports grounds), passive recreational space (parklands) and tree canopy cover is required.

In planning for green space, there should be an allocation of water to ensure that the services or outcomes to be provided by that space can be delivered in an ongoing sustainable way. Urban planning should identify and quantify the areas of public irrigated green space that are required to support the community. The provision of a sustainable water source for these areas should be part of the planning process.

It would not be expected that potable water was the major source of water for these areas. It may be recycled, stormwater, groundwater or desalination water. Whatever the source, security of supply is essential, because the services to be provided are essential.

Green space should not be considered as optional for a future city. If dry periods and drought occur, the maintenance of these areas is fundamental to the wellbeing and liveability of the city. It can be argued that the value of key green spaces is greater during periods of limited rainfall. Secure water sources should be obtained to ensure that core green space can be maintained.

Currently, Australia is using approximately 400 GL/year for all recreational/open space purposes. This includes all turf uses such as racing, golf, sports grounds and urban public landscapes. The represents 180 ML of irrigation water per 10 000 people, assuming a population of 22 million.

The volume of potable water currently used by councils in Melbourne to maintain open space is typically around 10 ML per 10 000 people (very approximate). The area dedicated to sports is typically within the range of 0.5 to 1.5 ha per 1000 residents. The City of Greater Dandenong has an open space target with a minimum area for sports of 1.5 ha per 1000 residents. It is important that some quantitative measures be incorporated in planning, so that there is a direct link between the services that are expected to be delivered for the wellbeing of the community and the amount of supplementary water required to achieve those outcomes.

Potential environmental impact of urban irrigation

There are significant issues associated with the use of urban water, and potable water in particular, to maintain turf and landscape areas. The use of water for these purposes removes this water from other potentially environmentally beneficial uses. The extraction of water for urban irrigation from groundwater potentially has a significant impact on the health and viability of groundwater reserves. The depletion of groundwater levels in Perth, Western Australia would be one such example. An associated issue is the degradation of the quality of the groundwater: many of these resources have increased in salinity over time. The urban areas of Wagga Wagga, New South Wales have shown significant degradation, in terms of salinity, over the past decades, as a result of extraction of water for irrigation.

Examples of environmental risks due to irrigation

The establishment and ongoing maintenance of an irrigated site can potentially have a negative impact on the environment in a number of ways. These include:

- leaching of applied fertilisers and pest and disease chemicals into groundwater and surface water courses
- chemical spills – a particular risk when site is hydrologically connected to water courses
- oil spills from machinery can contaminate soils and water bodies
- water extracted from the environment and used for irrigation water results in degradation of source (e.g. non sustainable river flows or reduced groundwater)
- applied irrigation water results in a rising watertable and salinity problems
- vegetation removal results in loss of biodiversity
- introduction of weeds and invading plant species
- soil degradation as a result of physical and chemical changes to the soil (e.g. compaction or sediment in runoff water)
- greenhouse gas contribution as a result of energy required for pumping and pressurisation of pipelines.

Table 2.6 outlines the level and range of greenhouse gas production associated with embedded energy in urban water. The source of energy used for pumping, the total pressure required and the efficiency of hydraulic systems influence the greenhouse gas emissions.

These calculations are based on a total hydraulic pressure of 50 m (500 kPa). Desalination, which requires high pressure to force solutions across membranes (reverse osmosis), greenhouse emissions will be much higher.

Table 2.6. Greenhouse gas emission factors and greenhouse gas production per ML of pumped water

State, territory or grid description	Greenhouse gas emission factor (kg CO_2- e/kWh)	CO_2 emission (kg CO_2- e) per ML (pressure 50 m)
New South Wales and Australian Capital Territory	0.89	173
Victoria	1.22	237
Queensland	0.89	173
South Australia	0.77	150
South Australia interconnected system in Western Australia	0.84	163
Tasmania	0.23	45
Northern Territory	0.68	132

Note: The CO_2 emission was determined using a pump efficiency of 70% and a pump delivery pressure (head) of 50 m
Source: adapted Commonwealth of Australia (2009) National Greenhouse Accounts (NGA) Factors

Management of irrigated space in the future

The urban water environment will continue to change and managers of open space will need to continue to be innovative and to find solutions that will allow the valuable services provided by these areas to continue.

The future contribution of irrigation to green space in towns and cities will depend on the adoption of the following:

- monitoring, evaluating and reporting water use
- emphasising irrigation performance and the attainment of high irrigation efficiency
- building human and technical capacity to achieve high irrigation efficiency
- managing sporting facilities, in terms of usage, within the context and constraints of water availability and the climate
- protecting environmental health
- accessing secure non-potable water supplies
- planning for climate change.

Water quality and the future

There will continue to be a shift from the use of potable water on urban landscapes to water of lower quality. Although the water may be of lower quality compared with potable sources, it will be suitable for the purpose, if managed appropriately. The key to the use of alternative supplies is the close ongoing management of water. The first step is a thorough assessment of the suitability of the water. If required, treatment processes are identified and management practices developed to allow the water to be used in a sustainable way.

The main quality issue, in terms of plant and soil health, is likely to be the amount of salts in the alternative supplies. The use of saline water has implications in many aspects of site water management. These include:

- monitoring of water source properties
- monitoring of site conditions and properties, including the soil
- the design and operation of the irrigation system
- staff expertise, skill and training
- site hydrology – surface and ground water protection and management.

The implications for the golf industry as a result of increased dependence on lower quality water (higher salinity) have been outlined by Neylan (2003) of the Australian Golf Course Superintendents Association (AGCSA) Technical Division. The following are the management systems that will need to be put in place to ensure that good quality golf courses can be sustained:

- introducing more salt-tolerant grasses
- applying more water to provide leaching
- constructing high permeability profiles
- applying gypsum applications
- increasing soil cultivation to improve permeability
- installing subsoil drainage to remove salts
- treating the water – acid and gypsum injection.

The use of potable water for urban horticulture

The value of green space in urban environments is high and should not be treated as a luxury or option in the planning of cities. In some situations, it may be appropriate to use potable water for urban horticulture. Clearly, the first option should be to use non-potable water, but if this is not available then potable may have to be used. The key considerations are that the value of the benefits produced by the water is high and that the water is used efficiently.

It is important to recognise the limitations of alternative sources:

- stormwater – sensitivity to rainfall, large storage areas or volumes required
- recycled treated effluent or industrial waste – water quality
- borewater – quality, degradation of the resource.

It is also important to recognise the need for 'essential irrigated green space'. This is urban green space that is essential to maintain the wellbeing of the community and protect the cultural and heritage values of the towns and cities.

Chapter 3

Water sources for irrigated turf and landscape sites

Water for irrigation

Securing a water source to supply irrigated sites requires detailed assessment of both the requirements and characteristics of the site and the suitability of the water. A long-term solution requires a match between the properties of the water source and the specific needs and characteristics of the site.

The site generally has requirements for:

- adequate water volume
- appropriate water supply conditions
- water quality suitable for sustainable use
- security of water supply.

The extraction of water from a source will have some impact on the condition or state of the source. These may be reduced resources, changed water properties or there may be some broader environmental impact. Each potential impact, which occurs as a result of the extraction of water, needs to be considered as part of the overall evaluation.

There are numerous potential sources of water for urban irrigation:

- potable (suitable for human consumption) mains supplies (reticulated systems)
- recycled or reclaimed water (treated effluent)
- stormwater (runoff from ground, paved areas roads, car parks, etc.)
- rivers, creeks/water courses
- groundwater
- rainwater (roof-harvested rainfall)
- greywater (bathroom, shower, in-house taps)
- sewer mining (water extracted from sewer main and treated locally)
- industrial water (water previously used as part of production or other process).

The particular characteristics of each water source needs to be considered when seeking a water source for irrigation. The following should be considered:

- volume available
- water quality – chemical, physical and biological
- water treatment requirements (if any)
- flow rate
- supply pressure
- cost
- availability – timing and duration
- reliability of supply
- conditions of use of water

- storage requirements
- human health issues
- regulatory considerations
- impact on environment
- license approvals and fees.

Potable and non-potable supplies

Potable water, which is available to the urban areas of Australia for the maintenance of landscapes, is a highly valuable resource. It is only in recent years of drought and water restrictions or periods of limited availability that many urban users have come to realise the true value of this resource.

The reasons why mains potable water has such a high value in the maintenance of urban vegetation include both hydraulic and water quality considerations. The high quality of mains potable water allows it to be used for urban horticulture without the quality constraints that are imposed by other sources of water that may contain contaminants, such as pathogens, sediments and salts.

Mains supply is available at most urban properties, pressurised and at flow rates to meet the basic irrigation needs of the site. The total volume of water available is generally not limited: it is only a matter of paying the cost of the water.

In the past, the supply conditions have generally meant that potable mains water was available at any time of the day and usually with good security. There are now often significant limitations during periods of drought and water restrictions.

Water delivered to a site under pressure from a reticulated mains supply is a significant benefit, providing the pressure meets the minimum hydraulic needs of the irrigation equipment. Pumping can be used to overcome shortcomings in supply pressure, but pumps are often not allowed to be connected directly to water authority mains and property service lines.

The accessibility of mains potable water means that there is generally no need for onsite storage: another significant advantage.

Water quality protection – backflow

A key factor in potable water mains systems is the protection of the high quality of the water supply. The underlying premise is that all water taken from the pipe network, at all delivery points, is safe for human consumption.

A potential source of contamination is the hydraulic interconnection between a pipe system with high-quality water and a low-quality hazardous source. This is referred to as a cross connection and can put the health of users at risk.

There are basically two conditions under which cross connection may occur:

1. **Back siphonage.** A low pressure occurs in the supply pipes and this results in siphonage of water back towards the supply point. This may occur if a pipe breaks or there are high extraction rates by firefighting pumps causing the water to drain out so that there is a partial vacuum created in that part of the system. The reverse in flow direction can potentially cause contaminated water to be conveyed to the supply side of the system.
2. **Back pressure**. In normal operation, the pressure on the supply side is higher than on the delivery side, so water flows towards the distribution system, where the pressure is lower. If the pressure for any reason is higher on the delivery side than the supply side, then water will flow back. This may occur if a pump or high pressure line is connected to the delivery side of the system.

Figure 3.1. The protection of potable water supplies, using backflow prevention devices (RPZ), is essential. Tapping points allow testing of the device.

Hazard rating

The categorisation of the risk of the system, to be installed, is made according to the following:

- high – potential to cause death
- medium – may endanger human health
- low – nuisance, but will not endanger human health.

The equipment required to be installed depends on the risk, the supply conditions, the design of the irrigation system and the nature of any chemicals that may be injected into the irrigation water (Figure 3.1). The risk that exists with irrigation systems is dependent on the layout of the system, the position of outlets relative to ground level and the use of pressurised devices, including chemical solution injection equipment. If the irrigation outlet is above the ground, then this reduces the level of risk, because there is effectively an air gap between the outlet and the ground. Pop-up sprinklers, which are installed at ground level, and the top of the sprinkler head may be submerged, are a higher risk.

Compliance with relevant standards and regulations is essential. The hazard rating is outlined in *Australian Standard* AS/NZS 3500.1. Plumbing and drainage – Water services and the performance and testing of backflow prevention devices are covered in AS/NZS 2845 (series) Water supply backflow prevention devices.

Irrigation systems that use some form of chemical injection (fertiliser, plant health chemicals, soil amendment, cleaning, etc.) pose a significantly higher risk.

The techniques available to prevent backflow include both mechanical devices and system layouts.

System design

The use of air gaps to prevent backflow means there is an air space at some point in the supply side of the system, which prevents transfers of liquid from the delivery side to the supply side. Filling of a tank through a riser, which has an air gap, is commonly used. This arrangement is referred to as break tank technique.

To ensure that the backflow device is going to provide effective protection, devices used for high and medium risk systems are required to be testable: the devices are fitted with connection points that allow the operation of the device to be tested. These devices need to be registered with the appropriate water authority. Testable devices are not required for low hazard systems.

Non-testable devices include:

- vacuum breakers
- double check valves.

Testable mechanical backflow prevention devices include:

- registered break tank (RBT) or air gap (RAG)
- double check valve assemblies (DCV)
- reduced pressure zone devices (RPZD).

As a general rule, where there is any risk of contamination, including because of backflow, RPZ devices will be required for potable water mains supplies. Details on specific plumbing compliance requirements for backflow prevention are available from local water authority websites.

In addition to the issue of compliance regarding backflow prevention, there are hydraulic considerations. Backflow-prevention devices cause a pressure loss and this needs to be allowed for in the hydraulic design and operation of the system.

Potable mains supplies – characteristics

The majority of urban irrigation is still carried out through connection to local water authority supply pipelines delivering potable water. Potable water is well suited for use as irrigation water. Drinking water needs to be chemically and microbiologically safe for human consumption and to meet high quality standards in terms of appearance, clarity and odour.

The primary consideration when investigating mains (potable) supply for irrigation are hydraulic aspects, including flow rate and pressure. The two key hydraulic performance characteristics are:

1. flow rate (L/min, L/s, m^3/h)
2. dynamic pressure (kPa) (pressure at operating flow rate).

The term dynamic pressure refers to the pressure in the pipeline when the water is flowing. The pressure that exists in a pipeline or other source, such as a pump, when there is no delivery flow from the source, is referred to as the static pressure. The static pressure is higher than dynamic pressure because, once the water flows, there are pressure losses in the pipe network.

The two factors that most strongly influence the supply available in the mains reticulation systems are the hydraulic conditions (pressure and flow) in the mains and the size of the service connection to the mains and associated headworks equipment. Limitations in any of these will constrain or limit the supply available to the site.

The size of the service line to the property, the valves, water meter size and type and, if a backflow prevention device is installed, will all impact on the flow to the property. The combination of each of these can result in significant pressure differential between the supply main pressure and the pressure available to the property internal irrigation supply line.

The size of the service line and the associated metering and control valves directly influence the supply available to the site. Service lines and meters have limits on allowable flow rates. In some urban sectors, rules are applied to determine the available supply conditions of a service. These are covered in Chapter 8, under 'Metered mains supply'.

Water supply pressure

The installation or presence of any valve or restriction, for example backflow prevention device, will reduce the delivery pressure available to a site. The complete hydraulic control manifold assembly needs to be assessed to determine the net pressure available to the irrigation system.

The availability of adequate pressure to operate sprinkler systems at optimum hydraulic conditions is essential. One area of concern to the irrigation industry is the risk of decline in supply pressure to a site. This decline may occur as a result of increased water demand in an area, changed plumbing (e.g. installation of backflow equipment) or as a result of planned reductions in pressure within the supply and distribution network. Some water authorities are reducing supply pressures to reduce the risk and occurrence of mainline pipe leaks. It is important that both the water supply authority and irrigation managers recognise the risk to the performance of the irrigation systems and provide supply performance protection in vulnerable localities.

Installation of in-line pumps is one strategy that is used to overcome low supply pressures. These installations, however, require specific approval from the water authority. In-line pumps can potentially have an impact on supply to other users in the network.

Benefits of mains potable supplies

Potable main supplies have the following benefits/characteristics:

- high-quality water (chemical, physical and biological)
- supply available under pressure
- available directly to the property
- on-site storage is generally not required
- relatively low capital cost to make the connection
- volume available is generally only limited by flow rate and cost (assuming restrictions are not in place).

Some of the limitations of mains potable supplies are:

- the maximum flow rate may be limited, resulting in less-than-optimum flows for some irrigation situations
- the supply is sometimes restricted to defined time windows (e.g. at night)
- protection is required to prevent cross connection in the form of backflow prevention devices. Pressure loss through these devices can impact negatively on the performance of the irrigation system.
- the cost of potable water can be very high for irrigation purposes
- the hydraulic supply conditions can be degraded, as a result of external factors, including property developments and increased demand.

Alternative water supplies for irrigation

A water supply alternative to potable water needs to be assessed in terms of both the hydraulic considerations and water quality.

A primary consideration in assessing alternative supplies is the issue of water quality. It is to be expected that alternatives to potable water, other than rainwater, will be of a lower quality. The other considerations of security of supply, access and cost. However, if the quality of the water is such that it cannot be used in the long term, in a sustainable way, then it is not a suitable alternative source.

Water quality

Irrigation considerations

Application of irrigation water to a site has an impact, not only on the vegetation, but also the soil. In some cases, it can have an impact on the irrigation hardware as well. The quality of the applied water is critical to any considerations of the suitability for irrigation and the management of water of a site.

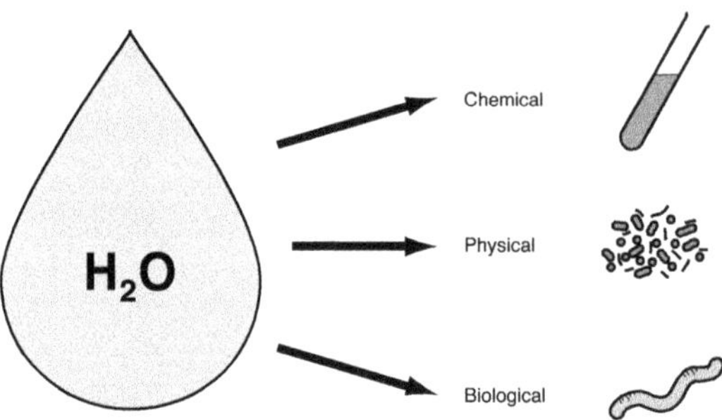

Figure 3.2. Water quality properties

The primary requirement of irrigation water is that it:

- is suitable for growing healthy plants
- is suitable for maintaining a healthy soil environment
- is suitable for using in water handling equipment, including pumping, treatment, control and application
- is safe for human health
- does not pollute or contaminate the environment, including water bodies.

When evaluating the suitability of water for a particular site, the assessment should consider not only potential short-term impacts, say 5 to 10 years, but also the medium to longer term: 20 to 50 years. Some soil problems, such as salinity and nutrient leaching, can take many years to show the effects of changes in water quality properties.

The term 'quality' refers to the characteristics or properties of the water. High-quality water is considered to exhibit properties that are within acceptable tolerances or ranges. Low-quality water exhibits parameter values that are outside the acceptable ranges.

Water is a very complex substance and can comprise many different elements in greatly varying levels or concentrations.

The quality of water can be considered using the following categories of properties (see Figure 3.2 and Table 3.1):

- chemical – constituent elements
- physical – including suspended particles
- biological – presence of organisms (e.g. pathogens).

Table 3.1. Potential impact of water quality on irrigation

Water quality property	Consideration – irrigation system
Chemical	Method of application • Selection of materials and equipment • Salinity management • Nutrient management • Fertigation equipment used • Runoff and drainage water impact on environment
Physical	Filtration requirements • Selection of emitter or irrigation outlet • Pump selection • Equipment selection e.g. temperature sensitivity
Biological	Filtration requirements • Water treatment process required • Disease management • Maintenance of irrigation system

The assessment of the suitability of the water source, as the supply of water for a particular site, soil type and crop, is the first step to be taken in a total water management planning process.

The quality of irrigation water has a direct impact on plant health, soil health, the environment and management practices appropriate to the site.

Detailed analysis of the properties of water proposed for use in irrigation, and the interpretation of the results by qualified experts, is important to the success of an irrigated site.

Water quality properties

The properties or quality of water can be assessed for irrigation purposes in terms of the following:

- pH (Acid or basic)
- salinity (electroconductivity: EC and total dissolved salts: TDS)
- specific ions that are potentially toxic (e.g. sodium and chloride)
- alkalinity (bicarbonate and carbonate)
- sodium hazard (proportion of sodium to calcium and magnesium)
- nutrients (e.g. nitrogen and phosphorus)
- micronutrients
- heavy metals
- hardness
- microbiological content
- corrosion.

pH: acidic or alkaline (basic)

The terms acidic and basic reflect the pH value of the water. The concentration of hydrogen ions in a solution determines the acidity or otherwise of the solution.

The pH scale is based on the negative logarithm of the concentration if H^+ ions. A solution with a pH of 7 is neutral. Solutions that have a pH lower than 7, and hence higher concentrations of H^+ ions, are acidic. Solutions that have pH values higher than 7, and hence lower concentrations of H^+ ions, are basic.

The acidity of irrigation water potentially influences plant growth, availability of nutrients, effectiveness of use of plant health chemicals, soil properties and the longevity of irrigation equipment. Water suited to irrigation is generally in the range of pH 6 to 8. High pH (in excess of 9) may indicate presence of bicarbonates and carbonates, which can cause calcium to precipitate and create hazards for the irrigation system (Yiasoumi *et al.* 2005). Low pH can reduce nutritional uptake by plants and also lead to soil acidity and corrosion of equipment.

Salinity

Water for irrigation may contain numerous salts including chlorides, sulphates, carbonates and bicarbonates of calcium, magnesium, potassium and sodium. Salinity refers to the concentration of all of the dissolved salts in the water or soil.

Salinity affects plant growth and soil health in a number of ways. The presence of salts reduces water absorbed through the plant root system. Salts may also adversely affect plants when there is direct contact between saline water and plant foliage. Plant response to salinity varies greatly. The relative sensitivity of ornamental landscape plants and grasses to salinity is presented in Appendix 5.

Salinity reduces the availability of soil water to the plants. Water moves from the soil reservoir through root hairs via the process of osmosis, which relies on differences in concentrations of sugars and salts between the roots and the soil to draw water into the plant. High soil salt concentrations reduce the rate of uptake of water. The soil may have ample volumes of water, but, because it is saline, this water is not available to the plants and so the plant effectively experiences 'drought' conditions.

Table 3.2. Units of electrical conductivity and salt concentration and their conversion

Units	Conversions		
Electroconductivity (EC)	**From**	**To**	**Multiply by**
deciSiemens per metre (dS/m)	dS/m	mS/cm	1
milliSiemens per cm (mS/cm)	dS/m	S/cm	1000
microSiemens per cm (μS/cm)	dS/m	EC units	1000
millimho per cm (mmho/cm)	S/cm dS/m	dS/m mmho/cm	0.001 1
Total dissolved solids (TDS)	EC (dS/m)	ppm	640
Parts per million (ppm)	ppm	mg/L	1

Salinity is typically measured with a meter that provides a reading of the electrical conductivity (EC) of the solution. A solution conducts electricity more readily as the concentration of dissolved salts increases. The recommended unit of electrical conductivity of a solution (EC_w) is decisiemens per metre (dS/m).

The actual amount of dissolved salts (TDS) in the solution is determined through weighing and is expressed as mg/L or ppm (parts per million). The relationship between the electrical conductivity (EC_w) reading of a saline solution and the mass concentration of salts can be calculated using the following expression:

$$EC_w \text{ (dS/m)} \times 640 = \text{TDS (ppm or mg/L)}$$

It should be noted that this conversion factor is approximate and needs to be adjusted for high salinity water. There are a range of units used for both electrical conductivity and salt concentration. A list of alternative units and conversions is given in Table 3.2.

The significance of the salinity readings should be appreciated. A reading of 1.0 dS/m is equivalent (approximately) to 640 kg of salt in a ML (1 000 000 L) of water. Therefore, an application of 100 mm depth of saline water (1.0 dS/m) on a hectare (10 000 m^2), represents a deposit of 640 kg of salts to that area. This represents approximately 5 tonnes on a sports ground of 1.5 ha over an irrigation season.

Table 3.3 defines the irrigation water quality classifications according to salinity concentrations.

Alkalinity

This is a measure of the capacity of the water to counter (buffer) changes in pH. The main components of water alkalinity are carbonates ($CaCO_3$), bicarbonates (HCO_3^-) and soluble hydroxides (OH^-). Alkalinity levels, expressed as mg/L calcium carbonate ($CaCO_3$) equivalent, of less than 90 mg/L are considered low risk for irrigation. The level of alkalinity can be

Table 3.3. Salinity classifications for irrigation water

Class	Classification	Electrical conductivity (EC) (dS/m)	Electrical conductivity (EC) (μS/cm)	TDS range (mg/L)
1	Low salinity	0.0–0.28	0–280	0–175
2	Medium salinity	0.28–0.80	280–800	175–500
3	High salinity	0.8–2.30	800–2300	500–1500
4	Very high salinity	2.30–5.50	2300–5500	1500–3500
5	Extremely high salinity	>5.50	>5500	>3500

Source: Australian Water Quality Guidelines for Fresh and Marine Waters, ANZECC, 1992.

adjusted or corrected through the injection of acid into the irrigation water. The origin of water largely determines its composition in terms of the carbonates and bicarbonates. Groundwater can be high in concentrations of these compounds.

Sodicity and the sodium adsorption ratio (SAR)

The sodicity of a soil refers to the relative proportion of sodium ions to the calcium and magnesium ions. Soils with high concentrations of sodium are more likely to cause a breakdown of the soil structure and the dispersal of fine soil particles. High amounts of sodium can have a negative impact on soil structure. Dispersive clays are a particular risk.

The dispersion of fine clay particles plugs the soil pores and decreases the opportunity for water to move through the soil. The net effect is reduced infiltration into the soil, reduced hydraulic conductivity within the soil and, is some cases, results in surface crusting. High concentrations of calcium and magnesium tend to counteract the negative impacts of soil dispersion.

The sodium adsorption ratio (SAR) provides a measure of the risk of soil dispersion. The risk is directly related to the soil type, existing hydraulic properties and the salinity of the water.

SAR calculation

The risks associated with the use of water with a high proportion of sodium to calcium and magnesium can be determined through the calculation of SAR.

$$SAR = \frac{Na^+}{\sqrt{Ca^{2+} + Mg^{2+}}}$$

where Na, Ca and Mg are expressed in milliequivalents per litre (me/L).

The calculation of SAR, which is a standard test procedure, should be completed by a test laboratory or water quality expert.

The SAR value needs to be considered in conjunction with the EC of the irrigation water. Generally SAR values of less than 3 to 5 represent no problems, but advice should be sought on management practices based on analysis of water analysis and understanding of the site soil properties and conditions (Neylan 2007). The management of turf sites, with irrigation water, that is high in sodium, is also covered in detail in the same paper.

Nutrients – Macronutrients and Micronutrients

Non-potable water sources, such as recycled and stormwater, are likely to contain nutrients that will be available for uptake by the plants. These main nutrients include nitrogen, phosphorus and potassium. Other nutrients include calcium, magnesium and sulphur. The water will also contain micronutrients, which are also available for uptake by the plants. These include boron, copper, iron, chloride, manganese, molybdenum and zinc.

The concentrations of each of these elements are required to determine the nutrient management practices to be adopted at the site. There are also potential soil and plant health issues if the concentrations of particular elements exceed specified limits.

The fertiliser program will need to be adjusted to accommodate nutrients available through the application of nutrient-rich recycled water.

Specific ions

Chloride

Chlorides are readily dissolved in water and are potentially toxic to plants. Plants are sensitive to high chloride concentrations, particularly if the foliage is wet by sprinkler or spray irrigation. Chlorides can also accumulate in plant leaves and exhibit scorching of foliage if levels are excessive.

The sensitivity of plants to chloride is plant specific. Chlorides in excess of 100 mg/L may be toxic to ornamental plants. Plants can tolerate higher concentrations of chlorides if uptake is via the root system rather than foliage exposure. Chlorides are readily soluble and mobile as water moves through the soil.

Sodium

The potential impact of sodium on soil structure is outlined earlier (sodicity). Sodium is also potentially damaging when applied directly to foliage if concentrations are in excess of 70 mg/L. High levels of sodium can also adversely affect root cells and root growth. Sodium, as with most elements, accumulates in the plant.

Sodium toxicity is a risk for intensely managed turfgrasses such as golf greens. Due to close and frequent mowing, the grass is stressed and vulnerable to excess sodium.

Boron

Boron is essential for plant growth. It is a micronutrient and usually present in very small concentrations. According to Harivandi (2000), turfgrasses can tolerate boron levels up to 10 mg/L, but some ornamental plants exhibit phytotoxicity levels to concentrations as low as 1 to 2 mg/L. Boron can be readily leached from soil by high quality water, such as rainfall.

Iron

The presence of excessive soluble iron and iron-loving bacteria in irrigation water can cause problems through the formation of sludge deposits and blockage of irrigation equipment including filters and orifices. Iron can also cause discolouration of surfaces, as the iron deposits remain after the water has evaporated.

Suspended solids

The amount of particles in suspension in irrigation water is an important indicator of the physical quality of the water. These include inorganic particles, such as clay and silt, and organic material, such as plant material, algae, bacteria, plankton and fine organic debris.

It is common for surface waters (e.g. rivers, lakes and stormwater), and sometimes recycled water, to have significant amounts of suspended particles. There are generally negligible amounts of suspended particles in mains water supplies. They can be introduced through pipeline failures and during construction works. Suspended solids can serve as carriers for other potentially harmful compounds, including pesticides.

Suspended solids present problems, such as:

- blockages – reduced effectiveness of irrigation equipment
- increased maintenance requirements, including servicing of filtration equipment
- a potential impact on the effectiveness of water treatment processes such as disinfection.

These particles (suspended solids) do not dissolve in water solution and so need to be removed by filtration.

Turbidity

The clarity of water is largely determined by the amount of suspended matter, such as clay, silt and fine organic material (e.g. plankton and algae), in the water.

The term to describe this property of the water is turbidity. It is a measure of the amount of light that is absorbed and scattered and does not pass through the solution. The measure of turbidity is the nephelometric turbidity unit (NTU), which is an optical measure. Ideally NTU should be less than 1.0 for drinking water and less than 5 for irrigation water, but this depends on the type of irrigation system in use.

Problems associated with high turbidity include blockages, sludge build up and staining. Treatment of turbidity is commonly achieved through the use of a coagulant, which causes the small particle to clump together (flocs) and then removed all following settling.

The effectiveness of disinfection techniques such as chlorination, ozone and ultraviolet (UV), is directly influenced by the turbidity (colloidal and particulate) of the water. Although clarification to less than 5 NTU is sometimes recommended for UV treatment, the water may need to be less than 2 NTU to achieve the required reduction in *Escherichia coli* (*E. coli*).

Hardness

The presence of calcium and magnesium carbonates in water can lead to the formation of precipitates and scale. This water is often referred to as being 'hard'. The presence of other elements, including iron, manganese, aluminium and zinc, can contribute to the hardness of water. Hard water is characterised by difficulty in causing soap to lather and producing scale in appliances. Water that originates in limestone regions and some borewater are often hard.

Hardness is measured in units of mg/L $CaCO_3$ equivalent. Water with less than 50 mg/L $CaCO_3$ is considered soft and water with more than 150 mg/L $CaCO_3$ is considered hard (Yiasoumi *et al.* 2005). Concentrations less than 100 mg/L are suitable for most irrigation uses.

One implication of irrigating with hard water is the potential risk of blockages from precipitates and the build up of scale in irrigation systems with small pathways, such as in micro-irrigation systems.

Microbiological

Pathogens

Water may contain organisms that cause disease and are a risk to human health. Any organism that causes a disease is classed as a pathogen. Faecal residues from humans and animals are particular sources of risk.

Plant specific pathogens include *Fusarium*, *Phytophthora* and *Pythium*. These are readily transferred in irrigation water, from diseased plants and soils to healthy plants. Conditions in irrigation water are well suited to many pathogens.

Organic content – biochemical (or biological) oxygen demand (BOD)

The presence of organic matter in water and the breakdown of this requires the absorption of oxygen as part of the process. There are many potential sources of organic materials in water, including decaying plants, animals, animal manure, sewage and wastes from industrial processes.

The level of oxygen demand to break down the organic material in water is referred to as the biological oxygen demand (BOD). Oxygen in water is also used up in chemical reactions and is referred to as the biochemical oxygen demand. Both sources contribute to the overall oxygen demand.

The main impact of oxygen demand is the health of aquatic life, because high BOD will deplete the oxygen supplies. A maximum BOD of 1 mg/L is considered appropriate by authorities for receiving waters.

Guide to preferred water quality parameter values for irrigation

Table 3.4 indicates the limits for the various water quality parameters that are not expected to present problems when used for irrigation.

Table 3.4. Guidelines for water quality for irrigation use

Water quality parameter	Preferred range or limit	Reference/source
pH	6.0 to 8.0	Various references[1]
Salinity – EC	<0.28 dS/m	ANZECC (1992)
Salinity – TDS	<175 mg/L	ANZECC (1992)
Alkalinity ($CaCO_3$ equiv.)	<100 mg/L	Handreck (2008)
Bicarbonate	<90 mg/L	Handreck and Black (2001)
Chloride (overhead sprinklers)	<100 mg/L	Handreck and Black (2001)
Sodium (overhead sprinklers)	<70 mg/L	Handreck and Black (2001)
Boron	<0.5 mg/L	Handreck (2008)
Sodium adsorption ratio (SAR)	<6	Neylan (2003)
Total Suspended Solids (TSS) (microirrigation drip systems)	<50 ppm	Burt and Styles (1994)

Note: The values shown are a guide only. Soil properties, plant response, irrigation method and composition of water influence appropriate parameter levels at a particular site.

1. pH recommended range varies depending on multiple water quality property, soil conditions and site requirements.

Corrosion

The corrosion of irrigation equipment is a water quality issue, in terms of operation and functioning of equipment, and the longevity of irrigation equipment. Corrosion of critical components, such as pump impellers and valves, can have a marked effect on the hydraulic performance of the system.

The risk of corrosion increases with decreasing pH. The hardness of water also influences corrosion. Soft waters are more prone to corrosion than hard water.

Water testing

When the potential broad implications of water quality are appreciated, it is clear that a detailed water analysis is warranted. There are many laboratories that can competently carry out water analysis. The laboratory should be National Association of Testing Authorities (NATA) certified.

The core irrigation water quality properties that should be tested are:

- pH
- salinity – as electrical conductivity (EC) and total dissolved solids (TDS)
- sodium and chloride
- bicarbonate
- alkalinity
- sodium adsorption ratio (SAR)
- total suspended solids (TSS).

This is not a comprehensive list. Each water source and site will have its own particular characteristics and requirements.

The laboratory analysis of water is only the first step in the water assessment process. Equally important is the interpretation of the results in terms of the suitability of the water for the particular crop, soil and site. Interpretation services are provided by some laboratories, consultants and government services. Generally, detailed analysis services are more available than interpretation services. It is important that interpretation be provided by appropriately qualified agronomists and experts.

A sample of a water quality analysis report, prepared by Australian Golf Course Superintendents Association (AGCSA), is presented in Appendix 7.

Recycled water supplies

Introduction

A secure recycled water source for irrigation has become the lifeblood for many turf and landscape sites. Access to a supply that is independent of rainfall and water restrictions is a very attractive proposition, in terms of the long term viability of the site. Recycled water (treated effluent) and reclaimed water are considered to refer to the same type of water in this book.

In general, the level of adoption of recycled water as a source for irrigation is low within the national context. The proportion of treated effluent water that has been used in recycling in Australia is relatively low, in the range of 12% (NPSI 2009). It is, however, increasing, with support from government. The green space sector – parks and sports grounds –, has, however, been a strong adopter of recycled water.

The overall water scene favours the use of recycled water. There is increasing pressure on potable supplies, the cost is increasing strongly and there is a growing view that potable water should not be used for irrigation.

Although stormwater is often considered, there are practical considerations in terms of the catchment yield, on-site storage requirements and the dependence on rainfall. For some open space managers, including those managing golf courses, bore water has become less attractive owing to lower quality and reduced yields. Overall the use of recycled water for irrigation makes sense.

Neylan and Peart (2005) reported that approximately 15 000 ML of treated effluent was used for the irrigation of approximately 3700 ha on golf courses. Some golf courses have been using recycled water for over 25 years (Figure 3.3). The use of recycled water to irrigate open space is generally well supported by the community. However, there are two main areas of concern: public health and environmental protection.

In summary, the reasons for the use of recycled water are:

- generally good security of supply reduces demand on potable supplies (an increasingly scarce resource)
- allows flexibility in water use

Figure 3.3. Recycled water, delivered by tanker, being used as emergency water for a golf course

Table 3.5. Sample (mean value) concentrations of chemical constituents found in recycled water

Parameter	Unit	Recycled water median values
pH	pH	7.9
Electrical conductivity (EC)	dS/m	1.3*
Total dissolved salts (TDS)	mg/L	675*
Sodium (Na)	mg/L	181*
Chloride (Cl)	mg/L	135*
Ammonium (NH_4)	mg/L	8.4
Total phosphorus (P_{tot})	mg/L	5.9
Calcium (Ca	mg/L	35
Magnesium (Mg)	mg/L	19
Total nitrogen (N_{tot})	mg/L	15.2
Sodium adsorption ratio (SAR)	$(mmolc/L)^{0.5}$	6

*Properties with higher than ideal values
Source: adapted from Stevens *et al.* (2008)

- in some cases, lower cost than potable supplies
- includes nutrients that can be used by plants
- it is only option because of regulatory constraints.

Composition of recycled water

The content of effluent and wastewaters generally is highly variable and hence the properties or qualities of the treated water are also variable. Although the various treatment processes modify and change concentrations of specific elements in the water, a range in values is to be expected. Effluent water properties are also expected to change over time.

Recycled water will generally contain:

- nutrients
- salts
- a range of chemical contaminants (e.g. heavy metals)
- pathogens (e.g. viruses and bacteria).

Note that anything that causes a disease is a pathogen. Pathogens include bacteria, viruses, protozoa, helminths and fungi.

Table 3.5 provides a sample of the chemical properties that may be exhibited in recycled water. It is important to note, that concentrations are highly variable, and this table is provided to show the ranges and relative importance of different components.

The parameter values recorded in the table indicate that the water quality is not ideal for irrigation, for example high EC, TDS, sodium and chloride. It can be used, but it needs to be well managed. It is for this reason that thorough testing should be undertaken and strategies developed, such as water treatment, to manage the water in a sustainable manner.

Recycled water and risk

The use of recycled water presents potential risks in terms of human health, horticulture and the environment.

Human health risks

The human health risks include illnesses and diseases, such as gastroenteritis, dysentery and various infections. Serious human illnesses such as hepatitis, meningitis can be caused by

Figure 3.4. Fine clay soil degraded as a result of application of saline recycled water

viruses in untreated water. Direct contact, consumption and inhaling aerosols (small droplets) are all potential contamination mechanisms. It is important to treat water to control pathogens to a level where the risk is eliminated or reduced to an acceptable level. Recycled water needs to be regularly tested for pathogens to protect people who may come into contact with it.

Horticultural risks – site, soil and plants

The risks to the landscape through the use of recycled or reclaimed water can be through direct contact with plant foliage, uptake by plants and modification of soil properties.

The horticultural risks may include:

- increased soil salinity
- degradation of high sodium soils
- excess growth, through nitrogen imbalance
- some nutrients and salts, such as phosphorus, chlorides and boron, may be available in excessive concentrations
- presence of pathogens.

The main water quality concerns for the golf industry, when using recycled water, are high salinity, sodicity and bicarbonates (Neylan and Peart 2005). Fine textured soils were identified as a particular problem, in terms of salts and sodium levels in irrigation water (Figure 3.4).

Environmental risks

These may include:

- increased salinity in waterways or ecosystems
- nutrient contamination of water bodies (referred to as 'eutrophication')
- damaging microorganisms released into aquatic ecosystems.

Assessing site for use of recycled water

The first stage of an assessment of the use of reclaimed water is a detailed consideration of the property and the site. A thorough understanding of the soils and geology, terrain, surface water flows and drainage, local water bodies, groundwater and the potential response of the vegetation to the use of the water is required.

The potential impact of the water on all areas needs to be considered. These include:

- vegetation
- soil

- water supply and irrigation system
- water storage
- environment
- users, visitors and staff Occupational Health and Safety (OHS).

Guidelines on use of recycled water

The use of recycled water for irrigation is strongly based on an assessment of the risks and adoption of a management approach to ensure that the water is suited to the situation. There are national guidelines on the use and management of recycled water: *The Australian Guidelines for Water Recycling. Managing Health and Environmental Risks, Phase 1* published by NRMMC, *EPHC* and NHMRC (2006).

States and territories (e.g. EPAs) provide advice at a more local level, and also in more detail. The compliance requirements are described in these guidelines, as are the classification of the recycled water, the irrigation methods, and the conditions of use and withholding periods. The classifications determined have important impact on the design of the system. For example, the requirement for a 4 hour withholding period, because of public access to the site, changes the hydraulic design dramatically. The delivery time available for irrigation may be halved.

The plumbing hardware requirements for the use of recycled water include identification with lilac/purple colouring: pipes, tape for standard pipes, headwork assemblies, sprinkler tops, valve boxes and water meters are required. The headworks assembly requires a backflow prevention device to be installed.

Signage is required to identify the nature of the water being conveyed and distributed and also to advise that it is not suitable for drinking. The relevant Australian standards are:

- AS/NZS 3500 *Plumbing and drainage* (Series). Standards Australia/Standards New Zealand, Sydney and Wellington.
- AS 1345-1995 *Identification of the contents of pipes, conduits and ducts.*
- AS/NZS 2845.1:1998 *Water Supply – Mechanical Backflow Prevention Devices.*

'Fit for purpose' approach and quality classifications

The quality of the water applied through the irrigation system should be suited to the vegetation and the site (Table 3.6). It is not necessary to use water of a very high standard, such as potable water, for most irrigation situations. This approach is referred to as 'fit for purpose'. It is, however, important that the properties of the irrigation water be suited to sustainable use in the long term.

The 'fit for purpose' approach has been an important shift in recycled water regulation. The new *Australian Guidelines for Water Recycling: Managing Health and Environmental Risks* (NRMMC, EPHC and NHMRC 2009) provide guidelines on the use of water of a particular quality being suited to a specific situation or application. This replaces the previous prescriptive approach, which defined classes of water and then identified categories of use that suited the particular class. State authorities still include classes (A, B, C and D) of recycled water in their local guidelines. Class A is the highest quality water and provides the greatest flexibility in use.

For the urban irrigator, it is most important to be aware of the national, state and local regulations that relate to the use of recycled water. The consequences – legal and financial – for failing to comply can be very serious.

Irrigating with recycled water

The use of recycled water can have numerous consequences, in terms of the design and management, of the irrigation system.

Table 3.6. Classes of recycled water – general description of treatment and irrigation uses

Classification	Treatment	Irrigation uses
Class A	Tertiary treatment with very low pathogen levels	Limited restrictions on use. Suited to food crops that are consumed raw. Some human contact (not consumption) possible. Unrestricted in irrigation method and public access. Suited to open space irrigation and to residential garden irrigation. Not suitable for drinking.
Class B	Secondary treatment. High degree of pathogen control, with some residual pathogen load.	Restricted use, withholding period (e.g.1 to 4 hours). Not for use on food crops consumed raw. Sprinkler systems suited with public access control. Buffer zones required.
Class C	Secondary treatment, some pathogen control or reduction. Significant pathogen load.	High level of management required. Restricted public access. Sprinklers with controlled public access (e.g. golf course). Wider buffer zones than Class B. Subsurface drip application removes public exposure risk.
Class D	Secondary treatment, low level of pathogen control	Very limited use on non-food crops (e.g. plantation timber, instant turf production or production nurseries). Restricted access.

Notes: It is important to refer to each state or territory classifications, water quality parameter specifications and associated use requirements. The requirements may vary from the general descriptions provided here.

Nitrogen, in its various forms, may be present in the range of 10 to 50 mgN/L in water. The delivery of a balanced nutrient program, which is essential for healthy plant growth, and preventing nitrogen losses, due to excess application of water, is one of the challenges of irrigating with recycled water. The irrigation system needs to apply water uniformly.

The salinity of recycled water is often high, relative to potable, and may be in the range of TDS 500 to 1000 mg/L or even higher. The long-term health of the soil needs to be considered. Drip systems, rather than sprinklers, are better for the application of saline water. Deep applications of water should be made, rather than shallow ones that can cause higher concentrations of the salts in the soil. Leaching, through application of water in excess of the plant evapotranspiration needs, is usually required to manage saline water.

Another area of concern is the potential impact of the recycled water on soil structure. The sodium adsorption ratio (SAR) provides a measure of the risk. Generally SAR less than 5 to 6 is not a problem, but soil and water testing at the site is the only way to really know.

The pH of the recycled water may have an impact on the availability of plant nutrients. In most cases, the pH is within acceptable range of pH 6 to 8. It may also affect the likelihood of precipitates forming, which can cause blockages. In these situations acid injection may be required.

Some of the irrigation risks associated with recycled water are in the functioning of the irrigation equipment: drip emitters in particular. Total suspended solids, total dissolved solids, iron, manganese and microorganisms are potential contributors to emitter blockages.

If sprinklers are used for the distribution of recycled water, spray heads that produce coarse drops should be used. Excess pressure on sprinkler nozzles produce fine drops, which are prone to wind drift and add to the risk. The buffer distance requirements, for example 50 m separation between sprinkler application and public use areas along boundaries, will be outlined in local guidelines.

The properties of recycled water may require selection of special materials for the irrigation system, such as stainless steel, so that they are compatible with the chemistry of the water.

The storage of recycled water, so that the quality is suitable for use, requires regular monitoring and management. The quality of stored water can have an impact on the treatment processes, the functioning of the irrigation system and the general health risks. Storage design, including depth, shape and aeration, can affect recycled water storage life. If algal blooms form, protection of the storage is required.

The suitability of the site for an earthen storage, in terms of seepage losses, needs to be investigated. A liner may be required, to prevent excess volume losses and eliminate the risk of contamination of the groundwater.

Provision may need to be made for disinfection equipment, such as chlorination to be incorporated into the water treatment processes. Provision may also need to be made for the blending or mixing of low salinity water, such as stormwater or potable water, with the recycled water to reduce the overall salinity of the water source. Treatment processes, except for reverse osmosis (RO), do not reduce the salt concentration in recycled water.

Irrigation systems (sprinklers and sprays) designed for recycled water

Sprinklers and spray irrigation systems that are to apply recycled water have special requirements. Although all irrigation should be efficient, there are additional considerations when recycled water – which has specific risks – is to be distributed and applied.

The core elements of a sound design approach are:

1. The system should be capable of meeting peak plant water demands and any leaching requirements.
2. Water should be applied uniformly because, in addition to efficiency considerations, there is the prospect of differential growth, due to the varying availability of nutrients.
3. Water should be applied effectively, with minimum wind drift.
4. Environmental sensors, including soil moisture and wind, should be incorporated into the control so that irrigation can be prevented following rainfall or under windy conditions.
5. The system should be designed and operated to avoid runoff. The soil should be managed to maintain adequate infiltration and adequate soil water-holding properties.

Details of best practice irrigation, using recycled water, are outlined in Chapter 7 'Best practice using recycled water'.

Surface water/stormwater harvesting and storage

Characteristics

The general nature of these systems is that relatively large areas of land are required for storage and treatment. Typically, the process involves diverting runoff to a collection point, physical filtration, biological treatment (vegetation and soil) and storage. The basic elements of the total system may consist of a sedimentation basin, a bioretention basin (provides flow moderation, nutrient uptake and pollution removal) and a storage pond or lake for the treated water.

A characteristic of natural treatment systems is that the effectiveness of the treatment is strongly influenced by the surface area of the basins, storage volumes, vegetation employed and local site conditions. The sizing of basins according to inflow rates, retention requirements, water properties and amounts is a critical design issue.

Stormwater supplies are generally characterised as being of moderate quality and available in relatively large volumes. The quality is strongly influenced by the nature and properties of the catchment and by the pattern of particular rainfall events.

Before use as irrigation supply water, stormwater requires some form of treatment. Removal of sediment is a key consideration when considering stormwater as a supply for irrigation, as is the provision of space for water treatment, transfer and storage.

Stormwater catchment yield

The water quantity available from stormwater is dependent on the catchment area, catchment runoff coefficient and rainfall pattern. In some circumstances, existing stormwater mains can be intercepted and used to provide supply for open space irrigation. Analysis of flows in these pipelines is required to gain an understanding of the contribution that could be made to the irrigation of a specific site.

Impervious areas, such as roads and car parks, provide a suitable catchments for the harvesting of stormwater (Plate 3.1).

The volume of water that can be harvested from a catchment is dependent on the:

- catchment area
- catchment topography
- catchment surface properties
- soil moisture level
- rainfall intensity and duration.

An estimate can be made of the potential volume of water that can be harvested using the runoff coefficient (ROC) (Table 3.7). The following expression is used:

$$V_{SW} = \text{Catchment/harvest area } (m^2) \times ROC \times P\ (mm)$$

where V_{sw} – estimated volume of water harvested over period; ROC – runoff coefficient; and P – rainfall for period (e.g. month)

Example – Stormwater harvesting

Catchment area: 1.5 ha
Catchment properties: 30% landscaped gardens (4500 m^2), 70% roads and car parks (10 500 m^2)
Location: Canberra
Rainfall: 60 mm (January)
Runoff coefficients (ROC): Landscape areas 0.2, Paved areas: 0.7
V_{sw1} = (4500 m^2 × 0.2 × 60) = 54 000 L
V_{sw2} = (10 500 m^2 × 0.7 × 60 mm = 441 000 L
Total volume (V_c) = 495 000 L = 495 kL

The estimation can be carried out on a monthly basis to obtain an annual estimate.

The proportion of this water that will be used in irrigation will depend upon the amount of storage available, treatment processes, extent of losses due to seepage and evaporation. Stormwater systems require detailed analysis of the hydrology to determine an effective design solution (Figure 3.5).

Table 3.7. Runoff coefficients for selected surfaces

Catchment category	Surface type	Runoff coefficient (ROC)
Building roof	Tiles	0.7–0.9
	Metal, corrugated	0.8–0.9
Paved areas	Concrete	0.6–0.8
	Bitumen	0.6–0.8
	Bricks, pavers	0.5–0.6
Ground/soil	Bare ground (flat)	0.0–0.30
	Bare ground (sloping)	0.2–0.3

Note: These values are a guide only. Site specific information is required to determine runoff volumes more accurately.

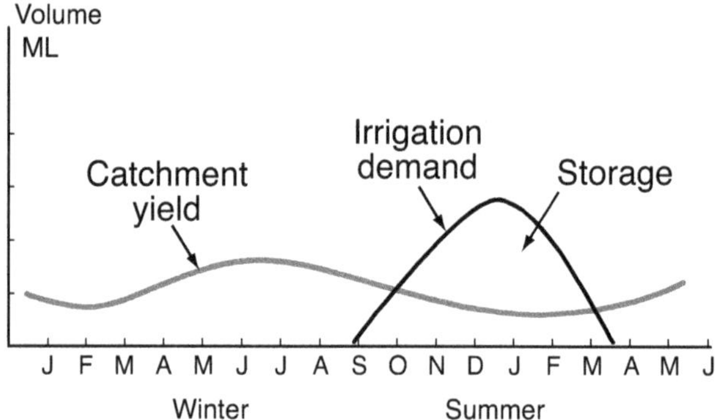

Figure 3.5. Stormwater yield requires significant storage to meet irrigation demand

The evaluation of the amount of water that can be harvested from urban catchments needs to include not only rainfall totals but also the pattern of rainfall. The intensity, duration and frequency all influence surface runoff quantity and quality. Design tools are available to aid in the thorough investigation of the potential yields for specific locations around Australia. An essential reference is *Australian Rainfall and Runoff*, published by the Institute of Engineers (1997). Another tool, MUSIC– developed by eWater Monash University, allows stormwater storage and treatment systems to be designed. The CRC eWater Toolkit can be found at www.toolkit.net.au.

The design of natural storage and treatment systems, including planning, technical aspects and implementation advice, is outlined in *Stormwater Biofiltration Systems, Adoption Guidelines* (FAWB 2009). There are numerous WSUD design resources available from capital city water authorities, including Melbourne Water and Sydney Water.

Stormwater quality

The quality of stormwater is highly variable and is dependent on the characteristics of the catchment. Table 3.8. provides a guide to water quality of stormwater. The potential impact of the pollutants, that may be found in stormwater, are detailed in Cross and Spencer (2009).

Table 3.8. Guide to water quality of stormwater

Parameter	Range/comments
Salinity – EC – TDS	Variable – generally suitable
pH	Acid/basic – generally suitable
Turbidity	Potentially high
Total nitrogen	Potentially high
Total phosphorus	Potentially high
Sodium adsorption ratio	Variable – generally suitable
Total suspended solids	Potentially very high. Large (gross pollutants) and small particles
Pollutants – metals	Zinc, copper, lead, cadmium- potentially high
Pollutants – organics	Petroleum hydrocarbons, herbicides, and pesticides potentially significant
Pollutants – microbes	Viruses, bacteria and protozoa including *E. coli*
Pollutants – toxic elements	Various toxins potentially significant

Note: This is a general guide only and each stormwater harvesting situation will be characterised by the properties of the catchment.

Figure 3.6. Bioretention filter (foreground) as part of stormwater treatment train for golf course irrigation (Royal Melbourne Golf Club)

There are numerous potential sources of pollutants in urban catchments: sediments, oils, heavy metals, microorganisms, pesticides and nutrients.

Water treatment options

Prior to planning and designing treatment processes, it is important to recognise the risks and understand the properties and characteristics of the various systems so that an effective and viable solution is determined for the site. Reference should be made to *The Australian Guidelines for Water Recycling: Managing Health and Environmental Risks (Phase 2) Stormwater Harvesting and Reuse* (NRMMC, EPHC and NHMRC 2009).

There are numerous water treatment processes to consider as part of a stormwater utilisation project. Target reductions for water quality improvement are required. Reducing suspended solids by 80% and nutrients, such as total nitrogen and phosphorus, by 45%, are typical.

The removal of large suspended sediment from stormwater is relatively simple. This can be done using traps, including gross pollution traps (GPT) and sediment ponds. The removal of finer sediment and treatment of the water to ensure appropriate properties is more challenging. The treatment process adopted is strongly influenced by the particle size removed (Table 3.9). Natural systems, such as wetlands, and mechanical systems are both used. Natural systems require considerable space and take some time to treat the water. Mechanical systems are space efficient (small space footprint), costly, provide processed water rapidly and usually have a significant greenhouse gas footprint, because they use electrical power (Figure 3.6).

The adoption of a natural system approach is generally best suited to situations where there is ample open space available for storage and treatment and the site is in close proximity to the area of use. Mechanical treatment systems require some space for storage of both the supply stormwater and the treated water, but much less overall than a natural system because the footprint of the mechanical treatment units is very small.

On-site tank storage

Both rainwater and stormwater systems usually require on-site storage. Provision for tanks, in terms of ground space, needs to be considered. The options are above-ground or in-ground storages. In-ground storage has the advantage of more readily accommodating large volumes, but the cost is significantly higher than above-ground storage (Figure 3.7).

Table 3.9. Stormwater treatment measures and particle size grading

Particle size range	Particle size grading	Guide to treatment measures
500 µm	Gross solids	Gross pollution traps (GTP)
125–500 µm	Coarse to medium sized particles	Sedimentation basins Grass swales and filter strips
10–125 µm	Fine particulates	Surface flow wetlands
0.45–10 µm	Very fine/colloidal particulates	Infiltration systems Subsurface flow wetlands
<0.45 µm	Dissolved particles	Subsurface flow wetlands

Source: adapted from MUSIC (2010)

The sizing of a stormwater storage for irrigation involves achieving a balance between the increasing size and cost of the storage and the increased reliability of supply (Figure 3.8). The expected reliability of the storage takes into account the irrigation demand pattern and the various climate and rainfall scenarios. The expected reliability of supply is partly dependent on the consequences if the storage fails. Tools such as MUSIC can be used to analyse stormwater storages.

The storage for irrigation involves large volumes. For example, to store a 4-week supply for a 1.5 ha (15 000 m^2) sports ground may require 900 000 L. A tank with a diameter of 8.0 m and a height of 2.0 m holds approximately 100 000 L. Nine of these would be required to hold this volume of water, so the footprint area required and cost would be significant.

Large underground storages, sometimes constructed using modular crates that allow flexibility in shape, are popular for stormwater applications (Plate 3.2). These storages may be 0.5 to 2.0 ML. Typically they are 1.5 to 2.0 m deep.

Figure 3.7. Modular concrete tanks provide large long-life underground storages (Source: Humes Water Solutions)

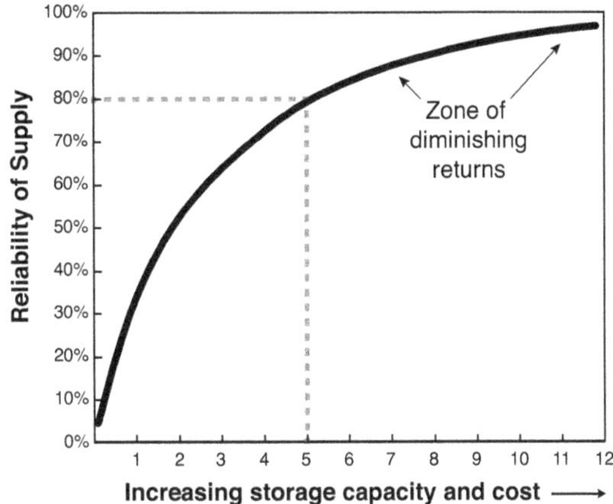

Figure 3.8. Increasing storage size increases reliability however the cost increase accelerates

Ponds and lakes

Characteristics

Large water bodies are a highly valuable asset in terms of potential water source and also as an environmental asset. The management of these water bodies, both in terms of water volumes and water quality, is becoming increasingly demanding, due to reduced catchment yields, changing climate (higher temperatures and evaporation) and increased awareness of the need to maintain the water body in a healthy condition.

The benefits of using water bodies include:

- potentially large volumes of water can be stored and made available
- optimum system design flow rate can be used (no flow constraints)
- they add to aesthetic value of the landscape.

The limitations of using water bodies include:

- a large area must be dedicated to water supply and storage
- natural water treatment systems require ongoing maintenance and management
- the water may contain iron that precipitates and causes contamination and blockage of irrigation systems
- the water quality may vary
- a risk of storage contamination with algae
- significant losses of water due to evaporation and, in some soils, due to seepage losses (unless impermeable liners are installed).

Management of open water storages

The maintenance of a water storage so that water quality is maintained and the formation of algae is prevented should be one of the aims of water management aims of the organisation (Plate 3.3). Algae affect irrigation system operation, aquatic life and storage amenity and aesthetics. Some blue-green algae, including Microcystis, are toxic to humans and also a risk to animals and fauna (Yiasoumi *et al.* 2005). Prevention of the problem is a much preferred option than treatment once the algae has formed. Both the design and management of the storage can reduce the risk of algae formation.

The storage conditions that favour algae formation are: medium to warm temperature (18°C to 30°C); nutrient availability; oxygen deficiency; sunlight and a stratified water column

Some design strategies to reduce the formation of algae include:

- The storage design should be deeper rather than shallower. This reduces the exposed surface area to stored water ratio and also assists in keeping storage temperatures lower.
- The creation of stagnant water should be prevented by encouraging water flow and mixing within the storage. This can be facilitated through mechanical mixing and aeration equipment. Encouragement of the movement and mixing of water, through recirculation, is beneficial in shallow (less than 2 m deep) storages.

Some management strategies include:

- Fertiliser rich runoff into the storage should be minimised or eliminated. Fertiliser composition, application rates, application techniques and irrigation practices should be carefully managed.
- Low storage levels at the time of high risk (summer) should be avoided (which is obviously difficult if water supplies are low).

Rivers and streams

Characteristics

Extracting water from flowing water bodies requires consideration of a broader range of factors because the source is an integral part of a much larger catchment. The health of the waterway and the protection of the local environment are key considerations. Approvals for extraction and overall water management that may have an impact on the waterway are the responsibility of a local catchment or water authority.

The benefits of extracting from rivers and streams include:

- large volume available
- water quality is generally reasonable
- pumps can be sized to achieve optimum delivery
- the water course acts as the storage.

The limitations include:

- suspended matter may cause blockages, if not filtered
- the water supply is potentially variable and water security uncertain
- flood events may damage the intake and pumping equipment
- access may be restricted or prohibited at times
- a license is generally required, regardless of availability of water.

Groundwater

Characteristics

The use of groundwater for urban irrigation has become increasingly popular in recent times as other sources, particularly potable, have become limited. Although some amount of groundwater is present at most locations, at some depth, the two key considerations are the quality of the water and the sustainability of extraction. It is important to recognise that groundwater is ultimately dependent on rainfall, whether in the short, medium or long term.

The regulation of groundwater is expected to increase in the future, because the protection of the resource is essential and, at many locations, the resource is threatened.

The benefits of using groundwater include:

- the access to water access is local (at the site)
- the water quality can be very good in some locations

- the extraction license fees are relatively low.

The limitations of using groundwater include:

- the water quality can be an issue, particularly if sustainable extraction rate is exceeded
- high salinity water is common and water quality can deteriorate over time
- the water may contain sand particles
- the water may contain iron with risks of precipitation and staining
- the discharge from a bore may be limited and this may require storage and re-pumping for some irrigation systems
- the position of the best water availability may not be convenient for the location of intended use
- the water may be deep below ground, which increases installation and pumping costs
- yields may decrease because of falling groundwater levels.

Rainwater harvesting

The harvesting of rainfall from the roofs of buildings should be considered as part of an overall water management strategy for a site. The amount of water that can be collected, stored and used is often relatively low compared with the amounts commonly required for irrigation. This means that it is important to consider the purpose for which rainwater is to be used. Rainwater is often better suited to be used for regular, non-seasonal demand, including washing of machinery, cleaning and plant protection spray solutions. The use of rainwater to flush toilets is another application, where there is a better balance between the water that can be collected throughout the year and the pattern of usage throughout the year. Under this scenario, relatively small storages are well used.

The roof area of a building is efficient in terms of harvesting of rainfall. It is estimated that in excess of 90% of the incident rain can be collected as runoff. The total area of roof is clearly determined by the size of the building.

Details on the amount of water that can be harvested from roof areas and the plumbing requirements to ensure reliable and safe operation are outlined in the *Rainwater Tank Design and Installation Handbook* (National Water Commission 2008).

As an example, if a building has a roof area of 60 m^2 and the rainfall in a month is 50 mm, then the volume of water that can be collected is:

$$\begin{aligned}\text{Harvested rainwater} &= \text{Rainfall} \times \text{Roof area} \times \text{Collection efficiency}\\ &= 50\ \text{mm} \times 60\ \text{m}^2 \times 0.90 = 2700\ \text{L}\end{aligned}$$

The amount of water that can be effectively harvested, over the period of a year, depends on both the pattern of usage, the pattern of rainfall and the size of the storage tank.

Greywater

Greywater as a source

Greywater refers to the wastewater that is available from laundries, showers, wash basins and the kitchen. The amount of greywater that is available is largely dependent on the number of occupants and the water pattern behaviour of the occupants. A key advantage of greywater, as an alternative water source to potable water, is the reliability of supply. Some water is available every day and the expected volumes can be predicted, with some reliability. This is not the case with rainwater.

Table 3.10. Typical constituents of greywater

Parameter	Unit	Greywater median values
pH	pH	7.9
Electrical conductivity (EC)	dS/m	1.3
Total dissolved salts (TDS)	mg/L	675
Sodium (Na)	mg/L	181
Chloride (Cl)	mg/L	135
Ammonium (NH_4)	mg/L	8.4
Total phosphorus (P_{tot})	mg/L	5.9
Calcium (Ca)	mg/L	35
Magnesium (Mg)	mg/L	19
Total nitrogen (N_{tot})	mg/L	15.2
Sodium adsorption ratio (SAR)	$(mmolc/L)^{0.5}$	6

Source: Stevens *et al.* (2008)

The quality of greywater is determined by the products used inside the dwelling and the water use practices carried out and so it is highly variable. There are risks associated with non-potable water supplies, including greywater, to be used for irrigation, where the quality of the water is much lower than a rainwater source.

Properties of greywater

The following are examples of the possible contents of greywater:

- bathroom water, including shower, bath, spa and wash basins – hair, soaps, shampoos, tooth paste, lint, body fat, oils and lotions, and cleaning products (and may also contain significant amounts of faecal matter)
- kitchen water – food solids, cooking oils, grease, detergents, and cleaning products in relatively large concentrations
- laundry water – soap, grease, lint, dirt and faecal matter (e.g. nappy waste).

Although kitchen wastewater is technically greywater, due to lack of suitability for direct use, it is usually not included in the greywater use category.

In terms of water quality and suitability for irrigation, greywater will commonly exceed ideal quality parameter levels or ranges, including pH, EC/TDS, TSS, sodium and phosphorus (Table 3.10).

Hazards and risks

The sources and potential hazards of greywater include:

- shower and hand washing basins – pathogens, salts, chemicals, water volume, physical contaminants
- washing machine – pathogens, nutrients, salts, chemicals, physical contaminants
- washing detergents – nutrients (in particular phosphorus) and salts
- cleaning products – toxic chemicals and physical contaminants
- bleaches – toxic chemicals.

The nature of greywater is that it potentially is a risk, in terms of human health, environmental health and the effective functioning of irrigation and water treatment equipment.

It is important to recognise the horticultural risks, such as plants that as sensitive to high salt levels and phosphorus. The circumstances of the use including method of application, such as wetting of foliage, need to be considered. Lists of plants sensitive to phosphorus (e.g. banksia,

Table 3.11. Greywater volumes estimated to be generated from a single person in Australia

	Volume used		
Wastewater source	**L/person/day**	**L/person/week**	**L/person/year**
Shower	56	392	20 384
Hand basin	6	42	2184
Laundry tap	7	49	2548
Washing machine	27	189	9828
Total – greywater	96	672	34 944

Note: Water available does not include water from kitchen including dishwasher
Source: adapted from RMIT University (2008)

grevillea and hakea) are presented in Stevens *et al.* (2008). High pH, which affects nutrient availability for plants, also needs to be considered.

Greywater systems

Greywater systems can be grouped according to two categories: treated systems and untreated systems. The type of greywater system that can be used to divert and use greywater depends on local authority regulations.

Broadly the techniques are to use a bucket to apply greywater directly (e.g. to garden) without treatment or to use a diversion type system. The following are the types of treatment used in various diversion type systems:

- primary treatment of screening and sedimentation – water distributed using a subsoil irrigation system
- secondary treatment including removal of solids, oils and grease and a degree of disinfection to reduce pathogens – surface and subsoil irrigation can be used
- tertiary treatment to Class A standard, where there are no restrictions on irrigation use.

Yield from greywater

Table 3.11 provides a guide to the yield that may be expected from a greywater system. These estimates will vary according to the type of appliances used and water use behaviour in the dwelling.

Greywater as an irrigation source

The regular supply of greywater means that it is well suited to irrigation in terms of maintaining irrigated areas. The main issue is the quality of the water and its management.

It is readily used at a residential property level, by using a bucket or temporary storage tank with pipe distributor, for example. In these cases, it is wise to modify the quality of the water as much as possible through the selection of environmentally friendly washing products (refer to the Landfax website: www.lanfaxlabs.com.au).

It is often a regulatory requirement that greywater can only be stored for 24 hours and then it must be disposed of, for example, to a sewer.

The soil on which the greywater is intended to be applied should allow good infiltration and have good water-holding properties. Soils that allow the water solution to rapidly percolate through are not well suited as the solution will readily move to lower soil layers and potentially contaminate groundwater.

It is important to only apply the volumes of greywater that will be taken up by the plants. Over application is likely to result in greywater polluting other areas, through deep percolation or surface runoff. Rainwater or potable water can be used to flush areas that have been irrigated using untreated greywater.

A significant constraint, in terms of supply, is the relatively small volumes that are available, compared with the volumes required for open space areas. Typically, only small areas can be maintained using greywater. For example, a household of three people could support a lawn area of approximately 40 m^2 if only shower greywater is used and the gross weekly application is 30 mm.

Other water sources

Industrial water

Many industrial processes require large amounts of water. In most cases, potable water is used. However, the quality of the discharged water is highly variable, in some cases, it can be toxic for landscape applications. Treatment processes are now available to modify the water properties to within acceptable limits. Technically it is possible to treat any quality water: it is the cost, often the capital cost, of treatment that determines the viability for a particular situation. The management of the waste from the treatment process is also an important consideration.

Sewer mining

Reclaimed water can be sourced from local sewer mains. This water is sometimes referred to as blackwater, because it contains sewage. This water would normally be transported to a water treatment plant, often located on the boundary or perimeter of the town or city. Extraction, treated and use at a locality within the urban area is a logical strategy, in terms of urban water management.

The local sewer main is tapped into and a local treatment plant installed to process the sewage. The treatment technology is generally some form of membrane system. Contaminated water can be treated to the point where the water quality is similar to potable water.

The following are the key characteristics of sewer mining systems:

- a suitable sewer main needs to be identified
- high cost is involved
- site storage is required
- high energy processes are used (which has greenhouse gas implications for both pumping and pressurised water treatment processes)
- after gaining access to the water source, it is necessary to obtain formal protection of the supply from the source.

The Pennant Hills Golf Club in Sydney has successfully converted from being strongly dependent on potable to now using water sourced from a sewer mining plant. The volume of water produced is 100 ML/year and the quality of the water is Class A (for further information go to the Sydney Water website: www.sydneywater.com.au and the AGCSATech website: www.agcsa.com.au).

Filtration and water treatment systems

Main water treatment processes

An understanding of the properties of water required for a particular irrigation situation is fundamental to good water management. These properties can be reasonably well defined in terms of the chemical, physical and biological requirements. The quality of the available water source may not meet the requirements in each of these categories. It is therefore necessary to modify the properties to achieve compatibility with the irrigation system and the required water quality of the site.

The main processes involved in modifying water are filtration and water treatment. Filtration is a process in which particles suspended in the water are removed by some form of

Table 3.12. Guide to filter mesh number and opening size equivalents

Screen mesh number	Opening size (microns)
10	2000 (2 mm)
20	850
40	425
100	150
200	75
350	45

physical barrier. The level of filtration ranges from coarse screens, which may trap plastic bottles, to membranes that can remove particles only a fraction of a micron in size. Water treatment includes biological, chemical, radiation and heat-treatment processes. Generally it involves targeting specific properties of the water. Water treatment is potentially quite complex, because several different processes may occur simultaneously, such as chemical reactions and precipitation. In order to achieve the desired water quality, knowledge of the condition of the water and how particular treatment processes function is required. Monitoring of the properties of the water is a key part of any water-treatment process.

The broad water treatment category includes disinfection. This involves processes designed to minimise the risk of disease to humans, plants and animals.

Filtration

Prevention of blockage of irrigation equipment is essential for reliable and effective performance. This is just one of the important roles of filtration. There are many techniques employed to remove particles from water. The suspended materials include silt and sand, organic matter, precipitates and bacteria, and also contaminants, such as oil. The trend towards microirrigation systems in recent years has meant that an increased emphasis is placed on the removal of fine suspended particles from irrigation water.

Filter screens and mesh number

The key performance feature of a filter is the size of the particle that it traps or blocks. This is largely determined by the aperture, or opening size, of the filtering barrier. The level of filtering that is achieved is described in terms of the mesh number: the number of openings per linear inch of filtering material. It is not an absolutely precise measurement, because the actual opening size depends on the thickness of the material used to construct the barrier or mesh. However, Table 3.12 is a good guide to the opening size associated with various mesh numbers.

The grade of filter (mesh number) and equivalent soil particle size is shown in Table 3.13.

Microirrigation blockages and water quality

Table 3.14 nominates the water quality parameters that should be monitored to reduce the risk of blockage and classifies the parameter values into three levels of concern.

Types of filtration equipment

The main types of filters used in irrigation are screen, disc, media, membrane and hydrocyclone or centrifugal filters. The first four are barrier-type filters and the last one uses the weight of the suspended particle and centrifugal force to separate these from the water.

In any filtration process, it is advisable to prepare the water for processing. The removal of particles from the solution places a sediment load on the filtering system. Regular cleaning and maintenance is required. Before water is filtered, it should be screened for floating debris and

Table 3.13. Classification of soils by particle size, with corresponding screen mesh numbers

Soil texture	Particle size range (mm)	Particle size range (microns)	Filter screen mesh number
Very coarse sand	1.00–2.00	1000–2000	18–10
Coarse sand	0.50–1.00	500–1000	35–18
Medium sand	0.25–0.50	250–500	60–35
Fine sand	0.10–0.25	100–250	160–60
Very fine sand	0.05–0.10	50–100	270–160
Silt	0.002–0.05	2–50	400–270
Clay	<0.002	<2	–

Note: Individual bacteria and viruses are smaller than clay particles
Source: Benham and Blake (2002)

large particles, and pass through a settling pond or screening process. Slowing the water down to the point where it is nearly stationary is very effective in allowing heavy particles, to settle out of the water. The size of the particle that can be carried in suspension increases as the velocity of water increases.

In some situations, it is recommended that a number of filters be used in series, so that the degree or fineness of filtration increases progressively through the filtration process. This reduces the load and maintenance of fine particle filters.

Filtration systems

Screen filters

The filtration principles of screen filters are very simple in that suspended particles are trapped on a perforated, flat or relatively flat surface. The minimum size of opening determines the size of particle trapped by the screen. The screen is typically arranged as a cylinder to provide a larger exposed surface area for the filter. The materials used in screen filters includes metals and plastics. Stainless steel is commonly used as the screen surface.

Screens are basically two-dimensional cleaning systems and have a relatively low surface area per unit of flow rate. For this reason, they have limited capacity to remove significant amounts of suspended material without the need for cleaning.

Screens are well suited to the removal of sand and grit and provide very effective filtration for microirrigation equipment. They are not suited to water containing organic (e.g. algae and small aquatic life) material, which can cause the filter to block very quickly.

Table 3.14. Water quality guidelines for clogging potential in subsurface drip irrigation systems

Parameter	Symbol/ abbreviation	Unit	Level of concern		
			Low	Moderate	High
Acidity/alkalinity	pH		<7.0	7–8	>8.0
Bicarbonate	HCO_3	meq/L	<2.0	>2.0	>2.0
Iron	Fe	mg/L	<0.2	0.2–1.5	>1.5
Manganese	Mn	mg/L	<0.1	0.1–1.5	>1.5
Hydrogen sulphide	H_2S	g/L	<0.2	0.2–2.0	>2.0
Total dissolved solids	TDS	mg/L	<500	500–2000	>2000
Suspended solids	SS	mg/L	<50	50–100	>100
Bacteria Count		No./L	<10 000	10 000–50 000	>50 000

Source: Burt and Styles (1994)

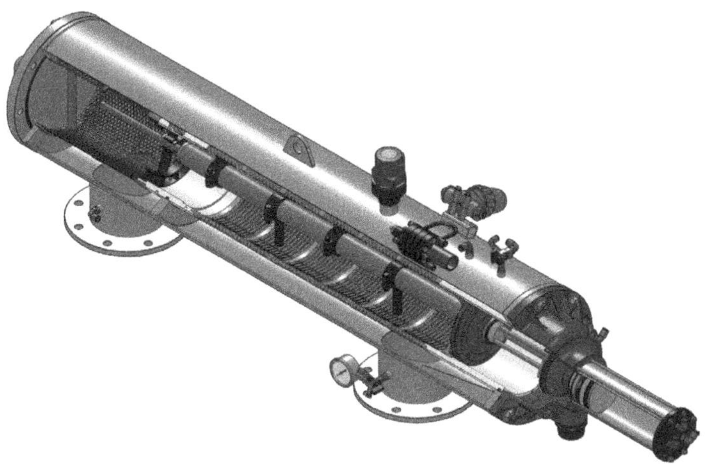

Figure 3.9. Sectionalised view of automatic self-cleaning screen filter (Source: Triangle Waterquip Pty Ltd)

There is a significant pressure loss when water flows through a screen filter. The rate of loss increases as the restriction, as a result of trapped particles, increases. Care needs to be taken to ensure this pressure loss does not become excessive.

There are numerous types of designs available to facilitate the cleaning of screen filters (Figure 3.9). They range from manual to fully automatic. The decision to clean is usually based on the pressure differential recommended by the manufacturer or when a certain time period has elapsed.

Disc filters

In this case, filtration is achieved through suspended particles trapped within narrow passageways formed between consecutive discs stacked on top of each other (Figure 3.10). The size and

Figure 3.10. Disc filter assembly providing protection for a drip system from suspended solids and incorporating chemical protection against root intrusion

configuration of the water passageways is determined by grooves in the surface of the discs. The discs are made of metal and plastic materials. The potential surface area available for trapping particles is much greater than with screen type filters.

The filter discs need to be loosened to allow trapped material to be removed. This is often done manually, but there are designs that allow automatic backflushing of this type of filter. Under some circumstances, cleaning is facilitated through the injection of acid into the irrigation water.

Disc filters are, in effect, a combination of screen and media type filters in the method of operation. They are well suited to microirrigation systems, where there are often requirements for relatively low flow rates and the water source contains both organic and inorganic material.

Media filters

The main difference between media filters and screen filters is that the media filtration is a three-dimensional process (Figure 3.11). Suspended material is trapped as the particle-laden solution moves through millions of sand particles. This method is well suited to both inorganic particles (e.g. sand and grit) and organic (e.g. algae). The technique is also well suited to remove large amounts of particles. Media filters can handle large amounts of suspended particles compared with screen filters.

The selection of the media sand is critical to the effective operation of media filters. The media sharp (e.g. sharp edges), size and size distribution are all critical.

Media filters need to be cleaned regularly to remove accumulated contaminants. Backflushing of the media (reversing of the flow through the media) is used to loosen the trapped particles and discharge to waste. The backflushing is undertaken based either on pressure differential across the media or at a set time.

Figure 3.11. Media filters provide three-dimensional filtering and effectively remove very fine particles for microirrigation systems (Source: Netafim Australia Pty Ltd)

Media filters need to be operated within specified flow rates and velocities and have specific requirements in terms of flow and pressure for backwashing. There needs to be a balance between effective cleaning and prevention of dislodging media and discharging some of the media with the backflush water.

Slow flow sand filters

These filters act as both physical separation devices and water treatment systems. The treatment system involves a relatively deep (in the vicinity of 800 mm to 1000 mm) layer of sand. Water moves slowly down through the sand layer, and pathogens, such as bacteria and fungi, are killed by microorganisms that have become established in the sand layer. It takes some time (up to several weeks) for the beneficial microorganisms to become established.

The sand is housed within a tank or similar enclosure. This is a natural treatment system, yet it is concentrated in terms of space. Slow sand filtration has relatively low treatment flow rates, so significant storage of treated water is required.

Membrane filtration

Physical separation of very small particles from water solution is possible using membrane technology. Filtration to fractions of a micron is possible. Table 3.15 outlines the various membrane technologies, pore size ranges and substances blocked.

Some of the characteristics of membrane filtration are the need to maintain the membrane in good condition: the membranes are relatively expensive and there is a significant energy cost associated with forcing the water through the small openings of the membrane.

Hydrocyclone or centrifugal sand separator

The water inlet to a sand separator is designed to create a high speed rotational flow. Due to centrifugal forces acting on the particles (sand) within the water solution, these particles more towards the outer edge of the rotating water. The separated particles are collected at the sides of the unit and the good quality water is discharged from the centre of the water flow.

These units are well suited to water sources with high coarse sand content, such as some groundwater supplies and stormwater catchment water sources. The operating principle is simple and removal of coarse particle size is readily achieved. Fine particles are not removed. Oil can also be removed using this principle.

Biofiltration

Natural water treatment systems provide not only direct benefits in terms of modifying the properties of the water, but often enhance the amenity of the area or landscape. They are reasonably effective at removing pollutants that arise from urban catchments, such as car parks, parkland and roads. Natural treatment systems involve a number of different chemical, biological and physical processes. In addition to multiple chemical reactions, there is often settling

Table 3.15. Membrane filtration terminology and contaminant blocking

Filtration type	Pore size range	Contaminant or pollutant blocked/removed
Microfiltration	0.1 to 1.0 microns	Parasites, *Giardia* cysts, suspended solids, some bacteria
Ultrafiltration	0.01 to 0.1 microns	Some viruses, some humic substances, colloids
Nanofiltration	0.001 to 0.01 microns	All viruses, some dissolved salts
Reverse osmosis	<0.001 microns	All dissolved salts, natural organic substances, pesticides.

of particles (sedimentation), filtration through sand beds and or vegetation and uptake on nutrients by appropriately selected aquatic plants.

The effectiveness of these systems is strongly dependent on the design, operation and management of the system. Provision needs to be made for regular maintenance to ensure ongoing effectiveness.

Selecting a filtration system

The size of the opening in the filtration process is a key value when selecting a system. Filtering guidelines are presented; however, manufacturer's recommendations should be followed:

- drip emitter – filter to one-fifth to one-tenth of the diameter of the emitter's smallest opening
- microspray, microjets and microsprinklers – filter to one-third to one-sixth of the emitter's smallest opening.

The following factors affect the selection of a filter:

- the pressure or head loss through the filter at design flow should typically be in range of 30 to 40 kPa
- the smallest size of particle to be filtered
- the sediment load should be readily managed by the filter
- the method of backflushing – manual, semi-automatic or automatic
- backflushing operational guidelines (e.g. backflushing that occurs at pre-set times or when the pressure differential exceeds a pre-set limit, such as 70 kPa, is often recommended)
- the management of filter waste
- the pressure rating of the filter housing and assembly.

Removal of total dissolved solids – reverse osmosis (RO)

The only techniques available to reduce the salinity of water are removal of salts through reverse osmosis (RO) or blending the saline water with low salinity water, such as potable water, rainwater or stormwater (depending on its source), to achieve an acceptable level of salts (Figure 3.12).

Figure 3.12. Reverse osmosis system (horizontal tubes) used to reduce salinity of irrigation water (Royal Melbourne Golf Club)

The process of osmosis works on the basis of water in solutions of low (salt) concentration moving to higher concentration. For example, water moves from soil of low concentration into the plant root system, which is at a higher concentration. In the case of reverse osmosis, pressure is applied to the water so that higher concentrated water is forced through a membrane barrier that has very small apertures. The size of the membrane openings is such that the dissolved salts are prevented from passing through, but (near) pure water will pass through.

RO requires pressure in the range of 200 kPa to 1000 kPa, depending on treatment requirements and design of the system. There is considerable energy used in water treatment through RO. The flow rates are not high. Maintenance of the membrane is important to ensure reliable and effective functioning of the unit. It is a very effective and reliable technology.

A byproduct of a RO system is that high salt concentrate solution is produced and needs to be disposed in an appropriate manner: it is potentially damaging to the environment.

Chemical water treatment

There are a variety of reasons why it is desirable to change the properties of water, in addition to the removal of suspended matter and reducing the salinity of the water. Some of the problems that may be encountered are the hardness of water, formation of scale or iron deposits and discolouration. The addition of chemicals is used to modify the water properties. There are also a range of treatment processes involving magnetism and electric currents that are promoted.

Adjusting acidity and basicity (pH)

Depending on the source of water, the pH may be neutral, acid or basic. For various reasons, it is often necessary to change the pH so that it is within an acceptable range to meet the water quality needs of the irrigation.

Acids, including phosphoric, sulphuric and nitric, are used to lower pH (e.g. prevent formation of precipitates in water). Compounds, such as gypsum ($CaSO_4$) and soda ash (sodium carbonate) (Na_2CO_3), are used to increase pH (e.g. improve soil structure).

Chlorination

Chlorine is injected into the irrigation water for several reasons, including as a disinfection technique and cleaning of microirrigation systems. Small concentrations of chlorine are used to protect the quality of water storages.

Chemical injection – caution

The injection of concentrated chemicals into the irrigation system presents a number of potential risks. There are risks to the plants or crop, the soil, the equipment, the environment and to humans. All risks need to be considered and assessed.

Disinfection systems of irrigation water supplies

For many reasons, water used for irrigation must be disease free. Water is a very effective carrier of disease and some water conditions promote the development and growth of disease organisms. Water-borne diseases are a risk to plants, environment and humans. Two important diseases to horticulture are *Fusarium* and *Phytophthora*. Both of these are readily imported in irrigation water.

Disinfection techniques

The options available for disinfection include:

- exclusion or removal of disease organisms by filtration
- heat treatment – kills organisms
- radiation (ultraviolet)
- chemical and oxidisation treatment

Filtration

The removal of pathogens requires filters that are very fine: membranes must block particles larger than 5 microns. This technique is referred to as microfiltration.

Heat

A water temperature in the range of 90°C to 95°C is required for effective disinfection of water. An exposure time greater than 30 seconds is generally required. This is an effective technique, but the management of the hot water is not as convenient as some other techniques and is also very energy expensive.

Radiation (ultraviolet)

Very short wavelength radiation, in the ultraviolet (UV) range, is effective in destroying many disease organisms that are a risk to horticulture. These include *Fusarium* and *Phytophthora*.

The radiation of a UV disinfection system must be able to pass through the water, so it is important to ensure the clarity of water to be treated. The presence of suspended particles can also reduce effectiveness through the shielding of disease organisms by the suspended particle. Pre-treatment of water for a UV system usually includes filtration to achieve the required clarity standards.

UV is effective, relatively harmless and suited to relatively high flow rates. UV lamps have a limited life. It is important that the lamp condition and accumulated hours of use are monitored to maintain effectiveness in treatment.

Chemicals and oxidisation

There are a range of chemical treatments suited to the disinfection of water borne pathogens. These include: chlorination, chlorobromination, chlorine dioxide and ozone. Each chemical treatment has its own characteristics, properties and management issues. Human safety in the handling of these chemicals is a key consideration.

Filtration for pumping equipment

The effective functioning of the irrigation system is initially dependent upon the water meeting the required physical standards. These will vary according to the irrigation method and the nature of the site. Microirrigation systems have more demanding requirements than sprinkler systems.

Before considering the filtering requirements of the inlet to the system, the physical properties of the water source should be assessed. Systems that will use water sources that typically have high levels of suspended solids, such as rivers, streams, lakes, ponds, reuse storage and stormwater storages, should be designed and managed to minimise the potential problems.

Strategies such as the provision for settling the larger particles, aeration to minimise risk of algal growth or build up of organic matter, minimisation of erosion particles entering the storage and the trapping of large debris (organic and inorganic) entering the water storage should be considered.

The suction side of a pumping unit is the source of many more operational problems than the delivery side of the pump. Flow restrictions and loss of suction more readily occur on the suction because of the low pressure in the suction line. The pump inlet should be positioned in clear water. The floating of the suction line is recommended for storages where the water levels fluctuate (Figure 3.13). Large screen areas are required to minimise the risk of blockages. Self cleaning is important to prevent flow restrictions to the inlet of the pump.

Figure 3.13. A floating suction line ensures that reasonable quality intake water is drawn into the pump

Fertigation

The process of delivering nutrients through the irrigation water is referred to as fertigation. Other chemicals, such as water conditioners and wetting agents, are also applied through fertigation systems. Basically fertigation involves chemical injection to complement irrigation management.

A key feature of fertigation is the capacity to deliver to the soil and root zone precise volumes and concentrations of chemicals. To do this, the delivery system – that is, the irrigation system – needs very uniform application and precise control. High standards of irrigation system design, operation and management are required.

The availability of nutrients, frequently in soluble form, means that the fertiliser is used more effectively than if it were applied in granular form. However, soluble nutrients are mobile and can be readily leached if excess water is applied.

The management of the fertiliser used in fertigation systems requires knowledge of the soil properties, nutrient requirements of the plants and the properties of the irrigation water. Agronomists can provide advice on appropriate rates to meet the nutrient balance requirements of the site. Because fertigation is more effective in nutrient application and use, the process is also more water efficient. Less water is required to achieve the required turf and landscape performance standards.

Fertigation systems range from single injection pumps and venturis for the incorporation of one compound to complex systems. Multiple solutions can be mixed in concentrations to suit the particular nutrient needs of the plants at any point in time (Figure 3.14). Fertigation control involves monitoring of input solutions and the distribution solutions.

Figure 3.14. Fertigation system incorporating triple solution feeds to provide on-demand customised nutrient mixes

Due to the risks associated with chemicals in the irrigation water, there is a need for operators to be aware of regulations that apply to the use of fertigation systems. This applies both to the personnel involved in operating and maintaining the site and the public who may be exposed through use of the site.

In terms of incorporation into the irrigation system, the point of injection must be positioned to ensure thorough mixing, and the whole process must be carefully controlled to protect the site, the environment and users.

Chapter 4

Irrigation methods

Overview of methods

Water application options

The aim of an irrigation system is to deliver water to the root zone of plants. It is best management practice to deliver the water effectively and efficiently. The method of water delivery to plants or crops has evolved over thousands of years. Today there are many techniques available, which are refinements of earlier irrigation methods. A core characteristic of current irrigation systems is the ability to deliver water to the plant in a controlled way.

The delivery of water to the plant can be achieved either from: above ground and above the plants or crop – often referred to as 'overhead'; on the surface or from below the surface. The oldest methods of irrigation distribute water as a surface flow, which is sometimes referred to as surface or flood irrigation. The method is also used to water broad area field crops, row crops and orchard trees, where small channels or furrows adjacent to the crop deliver water (Figure 4.1). Only small amounts of energy, and low pressure requirements, are generally required with surface irrigation methods.

Figure 4.1. Surface irrigation has been used for hundreds of years on urban trees

Application of water from above plants is a relatively recent development in the history of irrigation. Sprinklers and sprays were developed for irrigation in the 1930s in the USA. Because the water stream needs to be broken up using pressure, this method became categorised as a form of 'pressurised' irrigation. The application of water through the air in the form of aerosol drops presents challenges in terms of uniformity of application, due to the pattern of distribution, and the potential influence of the atmospheric conditions (mainly wind).

There is now a wide range of low flow rate irrigation technology, referred to as microirrigation, that delivers water to the crop or to individual plants on the surface or below the plant canopy. There are now also various forms of subsurface microirrigation in use. Microirrigation uses pipework to convey water to each plant, rather than using overland surface flow or sprayed droplets. This increased control of water distribution has important potential advantages in terms of efficiency. Although it operates on relatively low pressure, microirrigation is also a form of 'pressurised' irrigation.

The delivery of water from below the plant has been practiced for hundreds of years. One example is horticultural tree crops in the Mediterranean area, which were flooded from below as a form of sub irrigation. Watering from below is now commonly used in the nursery industry. More recent examples include the use of sand profiles on turf areas that are watered from below using the perched water table principle. In this case, however, the irrigation water is generally initially applied overhead using sprinklers.

A relatively recent development for urban horticulture has been the use of drip systems buried within the plant root zone. This technique is referred to as subsurface drip irrigation (SDI). The technique has been used since the 1960s in vegetable row crops in the USA, and is now being increasingly applied to amenity areas.

In addition to choosing the method of delivery of water, the method of control is also a major consideration. An important development in irrigation has been the adoption of automatic operation of irrigation equipment, which is commonly time-based control, but other feedback inputs, such as weather station data and environmental sensors (rain and soil moisture), are now being used. Automation is another aid in achieving greater precision in the delivery of water to the plant.

Main irrigation methods

The main irrigation methods are:

- surface/flood and furrow irrigation
- sprinkler and spray
- microirrigation systems including drip
- travelling rain gun irrigators
- travelling/mobile sprinklers
- boom irrigators
- centre pivots and linear move irrigators.

All of these methods can be categorised as 'pressurised', except for surface and furrow irrigation. The required operating pressure is important from the aspect of energy cost involved in applying water. High operating pressures result in higher energy expended per unit of water applied.

In this book, drip irrigation (surface and subsurface), microsprays and minisprinklers will be discussed in some detail.

Pressure categories of irrigation methods

Pressurised irrigation systems can be divided into low (0–200 kPa, medium (200–500 kPa) and high (500 kPa+) pressure methods. Low pressure methods involve microirrigation (see above). Medium pressure systems include: sprinklers; permanent (fixed) and semi-permanent (moveable) systems; manual move/portables systems, travelling booms and mobile (travelling) sprinklers. High pressure systems include large capacity sprinklers and travelling rain guns – soft hose and hard hose types.

Surface (flood) irrigation

Description and principles

Surface irrigation is the most common method of irrigation in Australia: it represents approximately 65% of all irrigation. Surface irrigation techniques have the advantage of low energy requirements and low cost, and low water quality, in terms of suspended particles, can be used.

The key disadvantages of surface application are the lack of precision in the delivery of water to the plant root zone, the uneven application due to bay lengths and varying wetting times. Successful surface irrigation is not achievable without detailed knowledge of the site soil properties, and understanding of the relationship between soil types, field slopes and irrigation run times. If these factors are not carefully considered, there is a likelihood of over-watering and through drainage, which will promote nutrient leaching, rising watertables and root zone saturation. Achieving high efficiency and uniform growth throughout the irrigated area is very challenging.

Characteristics of surface irrigation – benefits and limitations

The following is a summary of the typical characteristics of the surface method of irrigation:

- It is difficult to achieve precision application of water: the depth is variable.
- Evaporation losses are generally high.
- Labour input can be high.
- It may lead to salinity problems.
- Efficiency typically in the range of 60% to 70% (but may be lower).
- It is difficult to automate (although recent technology developments are improving this situation).

The benefits of surface irrigation include:

- It is suited to large areas.
- The cost of installation is low.
- Only minimal energy is required in delivering water to the field.

The limitations of surface irrigation include:

- It is difficult to achieve precise delivery.
- Water distribution and uniformity are strongly dependent on the soil type.
- It requires relatively flat ground.
- It is difficult to automate.

Sprinklers and sprays

Water distribution characteristics

Water distribution from sprinklers is often compared to rainfall. There are similarities, but the sprinkler technique has some important differences. One consequence of delivering water in a radial manner from a sprinkler is that the precipitation rate (PR) decreases as the distance from the sprinkler head increases. There is a much greater area to be covered by the water falling from the stream at the extremities of the wetted circle than there is close to the sprinkler head. This leads to a large variation in application depths.

Achieving uniformity in precipitation is a major issue with sprinkler irrigation systems. The circular pattern of coverage makes it difficult to achieve uniform application because considerable overlap is required to achieve even watering of the total irrigated area. Another important difference between sprinklers and rainfall is the greater range in droplet sizes from sprinklers and also droplet trajectory from sprinklers can be at oblique angles to the ground compared with some rainfall events. Although the average precipitation rate from sprinklers can be relatively low for example 5 to 20 mm/h – the instantaneous rate can be high as the water stream passes over a particular point on the ground during a rotation. It is the instantaneous rate and droplet size that influences the amount of soil erosion and compaction that may occur.

The breaking up of a water stream into droplets is a very simple process. It is achieved by providing an adequate flow and pressure through an orifice or nozzle. Achieving an even application of water to an area, using droplets, is a much more challenging task (Plate 4.1). There are numerous factors that work against the achievement of this goal. Some of these are:

1. The greatest amount of water per unit area is applied closest to the nozzle position.
2. A range of droplet sizes is produced, some small and some large. Small ones, closest to the sprinkler head, are prone to loss due to drift and evaporation.
3. Water distribution is generally in a circle, which is not well suited to continuous even coverage of a regular (rectangular or square) shaped areas.
4. Operation and effectiveness of delivery is sensitive to atmospheric conditions, such as wind and evaporation.

The pattern of distribution of water from a sprinkler depends on the:

- nozzle size
- nozzle design
- number of nozzles
- angle of nozzles – stream trajectory
- internal flow characteristics within the head
- rotation of nozzle assembly – speed and uniformity
- operating conditions – pressure and flow characteristics
- atmospheric conditions – wind, evaporation.

Sprinklers and spray heads

It is important to differentiate between sprinkler and spray equipment. In a sprinkler (also called rotors) system the nozzle assembly rotates and the water stream is continuously moving. In a spray system, the nozzle assembly is fixed and water is continuously applied to the same area.

Characteristics of sprinklers (or rotors) – benefits and limitations

The following is a summary of the typical characteristics of sprinkler and spray systems:

- All of the target area is wetted.
- The size of wetted area is medium to large.
- The precipitation rates range from low to high.
- The application is sensitive to wind.
- Evaporation losses can be significant.
- They can be used for microclimate modification – frost protection or crop cooling.
- They require a medium to high pressure supply.
- They wet the foliage, so the risk of disease is higher.

The benefits of sprinkler and spray systems include:

- They are suited to a wide range of topography and soil types.
- They allow controlled application.
- They are potentially highly efficient.
- The operation of system can be readily observed.
- They are suited to automation.
- They are suited to fertigation.

- They can be used for range of environmental modification purposes, including plant cooling (e.g. syringing of turf) and frost protection.

The limitations of spray and spray systems include:

- They have relatively high installation costs.
- Their performance is sensitive to atmospheric conditions of evaporation and wind.
- The wetting of foliage with lower quality water can cause plant damage and increase the risk of disease.
- They have reasonably high energy costs.
- There is a risk of damage to equipment in public areas.

Types of sprinklers and sprays

Sprinklers and sprays can be categorised according to the:

- operating duty (flow and pressure)
- mounting style (overhead or pop-up)
- type of drive (sprinklers)
- nozzle design and number
- materials of construction.

There is virtually an endless combination of the above characteristics available in sprinkler and spray products. These products can be used in permanent and temporary systems and mobile and fixed systems.

Sprinkler drives

The methods of drive used to rotate the nozzle assemblies include the use of the internal energy of the water flowing through the sprinkler. In the past, turbines operating gear trains, pistons and stainless steel balls impacting on anvils have all been used to drive sprinklers.

Other drives use the energy of the water stream discharging from the nozzle to drive the rotation (Figures 4.2 and 4.3). The impact drive, operating on a spring-loaded oscillating arm (also known or hammer drive), was the first developed and is still popular today (Figure 4.4).

In today's irrigation market, there are two types of drive: impact drive and gear drive.

Sprinkler performance

The key performance parameters of sprinklers are the:

- wetted area (represented by wetted radius or wetted diameter)
- discharge/flow rate
- operating pressure
- distribution precipitation profile.

The design of the nozzle assembly is critical to the performance of sprinklers. As nozzle orifice size increases, the required flow rate increases and the length of the stream and distance of coverage increases. To achieve optimum break-up of the stream, higher pressures are required as the nozzle size increases. Sprinkler performance is the outcome of numerous interacting hydraulic and design features (Figure 4.5).

The precipitation distribution profile of sprinkler heads is influenced by many factors. Each sprinkler nozzle has a limited range of pressures over which it will produce its optimum distribution profiles. Low pressures result in a donut shaped profiles and excessively high pressures result in non uniform distribution with higher precipitation rates close to the sprinkler head.

Table 4.1 provides a guide to the optimum pressures for various size nozzles.

Nozzle assembly – nozzle number and design

A sprinkler nozzle should give good break-up of the water stream over a defined range of pressures. The stream should break up along the whole length of the stream when operating under optimum conditions. Both distance of coverage and break up are important. Much effort goes

Figure 4.2. Sprinkler head rotation via oscillating impact arm ('knocker sprinkler')

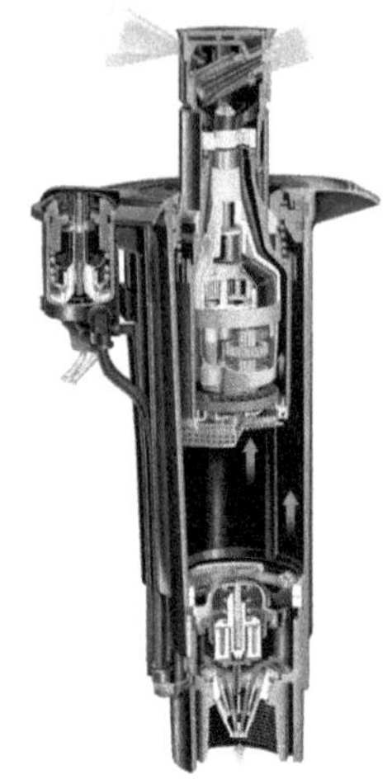

Figure 4.3. Water flow through turbine assembly provides continuous rotation of gear drive sprinklers (Source: Rain Bird Australia Pty Ltd)

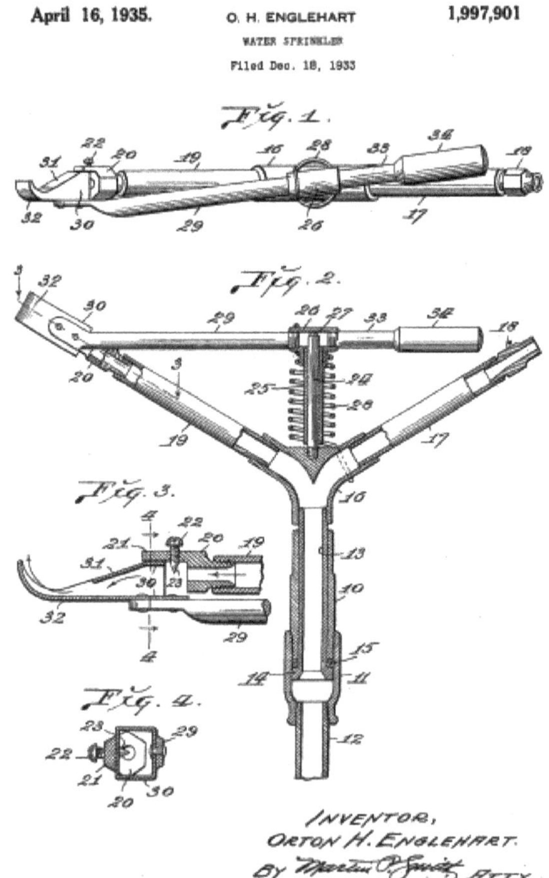

Figure 4.4. First sprinkler design patented in US, 1940 (Source: Rain Bird Australia Pty Ltd)

Table 4.1. Guide to operating pressures and flow rates for various nozzle sizes

Nozzle size (diameter of orifice)	Guide to optimum operating pressure (kPa)	Flow rate (L/min)	Wetted diameter (m)
3 mm	225	10	25
4 mm	275	17	30
5 mm	320	30	33
6 mm	360	36	35
7 mm	400	60	38
8 mm	415	80	40
9 mm	460	130	44
10 mm	510	160	50
12 mm	570	280	55

Note: This is a guide only. Refer to manufacturers' performance specifications for particular products.

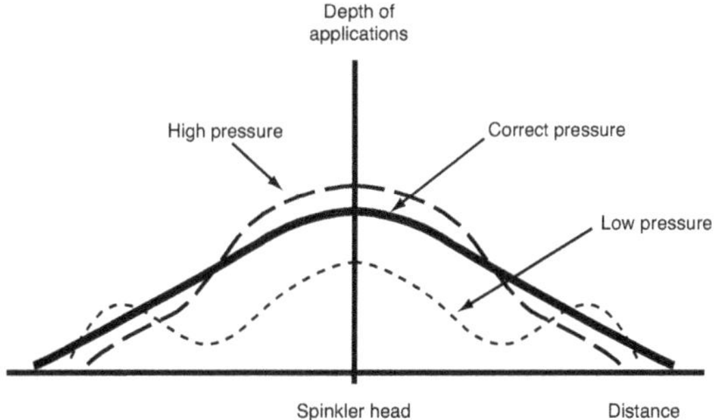

Figure 4.5. Sprinkler distribution profile variation with changing operating pressure

Figure 4.6. Multi-jet gear drive rotor designed for distance of coverage and a distribution profile that provides an overall uniform application

into the design of nozzles to achieve an optimum balance in the break-up properties of the stream (Figure 4.6).

Long tapering nozzles produce greater distance of coverage, but they may not possess the optimum droplet distribution that will allow high uniformity of application to be achieved. They are well suited to windy sites where distance of coverage is required.

Multiple nozzles are used on larger capacity sprinklers. The smaller nozzles cover the area close to the sprinkler head and larger nozzles produce the longer distances.

Nozzle design – nozzle angle

The optimum angle of trajectory of a water stream, to maximise distance, is in the range of 20 to 30 degrees. At low trajectory angles the distance is limited and at high angles the water stream is subjected to higher wind velocities and break up, so that coverage distance is limited.

The amount of distortion experienced by a sprinkler stream, as a result of wind, depends on the degree of break-up and the velocity of the wind. Naturally windy sites are a problem for any spray or sprinkler system because distribution patterns are readily distorted. Generally, high flow rates, lower pressures, larger droplets at lower trajectory angles are more suited to these situations.

There is generally a marked increase in wind speed with increasing elevation above the ground, and so the higher the water stream rises in the air, the greater is the exposure to higher wind speeds. For particularly windy sites, manufacturers offer a range of sprinkler with low trajectory angle nozzles (15 degrees or less) to minimise the effects of wind. The distance of coverage is, however, significantly reduced.

Sprinkler performance data

Manufacturer catalogues and websites provide performance specifications for particular products and models. The format for the presentation of this information varies, but it will generally appear somewhat similar to the data in Table 4.2.

The performance data show the increase in discharge rate as both nozzle size and operating pressure increase. There is also a marked increase in the precipitation rate. It increases 400% over the nozzle and pressure range. The precipitation rate increases rapidly with an increasing discharge rate, because the covered area only increases marginally. Each of the variables of nozzle size, operating pressure, layout pattern and spacing directly influence the

Table 4.2. Performance specifications for medium pressure sprinkler

Pop-up sprinkler (60 mm lift) performance						
Nozzle	Pressure kPa	Radius (m)	Flow (m^3/h)	Flow (L/min)	Precipitation (mm/h) ■	Precipitation (mm/h) ▲
1	344	8.8	0.16	2.7	4	5
2	344	9.1	0.20	3.4	5	6
3	344	9.4	0.27	4.5	6	7
4	344	10.4	0.36	6.1	7	8
5	413	11.6	0.45	7.6	7	8
6	413	11.6	0.61	10.2	9	11
7	413	12.2	0.77	14.0	11	13
8	413	12.8	0.89	17.4	13	15
9	413	13.7	1.25	20.8	13	15
10	413	14.3	1.86	26.8	17	19

Note: ■ square spray layout pattern
▲ triangular spray layout pattern

precipitation rate and the uniformity of the system. Selection of the right sprinkler nozzle combination requires consultation with irrigation system designers.

The pressure can be expressed in a number of different units including kPa, metres of water (m), pounds per square inch (psi), bars (atmospheric pressure) and centibars (cb). Pressure conversion tables are presented in Appendix 11.

The pressure nominated is usually the pressure at the base or inlet of the sprinkler. The pressure at the point of discharge at the nozzle orifice will be lower than the inlet pressure. The actual difference depends on the frictional characteristics of the internal flow pathway of the sprinkler. High flow rates through gear drive sprinklers can result in significant losses; for example, in the range of 20–50 kPa. However, if the system is designed to provide the stated pressure at the inlet, then the required performance of the sprinkler will be achieved.

The distance of coverage is presented as a radius in Table 4.2. It is also commonly presented as a wetted diameter (2 × radius). Caution is required in reading the charts to ensure that the radius or wetted diameter is being recorded.

The discharge or flow rate (Q) can be presented in several units, including L/s, L/min and m^3/h.

Sprinkler distribution profiles

Sprinkler manufacturers test all sprinkler heads under a range of conditions. In some cases, tests are carried out in-house and in other cases the heads are tested at independent test facilities, such as the Center for Irrigation Technology (CIT), Fresno, California. The main software package from CIT is 'Space Pro', which is available from Irrigation Australia Ltd (see the CIT Website: http://cati.csufresno.edu/cit/software).

Access to sprinkler performance data is available from manufacturers (Figure 4.7), irrigation designers or through ownership of the sprinkler performance data analysis programs.

The performance data reported by testing authorities, such as the wetted radius or wetted diameter, are determined by procedures outlined in American Society of Agricultural Engineers Standard ASAE S398 (ASAE 1999).

Turf sprinklers

Turf sprinklers have to operate in particularly demanding conditions. In addition to the need for high uniformities, the physical operating environment can be very harsh. The sprinkler body must be housed in the ground and the sprinkler head designed to raise to an elevated

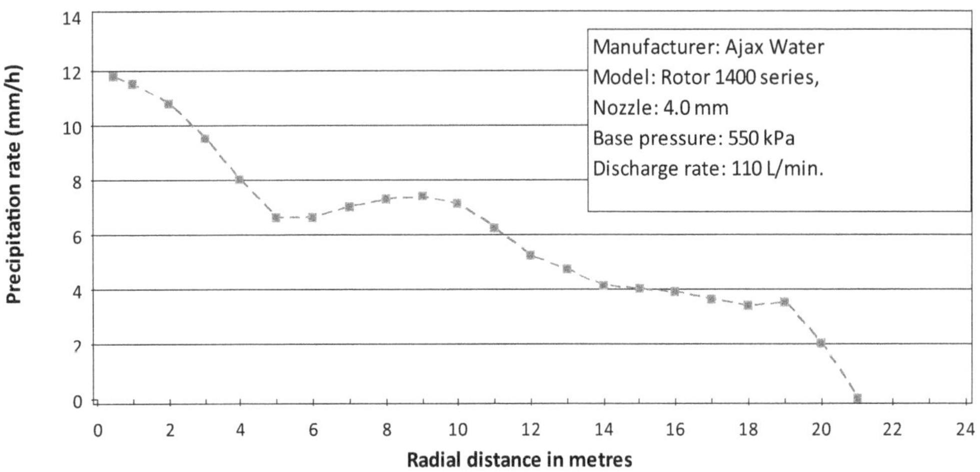

Figure 4.7. Typical sprinkler radial distribution profile (Source: Space Pro, CIT, Fresno)

position to effectively distribute the water. It must also withstand top loads (people and machines) and sand and silt may wash into the seals and working mechanism. Turf sprinklers are also exposed to a range of potentially damaging fertiliser and plant health chemicals.

There is a wide range of sprinkler options available for turf. The main areas of differentiation are in drive type, nozzle arrangement and materials used in construction. The majority of turf sprinklers use either impact or gear drive mechanisms. The key requirement of the drive is to provide reliable, even rotation at specified speeds, under varying environmental conditions (especially wind).

Sprinklers designed for turf applications must achieve high uniformity and water large areas. The uniformity of all sprinkler systems is very sensitive to the spacing of sprinklers. Uniformity data presented by Pair (1969) shows that, for sprinklers with triangular shape distribution profiles, the uniformity decreases markedly when the spacing exceeds the range of 65% to 70% of the sprinkler wetted diameter. The spacing of sprinklers so that the distance between the heads is 50% of the wetted diameter is commonly recommended for situations that require high uniformity, such as golf greens. This spacing is referred to as 'head to head' (Plate 4.2).

Due to the key role of the sprinkler in determining the overall efficiency and effectiveness of the system, continual monitoring of the performance and functioning of the sprinkler heads should be given high priority. Regular maintenance is essential.

Turf sprinkler selection checklist

The following features and characteristics of sprinklers should be considered before the selection of turf sprinkler heads.

1. **Performance** – required inlet pressure, effective wetted diameter, discharge rate, spacing and pattern for uniformity of application
2. **Drive mechanism** – suitability for particular application, rotation uniformity, rotation speed, wear characteristics and expected life under site operating conditions
3. **Materials** – body, nozzle assembly, cover, bearings and seals (e.g. plastic, stainless steel)
4. **Nozzle assembly** – design, number, smallest opening size (blockage risk) and stream trajectory angle
5. **Part-circle mechanism** – effectiveness, reliability and simplicity of adjustment
6. **Quality of manufacture** – degree of precision and finish (potentially dangerous sharp protrusions)
7. **Retraction method** – retraction force, sealing and effectiveness in excluding soil (sand) particles from interfering with sprinkler operation
8. **Operating height** – clearance of water stream above surrounding turf. (Total nozzle lift). Increased mowing heights and thatch require higher lift heads.
9. **Cover** – protection of sprinkler, protective safety cover and resistance to vandalism
10. **Filter** – inlet screen sufficiently fine to prevent blockage of smallest opening
11. **Anti-drain valve** – operating requirements (pressure)
12. **Serviceability** – ability to service and repair from the top
13. **Spare parts** – availability and cost.

The following sections compare the difference between sprinklers and sprays.

Shrub and lawn sprays

Sprays are an economic way of covering small areas (Figure 4.8). For example, if the required coverage is within the range of 2 m to 8 m, sprays are potentially well suited. Fixed nozzles of sprays apply water continuously to the wetted area.

Figure 4.8. Sprays are simple and reliable applicators

Sprays can be mounted above the plant or installed on telescopic risers so that the water can be applied over the top of the shrubs without the need for a tall, fixed and potentially unsightly riser.

Differences between sprinklers and sprays

Both sprinklers and sprays are commonly used in urban horticulture. There are key differences in performance and characteristics of these two irrigation techniques.

Sprinklers have the following characteristics:

- They produce a rotating stream, which requires some form of drive mechanism.
- They may require medium to high operating pressure.
- They have medium to high discharge rates.
- They have a large wetted diameter (up to 30 m).
- They have moderate precipitation rates (5–15 mm/h).

Sprays have the following characteristics:

- They produce a fixed stream.
- They require low to medium operating pressure.
- They have low to medium discharge rates.
- They have a small wetted diameter (up to 5 m).
- They have medium to high precipitation rates (20–50 mm/h).

Microirrigation

Description and principles

This method of irrigation is based on the delivery of water to individual plants or small groups of plants. The plants may be small, such as a seedling, or large, such as a tree.

Relatively low flow rates from individual outlets, called emitters, are characteristic of this method of irrigation. There are many different products that are available within this category. The use of plastic materials and low cost of manufacture means that there is a huge variety in designs and product options.

Types of microirrigation

Within the broad microirrigation category, there are numerous product groups. The main ones are as follows:

- point source drip emitter (individual drippers connected via microtube to supply line or attached externally to supply line), referred to as 'on-line'
- line source drip pipe (drip emitter incorporated in pipe/tube/hose), referred to as 'in-line' or dripline
- drip tape (line source) (drip emitter incorporated in wall of tape – collapsible pipe)
- microjets or microsprays
- microstream emitters (multiple small streams)
- microbubblers
- microsprinklers
- mini sprinklers.

Fundamental to a quality microirrigation system is the delivery of a precise quantity of water to the plant root zone. The actual quantity depends on the discharge rate (flow rate) of the emitter (microirrigation outlet) and the run time.

The discharge rate from an emitter is sensitive to variation for three main reasons: pressure variation, manufacturing variations in product dimensions and blockages and malfunctions.

Because microirrigation systems operate on low pressure (typically in the range of 100–200 kPa) small variations can be significant in terms of the correct functioning of the system. Relatively small variations in elevation along a lateral pipe, such as 2–3 m, can cause a significant variation in the flow.

Pressure variation can occur as a result of friction in pipes and fittings, elevation changes and supply pressure variation. Also, because flow pathways are small, any physical variation in the product dimensions can lead to significant flow rate variations. Product quality and correct operating conditions are critical to high efficiency with microirrigation equipment.

Characteristics of microirrigation – benefits and limitations

The following is a summary of the typical characteristics of microirrigation systems:

1. They use low discharge rates.
2. They require low operating pressures (typically 100–200 kPa).
3. They are potentially very efficient in application.
4. The soil properties determine the distribution from drippers.
5. Many items of equipment and components are required.
6. The level of maintenance is potentially high.
7. They require high-quality water.
8. They are of medium cost.

Benefits and limitations of microirrigation

As a result of these characteristics, the benefits of microirrigation include:

- improved effectiveness of water application – water is directed to the plant root zone (drip) or to the root zone of trees (microspray)
- reduced wetted area of ground (less evaporation and weed growth)
- decreased water loss and waste – low evaporation losses (in most cases)
- low energy usage because of the lower operating pressure.

The limitations of microirrigation include:

- a risk of emitter clogging, which can be caused by poor quality water supply, soil or roots
- many components in system, which require high maintenance and are vulnerable

- high-quality (physical) water required
- higher level of management skill required
- root zone wetting depends on the soil characteristics and outlet flow.

Drip irrigation

Description and principles

The underlying principle of drip irrigation is that water is delivered directly to the root zone of individual plants. Rather than wet the general area, the aim is to apply water to soil volumes with roots. To achieve this, the delivery flow rate is low. Low flow rates require small outlet orifice sizes and operation at relatively low pressure. This method of application allows soil moisture to be maintained at close to optimum levels, as required by the plant.

Characteristics of drip – benefits and limitations

The benefits of drip irrigation include:

- high efficiency of application
- lower energy costs
- relatively low installation costs
- minimal surface wetting – less weeds, usable surface/ground
- uniform application to individual plants is possible
- suited to fertigation
- precision application
- suited to the irrigation of narrow or irregular shaped areas.

The limitations of drip irrigation include:

- high-quality water required
- risk of plugging and damage
- soil water distribution strongly dependent on soil properties
- high maintenance requirements
- many components in system
- not well suited to seed germination and some plant establishment situations
- close spacing (long lengths) required for lighter soils.

Drip products

The role of the drip outlet, referred to as drip emitter, is to release water at a slow and controlled rate from a supply line (Figure 4.9). The key to the success of drip irrigation systems is the reliable discharge of very low flow rates, typically in the range of 1 to 8 L/h. Low emitter discharge rates are obtained using small orifice sizes and small flow pathways. The main risk in terms of reliability is blockage.

Drip emitters are used in a number of different supply configurations as follows:

1. Individual drippers supplied through small bore (diameter) tubing or attached directly, via a barbed connector, to the supply pipe. The supply pipe material is typically polythene (point source type).
2. Drip emitters incorporated into the supply pipe (dripline). Numerous designs are available, including emitter assemblies attached to the wall of the pipe and emitters designed to allow water to flow through the centre of the emitter (in-line source type) (Figure 4.10).
3. Drip tape (thin-walled pipe) where the drip emitter is incorporated into the wall of the tape (line source type).

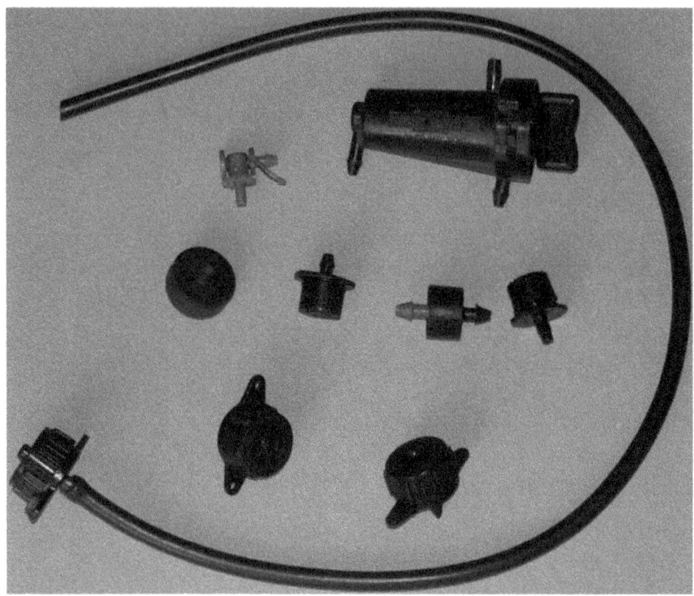

Figure 4.9. Examples of drip emitters designed to be supplied via mintube or fitted externally to pipes (on-line emitters)

There have been major developments in the design of drip emitters in recent years. The design of drip emitter to provide precise, low flow rate and reliable delivery operating in harsh environments with water of variable quality is challenging. Early designs were prone to plugging, variable discharge and temperature sensitivity (Burt and Styles 1994). Drip emitter design has improved markedly since then and with appropriately treated water can achieve high levels of efficiency and reliability of operation.

Drip emitter design has progressed from long flow path, small bore tubing (0.5 to 1.5 mm) to very compact, short flow path lengths (turbulent flow). The more recent emitters rely on turbulent flow, rather than laminar type flow, through the dripper. This type of flow is better able to handle any particles that may block the dripper. The emitter exponent value indicates

Figure 4.10. Exposed labyrinth passageway (turbulent flow) of emitter fitted to internal wall of dripline

Table 4.3. Emitter type and flow regime

Flow regime	Emitter exponent (*x*-value)	Emitter type
Variable flow path	0.0 0.3	Fully pressure compensating Partially pressure compensating
Vortical flow	0.4	Vortex emitters
Fully turbulent flow	0.5	Orifice flow, tortuous path
Mostly turbulent flow	0.6	Long or spiral path
Mostly laminar flow	0.9	Microtube
Fully laminar flow	1.0	Capillary flow

Source: Irrigation Association (1997)

the relationship between pressure and discharge rate from the emitter. High exponent values– for example, laminar flow – will increase in discharge rate strongly compared with turbulent flow type devices (Table 4.3).

A very significant development in drip emitter design has been in pressure compensation (PC). The amount of water discharged from an orifice increases as the supply pressure increases. Although some emitter designs employing turbulent and vortex flow exhibit reduced sensitivity to pressure variation, there is strong variation in flow as the pressure changes. As stated previously, there are numerous reasons why the pressure can change. The PC emitter usually operates on the basis of a flexible (e.g. rubber) diaphragm constricting the outlet opening as the pressure increases. The ability to achieve a constant discharge from a drip emitter over a range of operating pressures is a significant advantage in terms of achieving efficient irrigation.

Figure 4.11 shows the response of two emitter (nominally 2 L/h and 4 L/h) designs, non PC, to changes in supply pressure (Connellan and Denman 2009).

Drip emitter performance – coefficient of variation (Cv)

Some variation in the discharge from all manufactured irrigation outlets is to be expected. Due to the need for precision in delivery from drip emitters, it is important that variation for a

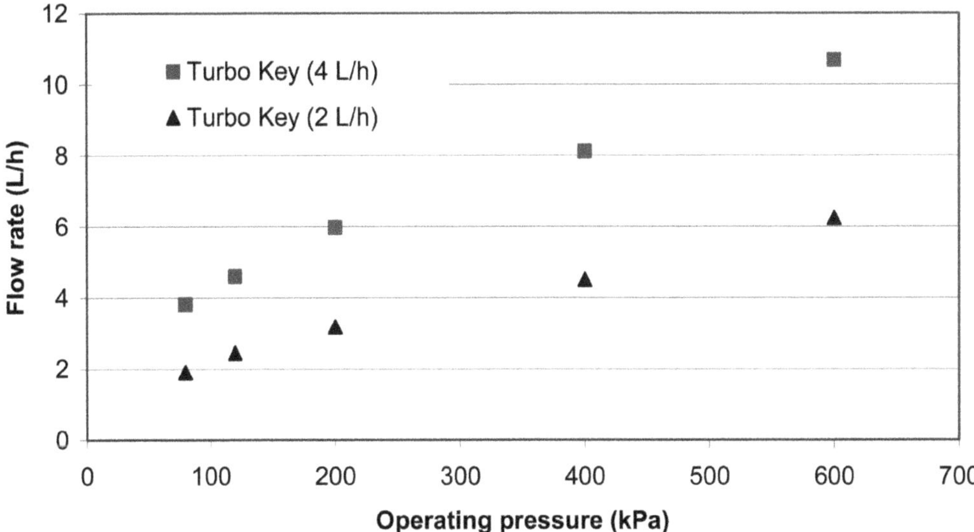

Figure 4.11. Emitter flow variation with increasing pressure (non pressure compensating emitter)

Table 4.4. Classification of emitter discharge performance and corresponding distribution uniformity

Classification	Emitter (Cv)	Distribution uniformity (DU_{Cv})
Excellent	<0.03	>96%
Average	0.03–0.07	91–96%
Poor	>0.07	<91%

Source: Harris (2006)

given set of supply conditions is within specified limits. The unit of measurement of variation is the coefficient of variation (Cv). Multiple emitters are tested and the flow rate variation is analysed.

The expression for the determination of Cv is:

$$\text{Cv} = \frac{\text{Standard deviation of emitter flow rates}}{\text{Average emitter flow rates}}$$

This test is usually carried out in a laboratory and is used by manufacturers and designers to determine the performance quality of the product. A separate test (emission uniformity, (EU) is used to test drip emitters in the field. (see Chapter 9)

Table 4.4 provides a guide to the values of Cv appropriate to three nominated performance classifications.

In order to achieve high uniformity of application, low variability in discharge from emitters is needed. Achieving a Cv less than 0.03 is therefore recommended.

Drip (surface) wetting patterns

The actual wetted soil volume by a drip emitter is highly variable. Both the diameter of the wetted area at any particular point in the soil profile and the depth of wetting can vary greatly. The rate of delivery, total amount delivered and the frequency of delivery can all influence the wetting pattern. The total volume of water delivered has a strong influence on the distance that the water moves in the soil.

The strongest influence on the wetting pattern is the soil type and properties. Only very general advice can be provided here. Soil testing at the proposed irrigation site is strongly recommended. In sandy soils, the wetting tends to be deeper than wider, but in heavy soils the wetting tends to be similar in width and depth, and in some cases wider than deeper (Figure 4.12 and Table 4.5). If there is an underlying relatively impermeable layer, then this will influence the wetting pattern and encourage wider wetting rather than deeper wetting.

The size of the root zone area to be watered and the soil type both strongly influence the number of drippers required for a particular situation. It is now common to water landscape

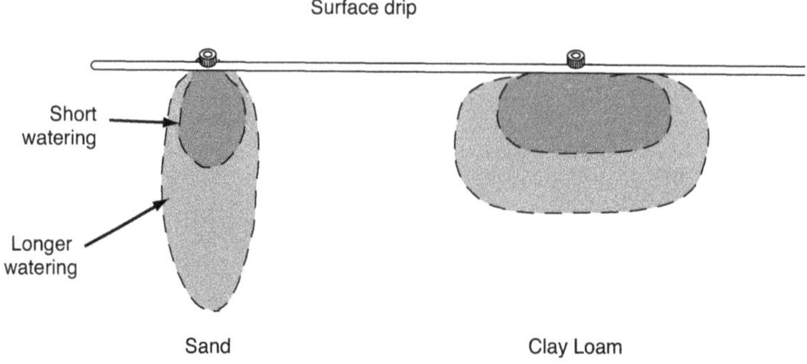

Figure 4.12. Representative wetting patterns of drip emitters on sand and clay soils

Table 4.5. Guide to wetted areas and depths for surface drip emitters

Property	Estimated wetted diameter and area		
	Sand	Loam	Clay
	Volume applied: 10 L		
Wetted diameter (m)	0.30	0.50	0.80
Wetted area (m^2)	0.07	0.20	0.50
	Volume applied: 50 L		
Wetted diameter (m)	0.50	0.90	1.60
Wetted area (m^2)	0.20	0.65	2.0

Note: These wetted diameters are estimates only.

Table 4.6. Guide to dripline spacing (surface drip systems e.g. garden beds) in various soil types

	Soil type		
	Sand	Loam	Clay
Emitter interval (S_{ei})	0.3 m	0.4 m	0.5 m
Dripline spacing – high uniformity (S_{ls})	0.3–0.4 m	0.4–0.5 m	0.4–0.5 m
Dripline spacing – medium/low uniformity (S_{ls})	0.4–0.6 m	0.5–0.7 m	0.5–0.7 m

Note: this is a guide only. A site soil test is required to determine emitter interval and dripline spacing.

areas so that the total area is watered with multiple driplines (parallel runs of pipe) spaced typically from 0.5 to 1.0 m apart, rather than individual drip units positioned adjacent to individual plants. These systems often replace areas previously watered with sprays.

The two key layout dimensions for these situations are the distance between emitters along the lateral pipe (emitter interval, S_{ei}) and the distance between lateral pipes (S_{ls}) or the dripline spacing.

Table 4.6 is a guide to dripline spacing for two levels of uniformity of coverage.

It should be noted that high uniformity spacing is used when near continuous wetting is required over an area (Figure 4.13). Medium uniformity spacing is used in situations, such as garden beds, where root systems of large plants do not require soil water to be uniform to exploit it.

Figure 4.13. Dripline installed in grid layout to provide near continuous wetting of garden area

Factors in the selection of drip emitters

The following is a list of the features and characteristics of emitters that should be considered when selecting a drip emitter:

- wetted diameter/radius/area
- discharge rate
- operating pressure
- application rate (mm/h)
- pressure sensitivity and limits
- flow variability (Cv)
- filtration requirements
- maintenance
- durability
- cost.

Subsurface drip irrigation (SDI) systems

Description and principles

The positioning of the drip emitter within the soil profile increases the potential efficiency of application because water is delivered within the soil volume containing the roots. Compared with a surface drip, this has the advantage that for plants with root systems at some depth in the soil profile, the soil wetting is not dependent on saturating the surface soil layer before water moves down to the main root area of the plant.

The release of water from within the soil allows water movement in all directions: upwards, laterally and downwards (Figure 4.14). The extent of movement depends on the hydraulic properties of the soil, the performance of the emitter and the design of the drip system (Figure 4.15).

The downward movement of water is mainly influenced by gravitational forces. The upward and lateral movement of water is mainly determined by capillary forces. Water movement in sandy soils, which have relatively large pore spaces, is mainly determined by gravitational forces. Water movement is predominantly downwards in sandy soils. In fine, heavier soils with significant clay content, water distribution is strongly determined by capillary forces, so there is a greater movement of water upwards and laterally.

Characteristics of SDI – benefits and limitations

These following benefits are listed in addition to those for drip and microirrigation:

- Water is delivered directly into plant root zone.
- Very high efficiency of application is possible.

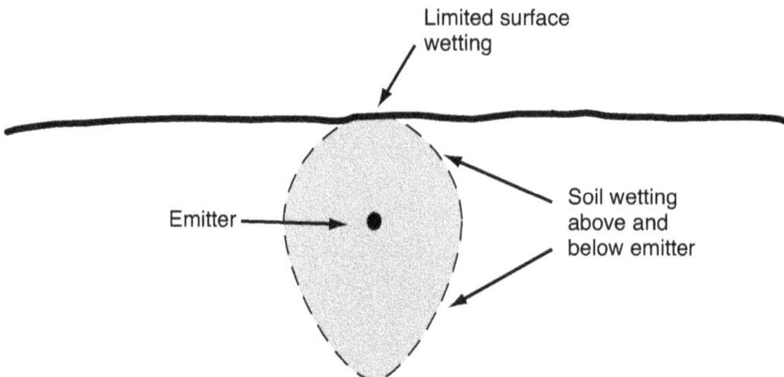

Figure 4.14. Water distribution from SDI emitter

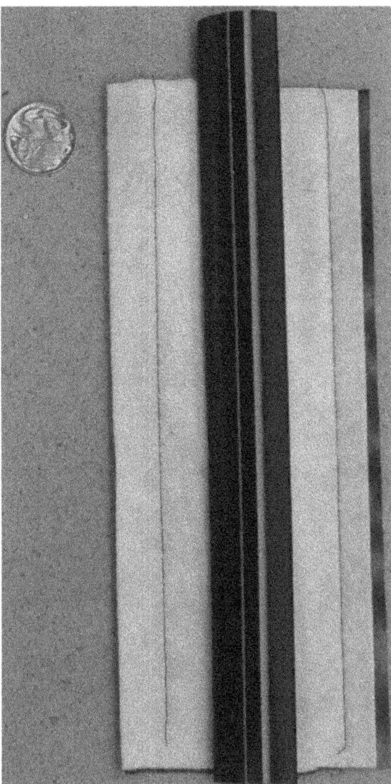

Figure 4.15. SDI drip product incorporating webbing and geotextile to assist in spreading water laterally and longitudinally

- No or minimal surface wetting means that the area is trafficable at all times.
- It allows the effective use of fertigation.
- A maximum soil volume is wetted from the emitter.
- Atmospheric conditions, such as wind or high evaporation, do not interfere with the effectiveness of application.

The limitations of SDI include:

- a reasonably high installation cost
- no flexibility in design once installed
- separate watering system likely to be required for seed germination
- root intrusion may be a problem
- requires specialised equipment installation (Figure 4.16)
- soil moisture monitoring is essential to check water distribution and system operation
- a risk of soil tunnelling in dispersive soils.

Performance of SDI

The ideal soil for SDI allows a balanced distribution of water upwards, laterally and downwards. Correct spacing and installation at uniform depth are critical to achieving high efficiency with SDI systems (Plate 4.3). Table 4.7 provides a guide to the distribution of water from SDI in a number of soil types.

Selecting SDI emitters

The performance requirements for all drip emitters are demanding. In the case of SDI emitters, there are additional considerations, because the device is operating within the soil environment.

Table 4.7. Emitter interval and spacing for SDI turf applications

Soil type	Clay	Loam	Sand
Dripper interval	400–500 mm	300–400 mm	250–300 mm
Lateral spacing	500 mm	400 mm	300 mm
Depth of dripline	150–200 mm	100–150 mm	50–100 mm

Note: These are a guide only.
Source: Adapted from Netafim (2007)

These include that the device should:

1. operate reliably and precisely in a potentially harsh (in physical and chemical terms) soil environment
2. be resistant to intrusion of roots
3. be resistant to ingress of debris through suck back
4. incorporate filtration to prevent blockage of emitters
5. use robust driplines to resist crushing from soil and growing roots
6. have an emitter capable of handling low quality water.

Porous pipe/weeping hose

This type of product is a flexible pipe or hose that is constructed from a rubber material that has many small holes over the whole surface of the pipe (Figure 4.17). It is designed to release water (weep or seep) along the entire length of the pipe. The diameter of the pipe is typically in the range of 13 mm to 20 mm.

The hose is designed to be positioned either on the surface or buried under mulch or in the soil. It is sold in lengths of up to 30 m. It is generally recommended to be operated at low pressure: less than 100 kPa (e.g. the recommended operating pressure for Aquapore is 70 kPa).

Figure 4.16. Tractor mounted dripline reels allow speedy and effective installation of an SDI system. Note the wheels used for soil compaction over the installed pipe.

Figure 4.17. Rubberised hose with micropores (porous pipe) designed to release water over the entire hose surface

Some key performance considerations are the discharge rate, the evenness of delivery along the pipe and the sensitivity of flow rate to increasing operating pressure. Also, the micropore material is prone to blockage if low quality water is used.

The delivery per unit length varies from around 47 L/h/m at 70 kPa to 95 L/h/m at 300 kPa for Aquapore. Tests on an alternative product, Ryset, showed an increase from around 30 L/h/m at 100 kPa to 91 L/h/m at 400 kPa, representing a tripling of flow rate over the pressure range tested. Control of the supply pressure is extremely important. Achieving efficiency in application requires controlled delivery along the length of the distribution piping. This is extremely difficult to achieve with the degree of variation exhibited with these types of products, if the pressure is not controlled.

In terms of uniformity of discharge along the length of the hose, both hoses showed higher delivery rates close to the supply or inlet end where the pressures are higher.

Microsprays

Description and principles

Microsprays are generally constructed using plastic and typically have small outlets in the range of 1.0–3.0 mm. They are designed to be used to water individual plants or groups of small plants. They are supplied using small diameter tubing (for example 6 mm) and often installed on risers.

Characteristics of microsprays – benefits and limitations

The benefits of microsprays include:

- low cost
- coverage can be targeted to individual medium- to large-sized plants (e.g. trees)
- a small area can be watered (wetted diameter up to 6 m)
- they are not dependent on soil properties for distribution
- you can readily observe the equipment operating.

The limitations of microsprays include:

- the fine spray is sensitive to wind
- the light construction is vulnerable to damage

Figure 4.18. Microjets under test at Burnley Campus, University of Melbourne, illustrating capacity for relatively high application rates and runoff

- a reasonable quality water is required
- they are potentially high maintenance.

Product description – microspray

The distribution of water from microsprays has the advantage of not being dependent on the properties of the soil. This applicator is described as an aerosol emitter, because the water is distributed through the air. They are very simple devices and are commonly constructed as single piece or two piece units.

The method of breakup of the stream usually consists of water being discharged from an orifice and hitting a surface designed to break up and distribute the stream. The water is distributed from a microspray as either a fan-type spray, fine streams or fingers of water (Figure 4.18).

Microspray typical performance data

As a category of irrigation equipment, microsprays tend to cover relatively small areas (only up to a few metres), have highly variable precipitation rates and can also have high precipitation rates (Table 4.8).

The orifice sizes typically range from less than 1.0 mm to 2.5 mm. The fan jet types typically cover a range of 1.0–2.0 m in wetted radius and the finger stream types cover a larger range of 1.0–3.0 m. The distribution pattern of microsprays is very irregular. There are large variations in precipitation rates over very short distances. The precipitation rate may change from 10–30 mm/h over a distance of 0.5 m.

Microsprays are suited to situations where radial coverage of 1 to 3 m is required. They are well suited to the watering of large plants, but the characteristics of non-uniformity and high precipitation rates need to be recognised. Also, the fine spray produced is prone to distortion in wind and operating at the correct pressure is essential to achieve reasonable efficiency (Table 4.9).

Table 4.8. Performance for various microspray nozzle sizes

Nozzle size	Operating pressure (kPa)	Discharge (L/h)	Diameter of coverage (m)
1.0 mm	100	40	5
	150	45	5–6
	200	50	5–7
	250	–	–
1.5 mm	100	75	7.5
	150	95	8.5
	200	110	9.0
	250	120	8.5
2.0 mm	100	120	6.5
	150	150	7.5
	200	175	8.0
	250	195	8.5
2.5 mm	100	220	11.0
	150	270	12.0
	200	315	12.2
	250	355	12.0

Microsprays are very sensitive to pressure variation. Higher pressures result in misting and water loss and low pressures result in very uneven applications.

Mini-sprinklers

Description and principles

These are low flow rate devices, which are constructed from plastic and incorporate a mechanism that allows the water delivery to rotate. The droplets are generally coarse and the radial distance of throw is in the range of 3 to 6 m. These features mean that they have low precipitation rates.

Characteristics of mini-sprinklers – benefits and limitations

The benefits of mini-sprinklers include:

- the low stream height results in lower wind losses
- reduced foliage wetting, because delivery can be under the canopy

Table 4.9. Precipitation and uniformity performance of microsprays

Microspray system precipitation rate and uniformity (DU) Microspray emitter tested: Antelco 1.0 mm, 180° Fan				
	Wind conditions – calm		Wind conditions – windy	
Test	Precipitation rate – average (mm/h)	Distribution uniformity (DU %)	Precipitation rate – average (mm/h)	Distribution uniformity (DU %)
100 kPa	19	35.7	13	66.7
200 kPa	31	62.2	–	–
400 kPa	38	70.8	24	77.8

Source: Connellan and Denman (2009)

Figure 4.19. Wide coverage at low application rates are a characteristic of mini-sprinklers

- well suited to a range of soil types
- low pressure, so less energy required for pumping.

The limitations of mini-sprinklers include:

- the foliage needs to be managed to allow uninterrupted distribution
- relatively light precipitation rates, so long run times required
- small size and construction – risk of physical damage.

Product descriptions – mini-sprinklers

The basic feature of irrigation devices containing the term 'sprinkler' is that they distribute water in a moving rotating stream. Two important consequences of this delivery technique is that water is distributed over much greater distances than sprays and the precipitation rate is comparatively low, often not more than 10 mm per hour.

This equipment is well suited to situations where wide distribution is required and soils are not suited to the use of drip emitters. Plants with large root systems, such as trees, are sometimes suited to this technique. The system suits soils that have both high and low infiltration rates.

There are many different designs employed in breaking up the stream and driving the rotation of the stream. Some units are specifically designed not to create excessive breaking up of the water stream and maintain large droplets, so that the distance of coverage is maximised. Mini-sprinklers tend to be available only in full circle coverage designs (Figure 4.19).

The performance of mini-sprinklers in terms of distribution profiles is usually irregular and they are sensitive to pressure variation. They are generally not designed to be overlapped in order to achieve uniform application.

Mini-sprinkler performance data

Mini-sprinklers typically have nozzle diameters in the range of 1.5 mm to 2.5 mm and an operating pressure in the range of 100 kPa to 200 kPa (Table 4.10). The maximum wetted radius is in the vicinity of 5 m to 6 m. The flow rate is in the range of 40–250 L/h.

Table 4.10. Example of mini-sprinkler performance

Nozzle	Mini-sprinkler performance specifications			
	Pressure (kPa)	**Flow rate (L/h)**	**Wetted radius (m)**	**Wetted diameter (m)**
1.6 mm	100	100	3.4	6.8
	150	120	3.9	7.8
	200	140	4.5	9.0
1.8 mm	100	125	3.9	7.8
	150	155	4.7	9.4
	200	175	5.0	10.0
2.1 mm	100	150	4.1	8.2
	150	185	4.9	9.8
	200	215	5.5	11.0

Performance summary of selected irrigation outlets

A comparison of the performance of microirrigation emitters is given in Table 4.11.

Table 4.11. Performance guide summary for selected irrigation outlets

Emitter/ applicator category	Emitter/applicator type	Flow rate (L/h or * L/h/m)
Drip products	In-line drip pipe, non-pressure-compensated (one dripper)	3
	In-line drip pipe, pressure-compensated (one dripper)	2
	Individual pressure-compensated dripper (2 L/h)	2
	Individual pressure-compensated dripper (4 L/h)	4
	Individual non-pressure-compensated dripper (2 L/h)	3(1)
	Individual non-pressure-compensated dripper (4 L/h)	6(1)
Microsprays and mini sprinklers	Microspray, half circle, 1.0 mm orifice	39
	Microspray, full circle, 1.0 mm orifice	40
	Mini sprinklers (2.3 mm orifice)	262
	Mini bubblers (set at 10 clicks)	76
	Shrubbler (set at 10 clicks)	22
Garden sprays and sprinklers	Border spray jet	403
	Pop up sprays	317
	Rotating stream lawn sprinkler	293
Hose products	Soaker hose (averaged for 7.5 m section length)	*100
	Porous pipe – Aquapore (averaged for 15 m section length)	*74
	Porous pipe – Ryset (averaged for 15 m section length)	*50
Hose-end products	Hand-held trigger – multi 'shower setting'	530
	Hand-held trigger – multi 'cone setting'	352
	Hand-held trigger – spray	1031
	Hose-end spray	983

Note: (1) Tested at typical residential pressure of 300 kPa
Source: Connellan and Denman (2009)

Table 4.12. Comparison of application performance of irrigation techniques

Category	Precipitation rate		Time for 10 mm application
	Category	Typical rate (mm/h)	
Sprays	High	30–50	12–20 mins
Sprinklers	Low to medium	15	40 mins
Microsprays	Medium to high	20	30 mins
Mini-sprinklers	Low to medium	8	1.25 hours
Microstream emitters	Medium to high	25	25 mins
Drip	Low to medium	5–15(2)	40 mins to 2 hours
Bubblers	Very high	50	10 mins

Notes:
(1) Time to apply 10 mm depth = (Depth (10 mm)/Precipitation rate) × 60 (min)
(2) The precipitation rate of drip systems will exceed the above rates if dripper spacing is close.

Comparison of performance of irrigation techniques

Each method of irrigation varies in application performance. Good irrigation practice requires defined amounts of water to be delivered. The operating time for each irrigation technique needs to be determined for each situation. Table 4.12 shows the impact of precipitation rate on run times of the various irrigation techniques.

Travelling sprinklers

Travelling rain guns (large sprinklers or rain guns)

Travelling sprinklers are sometimes used for the watering of sports grounds and parklands and also for turf production. Much of this equipment has been developed for the agricultural market. The high pressure rain guns used to apply the water are a feature of the large capacity units. This equipment has demanding supply requirements in terms of flow rate and pressure.

Although travelling sprinklers do provide flexibility in the use of equipment, the uniformity of application is generally significantly lower than fixed sprinkler systems. Also, the application depth is often low (less than 10 mm), due to the travel speeds and the wide spacing of each travel laneway.

The method of propulsion is from the water passing through the unit or an engine mounted on the machine. Water turbines are a common technique used to power a winch, which, in turn, pulls the irrigator carriage with rain gun along. An alternative design winds up the hose supplying the rain gun. In the latter case, the irrigator remains stationary during the application of water.

The supply hose to the rain gun needs to be both flexible and robust. Because of the need to keep the hose bulk to a manageable level, the hose diameter is limited. The friction loss in the hose can be significant.

The flow rate is typically in the range of 400–1000 L/min at an operating pressure of 500–800 kPa and the watering width is typically in the range of 70–100 m (Figure 4.20).

Characteristics of travellers – benefits and limitations

The benefits of travelling sprinklers include:

- lower capital cost compared with fixed systems
- the irrigator can be removed from site to site
- the system is suited to light applications on multiple sites.

Figure 4.20. Large coverage is achieved with travelling irrigator using high flow rate and high trajectory

The limitations of travelling sprinklers include:

- poor application uniformity in windy conditions
- high intensity and high energy usage– potential for soil damage (e.g. erosion and compaction)
- require a tractor or vehicle to move to next operating position
- the risk of theft or vandalism
- require high operating pressures.

Travelling boom sprinklers

These are similar to the gun-type systems, but they use a fixed or rotating boom with multiple sprinklers or sprays, rather than distributing the water from a single high pressure sprinkler (Figure 4.21). This method of application is significantly more uniform than rain guns. Water is supplied to the mobile unit through a hose. This type of irrigator is sometimes used on race-tracks, where the need for high uniformity and a reasonable depth of application is required.

Characteristics of boom irrigators – benefits and limitations

The benefits of boom irrigators include:

- The uniformity of application in wind is higher than for rain-gun systems.
- They use lower operating pressures than rain gun machines.
- The system can be removed from field to allow full utilisation of the area.
- The system can be used for multiple sites (but some restrictions due to large size).

The limitations of boom irrigators include:

- They generally require reasonably level operating areas.
- They can be difficult to move and accommodate, because of the size of boom structure and weight of the machine.
- The application depth is limited by the size of the supply line and travel speed of machine. Only light applications, less than 10 mm can be applied, if used on large areas.

Figure 4.21. Mobile boom type irrigator provide good uniformity using multiple medium pressure sprays (Source: Upton Irrigation)

Self-propelled small mobile lawn sprinklers

There are a range of mobile sprinklers supplied with hoses of diameter up to 25 to 40 mm. These are self propelled, using the rotation of a rotating nozzle arm to operate a gear mechanism and winch, to pull the unit along. Portability is the key feature of these units, but the risk of theft is a major consideration in public open space areas.

The flow rate is typically in the range of 20 to 60 L/min at an operating pressure of 300 to 500 kPa. The effective watering width is typically in the range of 20 to 30 m.

Characteristics of mobile lawn sprinklers – benefits and limitations

The benefits of mobile lawn sprinklers include:

- relatively inexpensive and portable
- reasonable uniformity of application
- flexibility in watering sites
- wide coverage.

The limitations of mobile lawn sprinklers include:

- the width of coverage is fixed
- some sensitivity to windy conditions
- require labour for set-up and shifting
- long operating times (over 12 h in some cases)
- potential target for theft
- require straight continuous travel path.

Manual and automatic systems

Comparison of systems

There is, and will continue to be, a trend towards automatic irrigation systems. There are sound reasons for this progression. Irrigating at night is recommended for efficiency reasons and this is very difficult to achieve with manual systems. Also, the cost of shifting the equipment is considerable and constrained by the availability of labour. Importantly, automatic systems operated according to precise scheduling programs are usually more efficient than manually operated systems, where the run time of equipment is sometimes determined by the work practices of the operator.

The following comments summarise the different features and characteristics of three types of irrigation systems: manually shifted sprinklers on stands; quick coupling valve sprinklers and automatic pop-up (fixed) sprinkler systems.

Manually shifted sprinklers on stands

The features and characteristics of manually shifted sprinklers on stands include:

- very labour intensive
- limited capacity to deliver water to the irrigated areas
- non-uniform application
- restricted to daytime use
- flexible in areas that can be watered
- low capital cost
- problems locating connection points at times.

Quick coupling valve (QCV) sprinklers

The features and characteristics of QCV sprinklers (Figure 4.22) include:

- reasonably large wetted area from each individual sprinkler
- prone to non-uniform application
- labour required for each set-up
- run time determined by availability of operator
- less expensive than in-ground pop-up sprinklers
- sprinklers and fittings prone to damage and wear due to use and handling
- risk of theft
- problems locating connection points.

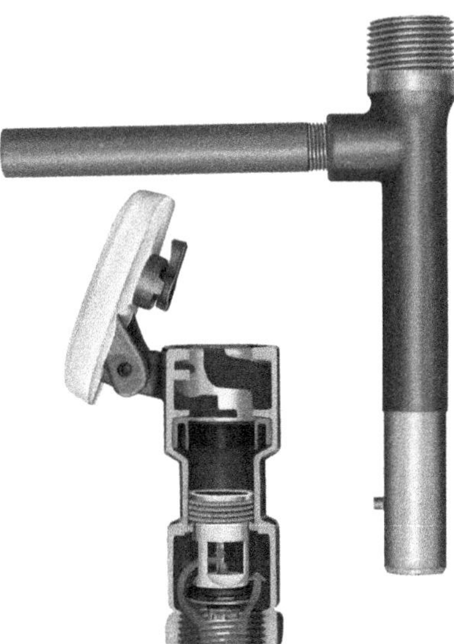

Figure 4.22. Quick coupling valves allow flexibility in water application for tasks such as watering in, plant establishment and topping up soil moisture (Source: Rain Bird Australia Pty Ltd)

Automatic pop-up (fixed) sprinkler system

The features and characteristics of automatic pop-up sprinklers include:

- potentially high uniformity of application and efficiency
- high capital cost
- capable of delivering a range of irrigation depths and timings
- maintenance of equipment is essential to maintain system effectiveness
- suited to night time watering and automation
- low labour input
- potential for over-watering.

Manual systems and people considerations

The following issues have been identified as factors or constraints that need to be considered when assessing the use of current manual systems in public open space areas.

1. Physically demanding
   - shifting hoses and sprinklers is very demanding during hot periods
   - occupational Health and Safety issues.
2. Workplace agreement/enterprise bargaining
   - conditions may limit work to daylight hours.
3. Night-time irrigation
   - potential safety issues for personnel attending equipment in the dark.
4. Labour requirements
   - manually shifting hoses is very labour intensive and time consuming
   - other responsibilities, including mowing, emptying bins, general park maintenance and administration duties, may determine when irrigation occurs.
5. Conflict in area use
   - day-time irrigation may conflict with use of area.
6. Safety
   - equipment (including hoses) potential risk (e.g. tripping) to users.
7. Maintenance
   - high maintenance requirements associated with manually operated equipment.

Efficiency of irrigation systems

Principles of irrigation efficiency

The overall aim of good irrigation is the delivery of water into the root zone without loss (Figure 4.23) so that soil moisture is maintained at a level that provides healthy plant growth. This aim can be considered in two parts: the effectiveness in application of water into the root zone (application efficiency) and the operation of the irrigation system to ensure that the right amount of water is delivered at the right time (water management efficiency).

The principles of irrigation efficiency are described in Chapter 1.

Components of irrigation application efficiency

The following are the potential ways in which water can be lost or wasted and so reduce the efficiency of application:

- atmosphere – evaporation losses and wind drift
- runoff
- non-uniformity of application
- interception by foliage
- deep drainage below the root zone

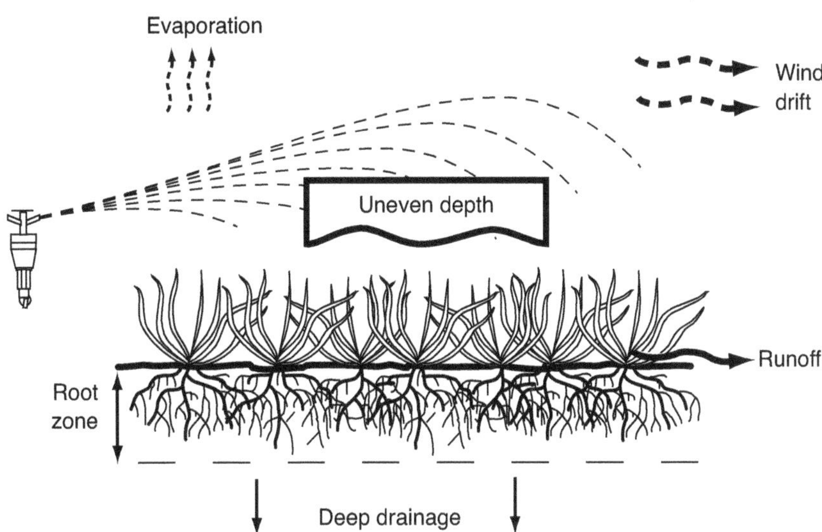

Figure 4.23. Potential sprinkler irrigation system losses

- missing the target area
- soil conditions that impede effectiveness of application.

The following sections discuss the factors involved and the relative importance of these, with regard to application efficiency.

Atmosphere – evaporation and wind

Water is lost to the atmosphere from moist and wet surfaces and from water droplets passing through the air. The factors that influence the amount of evaporation from sprayed droplets are:

- the size of the droplets
- the evaporation rate of the atmosphere
- the exposure time of the droplet in the air.

The evaporation rate is influenced, to some extent, by operational management decisions. Night-time watering has a lower evaporation rate than during day-time watering.

The droplet size is determined by the:

- nozzle orifice size
- nozzle geometry, design and materials
- operating pressure.

Droplet size generally increases as nozzle size increases, and as pressure decreases. In other words, large nozzles produce large droplets and small nozzles produce small droplets.

Table 4.13 presents the expected evaporation losses for three nozzle sizes, operating under a range of atmospheric conditions and operating pressures.

Evaporation losses in excess of 5% are very significant. The data show that during daytime operation, when temperatures are in the 30°C range, these losses can easily exceed this level. It is important to consult with irrigation designers in the selection of sprinklers and sprays that are going to be operated in conditions conducive to evaporation losses.

The main situations in which evaporation and wind losses occur are:

- daytime irrigation – higher evaporation rates
- high wind conditions – winds generally higher during the day than at night
- over pressurised nozzles – small nozzles in particular, such as microsprays.

Table 4.13. Evaporation losses of sprinkler irrigation/spray systems for different weather conditions

Nozzle size (mm)	Nozzle pressure (kPa)	Evaporation losses	
		Night-time 25°C/RH 30%/WS 5 km/h	Daytime 35°C/RH 10%/WS 5 km/h
4	200	3.5%	7%
4	275	5.5%	9%
4	415	7%	14%
8	275	1.5%	3.5%
8	415	4%	9%
8	550	5.5%	14%
12	415	2.5%	5%
12	550	3.5%	6%

Notes: Frost and Schwalen (1955) developed a nomograph, which has been used to determine the losses in the above cases. RH – relative humidity. WS – wind speed.

Droplet size is a critical factor in terms of potential evaporation losses. For sprinklers operating at approximately 300 kPa, the proportion of evaporation loss increases rapidly with nozzle orifice sizes smaller than 8 mm (Smajstrla and Zazueta 1994). These authors also point out that evaporation losses increase almost linearly from 1.4% at 150 kPa to 5% at 550 kPa. Note that this refers to evaporation losses only: wind drift losses can be much more.

Runoff

The situations in which runoff occurs include:

- when the system precipitation rate is higher than the soil infiltration rate
- when the irrigation system is operated for extended run times on fine soils (e.g. loam and clay)
- where there are compacted soils
- on sloping sites.

Non-uniformity of application

The main source of losses associated with non-uniform application or delivery is that part of the irrigated area will need to be over-watered to ensure that the low precipitation areas receive adequate water. This characteristic applies to both sprayed (sprinklers and sprays) and drip systems.

Interception by foliage

The main sources of interception losses are:

- foliage obstructing spray distribution in garden beds
- pop-up sprinklers and sprays set too low, so that surrounding grass/turf prevents clear delivery of water.

Deep drainage

The main sources or situations in which losses associated with deep drainage occur include:

- light, free draining soils (e.g. sand) through which the irrigation application will readily percolate
- excessive run times
- irrigation methods, such as surface irrigation, on relatively light soils
- irrigating soils which are already at maximum water-holding capacity.

Figure 4.24. Combined characteristics of fine droplets and poor application control can lead to significant runoff with sprays

Missing the target

Reasons for water not reaching target area include:

- overspray – the radius of spray is greater than the distance to be covered (Figure 4.24)
- poor adjustment – part-circle sprinklers and sprays
- significant wind drift.

Surface and soil conditions

A number of soil and surface conditions can have a negative impact on efficiency of application (Fig 4.25). These include:

Figure 4.25. Build up of organic matter in turf (thatch) reduces penetration of rainfall and irrigation

Table 4.14. Guide to scale of losses that may be experienced with various irrigation techniques

Loss category	Sprinkler /sprays	Microspray	Drip – surface	Drip – subsurface	Comment
Spray droplet evaporation	Medium	High	Low	None	Depends on droplet size and atmosphere evaporation
Wind drift	Medium	High	None	None	Droplet size and wind speed
Surface evaporation	Medium	Medium	Low	Very low/ None	If mulch, very low
Runoff	Low/medium	Low/medium	Very low/ none	None	
Deep drainage	Low/medium	Low	Low	Low	Soil and application depth dependent
Missing the target	Low/medium	Low	Low	Low	If good design and good maintenance

Note: This does not include losses due to incorrect scheduling of irrigation.

- hydrophobic (water repellent) soils
- thatch – some of the applied water is captured in the thatch and re-evaporates into the atmosphere
- mulch that intercepts irrigation application.

Soil conditions, such as compaction, may also have an impact on efficiency of application.

Table 4.14 shows the relative importance of the various potential sources of loss associated with selected irrigation methods.

Table 4.15 provides a guide to the relative importance of the various factors than can influence efficiency. In this case, the data refers to agricultural sprinkler systems, but the range in values indicated is applicable to urban situations.

Summary guide to efficiency of irrigation methods

Table 4.16 summarises the ranges in efficiency for various irrigation methods that have been obtained from a number of sources. The important aspect to consider is the relative efficiency of the various methods.

From the previous discussion, it can be seen that numerous factors can have an impact on the efficiency of application at a site. The importance of many of these can be significantly diminished through correct equipment selection, competent design and good management.

Table 4.15. Guide to potential losses in sprinkler irrigation systems

Source of loss	Range	Typical value
Uneven application (non-uniform distribution)	5–30%	15%
Excessive application depth	0–50%	10%
Surface runoff	0–10%	<5%
Interception by plants	0–3%	<2%
Blown away by wind	0–20%	<5%
Evaporation in air	0–10%	<3%
Watering non-target areas	0–5%	<2%
Leaking pipes	0–10%	<1%

Source: adapted from Aqualinc Research Ltd (2006).

Table 4.16. Guide to application efficiency (E_a) for various irrigation methods

Irrigation method	Applicator efficiency range	Average E_a efficiency
Sprinkler – night	65–80%	75%
Sprinkler – day	60–70%	65%
Spray	65–75%	70%
Mini-sprinkler	65–75%	70%
Microspray	60–70%	65%
Drip – point source	85–95%	90%
Dripline – line source	75–90%	85%
Dripline under mulch	85–95%	90%
Subsurface drip (SDI)	85–95%	90%
Surface/flood	50–70%	60%
Travelling irrigator	55–70%	65%

Note: Many factors have an impact on the application efficiency of an applicator, including uniformity, runoff, wind drift and evaporation. These are generalised groupings to reflect differences between various types of applicators.

Some examples of poor irrigation selection

The following are examples of irrigation systems where there is not a good match between the performance characteristics of the technique, and the site conditions:

- microsprays and low precipitation rate sprinklers and sprays on mulched areas
- SDI on sand (ineffective coverage)
- sprays on wind prone sites (atmospheric losses)
- sprinklers and sprays used in daytime
- sprays (high precipitation rate) on sloping sites (runoff)
- sprays on low infiltration rate soils (ponding and runoff)
- driplines in low density (sparsely planted) garden beds
- sprays positioned at ground level in garden beds (distorted distribution)
- sprinklers operating below optimum pressure (non-uniform application)
- Non pressure compensating (PC) drippers operated on water supplies not pressure regulated.

Selecting an irrigation method – checklist

A broad range of factors need to be taken into account when selecting an irrigation method. These include:

- plants to be grown – requirements for healthy growth, landscape design and water use characteristics
- soil and topography – soil water properties and elevation changes
- climate – evaporation rates, wind, rainfall
- water supply – quality, amount and delivery rate and frequency
- labour requirements – skill, amount and availability requirements
- financial resources available to replace asset and maintain system in effective operating condition
- automation capability
- fertigation (injection of chemicals) capability
- energy/power available for pumping
- regulatory requirements in terms of types of systems allowed
- environmental conditions – impact and regulations

- maintenance machinery and equipment requirements
- cost – capital and operating.

Summary – key issues in selecting an irrigation method

1. Understand the properties of the site.
2. Identify the potential efficiency constraints of the site.
3. Select a method that maximises application efficiency.
4. Select a product with performance specifications that meet the site requirements.

Web resources – irrigation systems and equipment

Product directory

Irrigation Australia Limited (IAL) Irrigation Online Products and Services Directory. This includes a full listing of companies and organisations involved in manufacture, supply, consulting and education for the irrigation industry.
Website: www.irrigation.org.au/irrigationdirectory.htm

Major irrigation manufacturers

Hunter Industries	www.hunterindustries.com/
Irrigation Water Technologies (IWT)	www.iwt.com.au
Nelson	www.nelsonirrigation.com/
Netafim	www.netafim.com/
Rain Bird	www.rainbird.com/
Toro	www.toro.com/irrigation/allirrigation.html

Chapter 5

Plant water use and irrigation budgets

Importance of knowing landscape water use rate

Knowing the amount of water used by plants is fundamental to the sound management of landscapes in the urban environment. Without this knowledge, irrigation management is a matter of guesswork, rather than informed decision making. Although technology can assist in improving the quality of irrigation decision making, there is still the need for the site manager to be informed about the amount of water used, the factors influencing plant water use and the landscape performance of the site.

The reasons for needing to know the amount of water required are numerous. They include:

- The design of an irrigation system requires the designer to match the delivery capacity of the system to the peak water demands of the plants.
- The scheduling of the irrigation system, so that the correct amount of water is applied at the right time, depends on knowing the water demand. Irrigation should be scheduled according to a water budget.
- The evaluation of the overall effectiveness and efficiency of an irrigation system can only be carried out by comparing the actual water supplied to the site with the amount required by the plants.
- The selection of water storage capacities requires the assessment of weather data, including rainfall, and the estimation of water volumes expected to be used over the irrigation season.
- The economic analysis of urban landscapes depends on knowing the amount of irrigation water that needs to be supplied and the cost of this water.

Knowledge of water consumption is not an option: it is a necessity for good site water management.

Under irrigation water volume allocation regimes, which are becoming more common, it is also essential to know the total areas that can be watered, when they can be watered and the quality of the landscape that can be maintained. This is a core component of urban water resource planning.

Plant water use

Water movement through the plant

The movement of water from the soil, through the plant to the atmosphere is a complex process. Soil water enters the roots, through the process of osmosis, and travels through the xylem up the stem or trunk and eventually exits the plant through stomata in the leaves. The majority of

Figure 5.1. Leaf size, number and total area all influence water use rate

stomata are located on the underside of leaves. Plants vary in the manner in which water enters, moves through and exits the plant.

The level of demand placed on the plant to transpire water by the evaporative conditions in the atmosphere is influenced by air temperature, solar radiation, relative humidity and wind. These vary greatly from day to day, during a particular day and seasonally.

The availability of water in the soil, particularly with regards to the forces that need to be overcome to allow movement from the soil into the roots, also varies considerably.

Taking into account all of these potential variations, the actual water use, at any point in time, is obviously dependent on numerous plant, soil and atmospheric factors.

Plants can be broadly grouped according to being low, medium or high water users. Examples of high water users include hydrangeas, rhododendrons, birch trees, cherry trees and grasses, such as ryegrass and fescue. Examples of low water use plants include correas, dianellas (Australian native flax), New Zealand flax, grevilleas, boobialla, native rosemary and yucca.

A more comprehensive list of plants and water use rates is presented in Appendix 4.

An important plant categorisation that is relevant to urban water management is the designation of C3 and C4 plants. This is particularly important for turfgrasses: the largest plant group that is managed and irrigated in urban areas.

The plant size, leaf area, plant architecture, stomata number and geometry and the plant physiology involved in transporting solutions through the internal structure of the plant also vary (Figure 5.1).

The conversion of carbon dioxide to sugars using sunlight is the basis of photosynthesis. The processes and pathways by which this occurs vary between different plants. Plants that have adapted to warmer climates (C4 plants) are different to those that have adapted to colder climates (C3 plants). C4 plants incorporate mechanisms, including stomata number, geometry and behaviour, that limit the rate of water transpired and are more efficient in terms of photosynthesis and water use. C3 plants have higher rates of water use in warmer climates. During colder periods, the C4 plants can become dormant, whereas the C3 plants will continue to

Table 5.1. Turfgrass species: C3 and C4

Main turfgrass categories	
Cool season species (C3)	**Warm season species (C4)**
Kentucky bluegrass	Buffalo grass (St Augustine grass)
Tall fescue	Couch grass
Hard fescue	Kikuyu grass
Perennial ryegrass	Zoysia grass
Annual ryegrass	Seashore paspalum
Creeping bentgrass	Carpet grass
Wintergrass	Centipede grass

thrive. The higher water use efficiency of the C4 plants is a valuable asset for irrigated turf. Table 5.1 lists the grass species according to the two plant categories.

Transpiration and evapotranspiration

The process of water evaporating from within the plant and exiting through stomata is called transpiration. Water loss from a plant is mainly, but not exclusively, from the leaves. Water also evaporates from the soil around the plant and between plants. The sum of the water lost from the soil surface (evaporation) and water used by plants (transpiration) is called evapotranspiration (ET). Evapotranspiration is commonly expressed as millimetres per day (mm/day).

It should be noted that there are several ET terms in use. The term ET_c is used to designate the rate of water use by a plant or crop. The subscript defines the particular use of the term. For example, ET_L refers to the evapotranspiration rate of the landscape. When there is no subscript, then ET is referring to water use by plants generally and not specific to any plant group.

Factors influencing plant water use

There are numerous factors influencing the plant water use rate. The major ones are the:

- plant species
- stage of growth
- weather – the evaporative demand
- available water in the soil
- water quality – salinity.

Climate factors influencing evaporation rate

The main driving force of plant water use is the evaporative demand or the evaporation potential of the atmosphere surrounding the plant foliage (Figure 5.2). The following climatic factors directly influence the level of evaporative demand and hence the rate that plants use water:

- solar radiation – provides the energy for evaporation
- air temperature – the evaporation rate increases with temperature
- relative humidity – indicates dryness of the air
- wind speed – wind increases the rate of water use by the removal of water molecules.

Techniques to estimate water demand

Observation and measurements

The techniques available to estimate the amount of water required by a plant include using assessments of the plant, the soil and the weather.

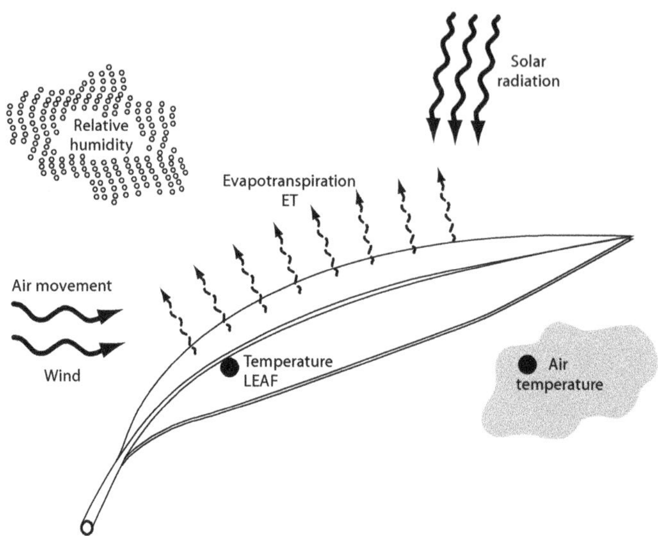

Figure 5.2. Climate factors influencing the plant transpiration rate

A very simple or basic technique is to visually observe the condition of the plant and assess if it is stressed. Some of the visual indicators of plants deficient in available water and stressed are leaf curl, wilting, drooping branches and leaf fall. Plants with large or fleshy leaves are often used as indicator plants. Indicator plants can be used to provide a visual alert (e.g. through drooping) as an early warning that the soil moisture is reaching low levels. Examples include primrose, hellebores, plectranthus, hydrangeas, dogwood and impatiens. It is important that the indicator species selected is suited to the site and appropriate in response to reflect the soil moisture requirements of the main plants.

The use of instruments allows more precise estimates to be made. One technique is to measure leaf temperature, using hand-held infrared thermometer, and determine the level of water stress, which is indicated by how much higher the temperature of the leaf tissue is than the surrounding air temperature. The elevated temperature indicates that the plant does not have enough water available to the leaves for natural cooling of the leaf. The water use rate of trees can be measured using sapflow meters.

Soil moisture measurements can also be used. The most accurate and reliable method of measuring soil moisture is to take a soil sample, weigh it, dry it and then weigh it again. This provides an accurate measure of the actual amount of water in the soil. This is called the gravimetric method. It is, however, labour intensive and time consuming and not practical as a regular monitoring method for plant water use.

A scientific-based technique is to estimate plant water requirements using climate data. This is based on measurement of evaporation or by calculating evaporation from local climate parameters. The actual evaporation rate is available from the Bureau of Meteorology, private climate data providers or from a local on-site weather station.

Water demand estimation using climate data

A generalised equation used for the determination of plant water requirement is:

$$ET_c = \text{Multiplier or adjustment factor} \times \text{Evaporation reference (mm)}$$

where:

ET_c – Plant or crop water use.

Figure 5.3. The amount of water lost from the pan daily is a measure of the evaporation rate

The calculation of water requirements and irrigation generally is expressed as a depth, in millimetres. There are numerous terms used to estimate plant water use rates. The above expression, allows water use to be estimated using a local evaporation rate and adjustment factor to reflect actual water use by the plant.

Evaporative pan estimation technique

There are two evaporation references in use: evaporation from evaporation pan (E_{pan}) (Figure 5.3) and reference evapotranspiration (ET_0).

The first is the reading obtained from the evaporation pan, which is the main evaporation instrument that has been used by the Bureau of Meteorology (Figure 5.3). The ET_0 reference will be increasingly used in the future.

The multiplier factor, in the general expression above, needs to be defined, for each type of evaporation reference that is used to estimate the crop water requirement. When the multiplier factor is used, in conjunction with evaporation pan data, it is called a crop factor (CF) and when it is used in conjunction with a reference crop evapotranspiration (ET_0) it is called a crop coefficient (K_c).

The water used by a particular crop or plant, measured in millimetre depth of water, can be determined using the following equation (Figure 5.4):

$$ET_c = CF \times Epan$$

where:

ET_c – evapotranspiration by a plant or crop, which can be expressed per day, week, month or year

CF – crop factor. The proportion of water used by the crop/plant compared with the water evaporated from a Class A evaporation pan.

E_{pan} – depth of water evaporated from the Class A evaporation pan (mm). The evaporation pan reading is taken daily at 9 a.m. each day.

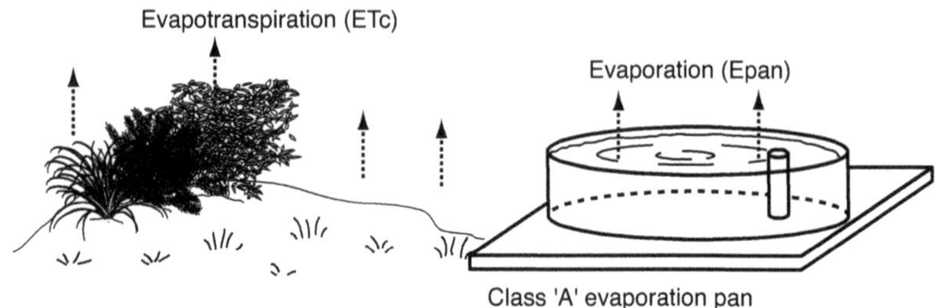

Figure 5.4. Evaporation (E_{pan}) and crop factor relationship

The values of crop factor typically range from 0.2 to 0.8. A crop factor of 1.0 would represent a plant that uses water at the same rate as water evaporates from the open water surface of the evaporation pan. This is a relatively high value for a plant group. Sugar cane, a C4 plant, has a peak CF value close to 1.0. Plants generally use water at a rate less than this. In some extreme cases it is only 20% (or CF = 0.2) of this rate.

Table 5.2 provides a guide to crop factor values for a range of plant types and also plant condition.

Evaporation pan example – plant water use

Turfgrass species: warm season grass – kikuyu, strong growth
Crop factor (CF): 0.5 (refer to Table 5.2)
Evaporation: 8 mm/day (assumed reading from Class A evaporation pan)
ET_C: $(0.5 \times 8) = 4$ mm/day

Reference evapotranspiration (ET_o) technique

An alternative technique to E_{pan}, to estimate plant water use (ET_c), is to use the evapotranspiration rate of a reference crop (ET_o). The reference crop or plant is selected to be a high water user, so other species will use water at a lower and proportional rate to the reference value.

Table 5.2. Crop factor (CF) values for a range of plant types and growth performance levels

Plants	Growth performance	Crop factor
Vegetables, fruit trees	Ample produce	0.85
Lawns, cool season grass (e.g. fescue, ryegrass)	Lush	0.8
	Strong growth	0.7
	Moderate growth, just acceptable	0.65
Lawns, warm season grass (e.g. couch, kikuyu)	Lush	0.6
	Strong growth	0.5
	Some growth, just acceptable quality	0.3
Ornamentals	Establishment, vigorous growth, large leaved	0.8
	Strong growth	0.65
	Moderate growth, somewhat drought-tolerant plants	0.4
	Low growth rate acceptable, plants drought tolerant	0.2–0.4
	Minimum; desert plants	0.1–0.15
Vegetables	Strong growth, fruiting	0.7–0.9

Source: Adapted from Handreck (2008)

The method for the determination of ET_o has developed over time. Experience with the initial Penman estimation in FAO – 24 Report (Doorenbos and Pruit 1977) showed that it tended to overestimate ET_o. Although there were other models including Blaney-Criddle and Thornthwaite and Radiation, these did not provide the required level of accuracy and consistency. The methodologies for the determination of ET_o were revised in 1990 and the current Pennman-Monteith method was adopted.

Reference evapotranspiration (ET_o) is defined as 'the rate of transpiration from a hypothetical reference crop with an assumed crop height of 0.12 m, a fixed surface resistance of 70 sec/m and an albedo of 0.23, closely resembling the evapotranspiration from an extensive surface of green grass of uniform height, actively growing, well-watered and completely shading the ground'.

In addition to changes in climate parameters used, a hypothetical reference crop, rather than a living short green crop that was not species specific, was adopted. This provides a more stable and predictable reference for the determination of crop coefficients. Details of the methodologies are outlined in FAO Irrigation and Drainage Paper No. 56 (Allen *et al.* 1998).

It is important that a source of ET_o data be accessed from Bureau of Meteorology (BOM) or local weather station. ET_o provides a reference evaporation value of sufficient accuracy to make sound irrigation management decisions.

There is a range of evaporation data available. The potential evaporation, the maximum that can occur at a locality (various definitions), is available as Potential Evapotranspiration (PET). This is a calculated value using historical climate data. PET represents the evaporation from water bodies (e.g. small water storages or dams) and is a valuable water resource management tool. E_{pan} evaporation, which is measured, not calculated, is also a potential evaporation rate.

Reference ET_o, which can be determined locally and in real time, is used throughout this publication. It can also be estimated using E_{pan} data and pan coefficient (K_p) (Appendix 3.4).

The method for the determination of ET_o, crop coefficients and PET continue to be developed and refined (Dodds *et al.* 2005).

A local weather station (e.g. on site), which accurately records the various climate parameters, is required to record readings (Figure 5.5) typically on an hourly basis (or more frequently), if required. The data relate to the conditions that are being experienced at that time. A data processing unit, either incorporated into the weather station or in a separate computer, is required to convert the readings into a calculated value of reference ET_o.

In the case of the E_{pan} the reading is taken at 9.00 am so the values relate mainly to the previous day. The accuracy of the readings, the proximity to the irrigated site and the currency

Figure 5.5. Weather station with sensors to provide frequent ET_o estimations

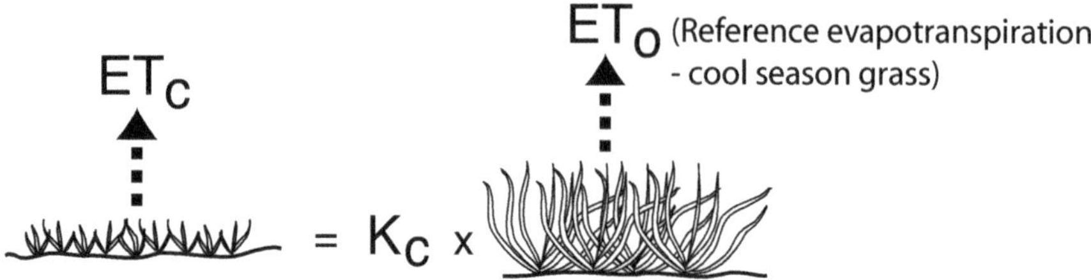

Figure 5.6. The reference evapotranspiration (ET_0) and crop coefficient expression

of the readings are all important factors favouring the use of ET_o over E_{pan} in the real time assessment of plant water needs.

The following climate parameters need to be monitored and accurately recorded, by a weather station, in order to obtain accurate ET_o values:

- air temperature
- solar radiation
- relative humidity
- wind speed
- wind run (km).

Calculating ET_c using the ET_0 formula

ET_c can be calculated using the following equation (Figure 5.6):

$$ET_c = K_c \times ET_o$$

where:
ET_c – Crop evapotranspiration rate (mm)
K_c – Crop coefficient
ET_o – Reference evaporation rate (using hypothetical reference crop)

Evaporation references and sources

Evaporation pan reference data (E_{pan})

The most commonly available evaporation data are available from the Bureau of Meteorology (website: www.bom.gov.au). Long-term historical data are available in the section 'Climate Averages'. Evaporation pan data are, however, only available at limited number of weather stations around the country. Proximity to an official weather station is a consideration for urban water managers.

Table 5.3 is an extract from Adelaide (Kent Town) showing the presentation of average daily evaporation for that locality.

ET_0 reference sources

ET_o data are generally calculated from local weather stations. The data are not currently readily available at each location in Australian urban areas. A valuable weather service in the USA is the Californian Climate Irrigation Management Information Service (CIMIS), where ET_o data are available as a public service, for all locations in that region.

Private irrigation management services, such as Micromet Pty Ltd, use local ET_o data in developing site specific irrigation schedules. ET_o data can be obtained from Bureau of Meteorology website: (www.bom.gov.au/watl/eto).

Table 5.3. Daily (average) climate data for Kent Town, SA

Statistics	Jan	Feb	Mar	Apr	May	Jun	July	Aug	Sep	Oct	Nov	Dec	Annual	No. of years	Dates
Mean daily wind run (km)	263	244	227	203	191	205	220	244	261	273	275	271	240	31	1977–2008
Max wind gust speed (km/h)	80	93	97	78	84	89	84	93	93	89	97	102		30	1997–2008
Mean daily sunshine (hours)	10.4	10.1	8.7	7.2	5.5	4.6	4.8	6.1	6.8	9.3	9.1	9.5	7.6	32	1997–2008
Mean daily solar exposure (MJ/m^2)	27.5	24.5	19.9	13.7	9.5	7.5	8.4	11.7	15.6	20.5	24.8	26.6	17.5	19	1990–2008
Mean number of clear days	11.7	12.5	11.3	8.1	4.7	3.5	3.9	5.2	6.2	6.5	6.9	8.0	87.5	30	1977–2008
Mean number of cloudy days	7.4	6.6	9.2	11.3	15.6	15.9	15.3	13.9	13.2	12.4	10.7	10.9	142.2	30	1977–2008
Mean daily evaporation (mm)	7.2	6.7	5.0	2.1	1.8	1.4	1.5	2.1	3.1	4.4	5.7	6.5	4.0	32	1997–2008

Source of data: Bureau of Meteorology, Climate Averages, Kent Town, Station Number: 023090

Evaporation pan coefficient

The rate of water evaporation (mm/day) from an evaporation pan is higher than the evaporation rate from a high water use grass. This means that the value of the multiplier in each of these equations must be different so that similar crop water use estimate values are determined. The pan coefficient (K_p) is used to determine ET_o using evaporation pan data: $ET_o = K_p \times E_{pan}$

Although an approximation, K_p 0.8, is used here, the actual value depends on several factors, including the siting of the pan and weather conditions. Procedures for determining more accurate values of K_p are outlined in FAO *Crop Evapotranspiration – Guidelines For Computing Crop Water Requirements* (Allen *et al.* 1998).

CF values are lower – typically in the range of 20% lower – than K_c values. For example, a plant species with a CF of 0.65 is the same as a K_c of 0.8.

Comparing E_{pan} and ET_o

Both measures of evaporation have a role in water management. The E_{pan} data have provided a good understanding of this key climate parameter for the past 50 years.

As a weather instrument, the evaporation pan is prone to some inaccuracies. Local microclimate conditions, wind turbulence, pan contamination, thermal behaviour of the pan storage water and manual reading all contribute to some degree of inaccuracy. E_{pan} data are generally not available a local level.

Accurate measurement of all climate parameters used to calculate ET_o is essential.

ET_o weather stations

The ET_o expression

A weather station, designed to generate accurate ET_o values is a precision instrument. There are numerous readings that need to be accurately obtained, and sophisticated mathematical modelling is required, to calculate the reference evapotranspiration rates for that time of day. The mathematical calculation of the water use by a plant is a complex process. The water demand or evaporative demand depends upon the solar energy incident on the plant (and soil), the heat energy in the air, the speed of the wind, the pressure on the water molecules to migrate from within the plant structure to the outside air and the wind speed.

To achieve this, the following parameters need to be accurately measured: solar radiation, air temperature, relative humidity (used for vapour pressure deficit calculation), wind speed and wind run.

Data accuracy requirements

Standards in instrumentation for weather stations are set by the Bureau of Meteorology and compliance with these standards is strongly recommended. The accuracy of a reading can be influenced by a number of factors, including the:

- sensitivity of the sensor to changes in climate parameter
- operating range of sensor
- accuracy of calibration
- stability of the sensor signal (drift characteristics)
- quality of weather station site in representing local conditions
- micro representation of sensor installation (tower/station)
- maintenance of the sensor (dirt, calibration, etc.).

Table 5.4 outlines the accuracy required in the various sensors used to measure the weather parameters in a weather station.

Table 5.4. Accuracy of weather station sensors

Sensor	Range	Unit	Accuracy
Solar radiation	–	W/m	3%
Air temperature	–25 to +60	°C	0.3°C
Wet bulb temperature	–25 to +60	°C	0.3°C
Relative humidity	2 to 100	%	3%
Wind speed	1 to 90	m/s	1 m/s
Wind direction	0 to 360	(°)Degree	5°
Rainfall	0 to 999.8	mm	2%

Source: Bureau of Meteorology (2005)

When purchasing a weather station it is important to know the method of determination of ET_0 and its compliance with Penman-Monteith methodology and also instrumentation compliance with recognised industry standards.

Siting of weather stations

The accuracy of the calculated ET_0 reading depends not only on the accuracy of the sensors, but also the siting of the weather station. The key consideration is to position the weather station so that the measurements are representative of the general weather conditions at the site. Both the location of the weather station and the position of the instruments, within the station structure or tower, are important. Weather stations should be located in open areas and not be affected by nearby buildings, trees and surfaces, such as bitumen roads, car parks, paved areas, that may produce unrepresentative readings.

Wind is an important consideration. It is commonly recommended that the weather station be located at a minimum distance of 10 times the height of nearby buildings or trees.

The solar radiation sensor (solarimeter or pyranometer) needs to have clear access to the sky, throughout all daylight hours.

Some siting problems that occur include:

- The solar radiation sensor is shaded for part of the day by adjacent buildings or trees.
- The site is close to, or on, a hard surface area.
- The sensors are sited on top of buildings, which often increases wind speeds and reduces rainfall. The maximum recommended height for rainfall sensing is 2.0 m.

Maintenance and calibration of weather station instruments

The change in sensor signal output over time (drift) is another important consideration. Relative humidity and solar radiation sensors are prone to drift and require regular recalibration.

Over time, inaccuracies are expected in an instrument array such as a weather station. In addition to the risk of component failure, there will be dirt accumulation, which will affect the output signals of some instruments and sensors to such an extent that they will need to be recalibrated.

The removal of dust on a regular basis from solar radiation sensors and leaves and debris from tipping buckets are two key maintenance considerations. Weather station sensors and tipping bucket rain gauges should be checked and cleaned on a monthly basis as part of a regular irrigation maintenance program.

Calibration of the sensors should be carried out at least every 12 months by an authority or organisation certified or accredited to perform this task.

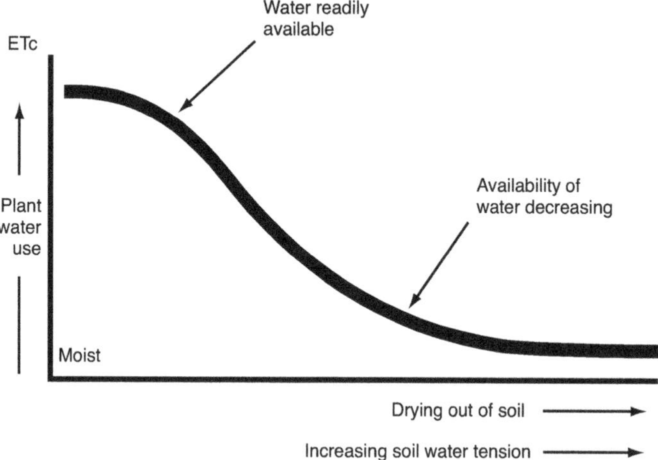

Figure 5.7. Plant water use decreases as available soil moisture decreases

Plant response to available water

With water readily available in the soil, plants will grow more rapidly than if water is not as plentiful or cannot be readily extracted. The use of high crop factor and crop coefficient values results in higher water use estimates and applications of water and so encourages stronger growth and increased water use. In using crop factor and crop coefficient values, it is important to recognise that, if the soil is relatively dry (water not readily available), then the water use may be less than the value estimated (Figure 5.7).

Crop factors and crop coefficients

Determining the crop coefficient

To make an estimation of water use by vegetation two sets of information are required. The first is a value of reference evaporation for the site and the second is the multiplier factor (crop factor or crop coefficient). The value of the crop factor or crop coefficient for a particular crop or plant has generally been developed from field trials. The amount of water used by the vegetation is typically measured using a lysimeter, which precisely weighs the amount of water used by the plant and evaporated from the soil, over a defined period: day, week or a month. The measured amount of water used is then used to determine a value for the crop factor or crop coefficient.

In the case of the crop coefficient (K_c), the value is determined using the following expression:

$$K_c = \frac{ET_c}{ET_o}$$

The value of the crop coefficient is the proportion of water used by the crop (evapotranspiration) to the predicted water used by the reference crop (ET_o). The coefficient value determined reflects the water use characteristics of the plant at that stage of growth and at that time.

Crop coefficient curve

The rate at which a crop or plant uses water varies with the stage of growth of the plant. The way in which the value of K_c (and CF) varies with the time or growth of the plant is described by the crop coefficient curve (Figure 5.8).

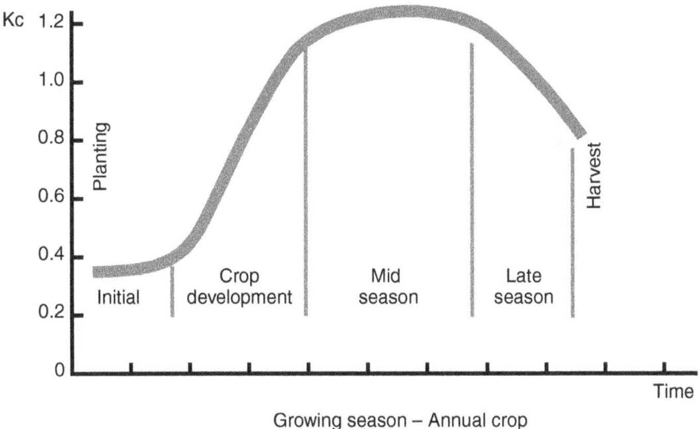

Figure 5.8. Typical crop coefficient curve for growing season of crop/plant

In agriculture a range of K_c values are used to reflect the different water use rates of a growing crop. K_c values are allocated for the initial (K_{ci}), mid-season ($K_{cm\text{-}s}$) and end of crop (K_{ce}) growing stages (Allen *et al.* 1998).

The change in water demand for a fruiting crop, such as tomatoes, is much more marked or variable than it is for a perennial plant, such as a woody shrub. The value of the crop coefficient for a fruiting crop can change markedly in a number of days. The rate of change in water use of perennial plants, used in amenity horticulture, is not as rapid.

Often, in amenity horticulture fields, crop coefficient and crop factor tables present only one value for a species. This is acceptable to gain an appreciation of the maximum amount of water that is required, but the variability that can occur should be appreciated.

The nature of many landscape plantings is that they contain a mixture of plants with different water requirements and different growth patterns. Although the demand for water is relatively higher in summer, plants still use water over the cooler winter months. Figure 5.9 illustrates how the demand pattern may be expected to vary over the full year.

Crop coefficients for ornamental landscape plants

There is only limited information available on water use coefficient values for landscape plants. While agricultural crops have been thoroughly researched, the same quality of data is not available to growers of ornamentals.

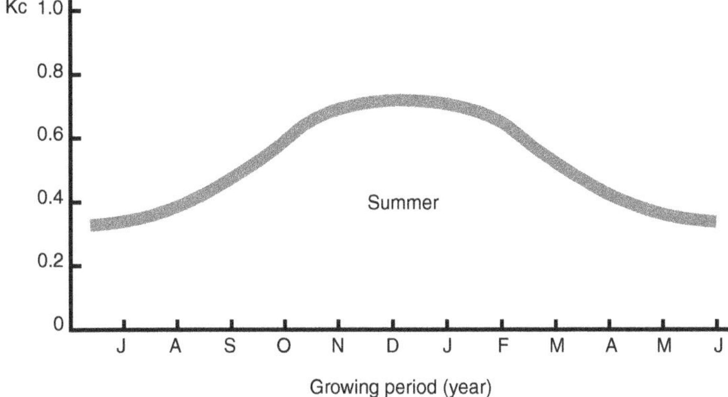

Figure 5.9. Crop coefficient curve for mixed planting landscape (a garden bed) from July to June

Table 5.5. Irrigation requirements for selected landscape plants[1] – results of trial by Pettinger and Shaw (2004)

Common name	Scientific name	Percentage of ET_0	Water use rating[2]
Boobialla (creeping)	*Myoporum* 'Pacificum'	0–36%	Very low
Correa (coastal)	*Correa backhouseana*	18–36%	Very low
Cassia (silver)	*Cassia artemisioides*	0–36%	Very low
Hawthorn (Indian)	*Rhaphiolepis indica*	18–36%	Low
Hibiscus (rose of China)	*Hibiscus rosa-sinensis*	40–60%	Low
Ivy (English)	*Hedera helix*	20–30%	Very low
Lantana (trailing)	*Lantana montevidensis*	18–36%	Very low
New Zealand flax	*Phormium tenax*	18–36%	Low
Pittosporum (Japanese)	*Pittosporum tobira*	18–36%	Very low
Strawberry tree (compact)	*Arbutus unedo*	18–36%	Very low

Notes:
1. This is selection from a trial involving more than 30 species.
2. Adapted to include water use rating adopted in crop water estimates in the Burnley Plant Directory.

An important resource is the *Water Use Classification of Landscape Species (WUCOLS): A Guide to the Water Needs of Landscape Plants* (Costello and Jones 2000), which lists the water use rating of many species. This is a comprehensive guide, but this information has limitations due to the differences in climate regions of California on which it is based.

The document also provides a guide to the crop coefficient values, for each of the water use categories. These are:

Low water use – K_c 0.1–0.3, Average K_c 0.3
Moderate water use – K_c 0.4–0.6, Average K_c 0.5
High water use – K_c 0.7–0.9, Average K_c 0.8

Plant directories, such as the Burnley Plant Directory, University of Melbourne, allow water use ratings for selected landscape plantings to be obtained. The website www.SmartGardenWatering.org.au allows users to access a modified version of this database.

In a trial, carried out by Pettinger and Shaw (2004), over 30 landscape plants were grown under a number of irrigation regimes to establish an irrigation amount required to achieve acceptable performance. The proportion of ET_0 for 10 of these plants is presented in Table 5.5.

These results show that many (9 out of 10) ornamental species maintain their aesthetic and functional value when irrigated at 18% to 36% of ET_0 (Plate 5.1).

Crop factors and coefficients for turf

CF values for turf are presented in several references, including Handreck and Black (2005). Generally, turfgrasses are grouped into cool season (C3) and warm season (C4 plants) categories. The CF values for warm season grasses are generally lower (0.2–0.7) than for cool season grasses (0.6 to 0.9).

The data presented in Table 5.6 illustrate the lower water requirements of warm season grasses. The variability in water use within the broad grass categories is also evident between the different grass species (Appendix 4.5.2). In trials conducted in Western Australia, Short and Colmer (1999) showed that warm season grasses can be maintained in a healthy growth condition with irrigation rates in the range of 50–60% of E_{pan}, compared with 80–100%, for cool season grasses. They found when C3 grasses, fescue and ryegrass, were watered at 40% of E_{pan}, turf colour (greenness) was significantly reduced. At this rate of water availability, the photosynthetic rate was not adequate to maintain adequate chlorophyll (green pigment).

Table 5.6. Guide to comparative ET_c rates for the major turfgrasses

Relative ranking	ET_c rate (mm/day)	Cool season turfgrass species	Warm season turfgrass species
Very low	<6		*Buchloe* spp.
Low	6–7		*Couch grass hybrids
			Centipede grass
			Couch grass
			Zoysia grass
Medium	7–8.5	Hard fescue	Bahia grass
		Red fescue	Seashore paspalum
		Chewings fescue	St. Augustine grass
High	8.5–10	Perennial ryegrass	Carpet grass
			Kikuyu grass
Very high	>10	Tall fescue	
		Creeping bentgrass	
		Wintergrass *(Poa annua)*	

Notes:
1. The daily water use rates relate to grass species grown in their respective climatic regions of adaptation and preferred cultural regime.
2. There is significant variability within and between species.
3. These data provide a guide to relative water use rates only.

Source: Neylan (2007)

Table 5.6 shows the relative rates of water use for turfgrasses. These values, which were based on USA climate conditions, are for high quality, strong growth conditions.

Further detailed water use rates for turfgrasses are presented in Table 5.7. These values have been published by the Center for Irrigation Technology, University of California, Fresno (CIT 2011). They represent values required to achieve vigorous growth and assume that water is readily available. These values demonstrate the way in which water demand varies throughout the year and varies between species.

The use of a single K_c or CF value for grass in water estimation calculations will result in an over-estimation of the requirements throughout the full growing season because the value for a particular species changes throughout the growing season.

Table 5.7. Monthly crop coefficient (K_c) values for selected turfgrasses

Grass	Jul	Aug	Sep	Oct	Nov	Dec	Jan	Feb	Mar	Apr	May	Jun
Bentgrass (CS)	0.61	0.64	0.75	1.04	0.95	0.88	0.94	0.86	0.74	0.75	0.69	0.60
Bermuda (WS)	0.55	0.54	0.76	0.72	0.79	0.68	0.71	0.71	0.62	0.54	0.58	0.55
Bluegrass (CS)	0.61	0.64	0.75	1.04	0.95	0.88	0.94	0.86	0.74	0.75	0.69	0.60
Kikuyu (WS)	0.55	0.54	0.76	0.72	0.79	0.68	0.71	0.71	0.62	0.54	0.58	0.55
Paspalum (WS)	0.55	0.54	0.76	0.72	0.79	0.68	0.71	0.71	0.62	0.54	0.58	0.55
Ryegrass (CS)	0.61	0.64	0.75	1.04	0.95	0.88	0.94	0.86	0.74	0.75	0.69	0.60
St. Augustine (Buffalo) (WS)	0.55	0.54	0.76	0.72	0.79	0.68	0.71	0.71	0.62	0.54	0.58	0.55
Tall fescue (CS)	0.61	0.64	0.75	1.04	0.95	0.88	0.94	0.86	0.74	0.75	0.69	0.60
Zoysia (WS)	0.55	0.54	0.76	0.72	0.79	0.68	0.71	0.71	0.62	0.54	0.58	0.55

Notes:
1. Key: CS – Cool season grass, WS – Warm season grass
2. The values have been transposed to reflect the seasonal months of the Southern hemisphere.

Source: Adapted from CIT (2011)

Table 5.8. Example of seasonal crop coefficient values used in South-Eastern Australia

	Crop coefficient (K_c)				
Vegetation category	**Summer**	**Autumn**	**Winter**	**Spring**	**Summer**
Turf (warm season) – medium quality	0.40	0.30	0.05	0.30	0.40
Turf (warm season)– low quality	0.30	0.25	0.05	0.25	0.30
	Landscape coefficient (K_L)				
Garden bed – mixed plantings – high quality	0.60	0.45	0.30	0.45	0.60
Garden bed – mixed plantings – medium quality	0.40	0.30	0.20	0.30	0.40
Garden bed – mixed plantings – low quality	0.30	0.25	0.15	0.25	0.30

Table 5.8 presents examples of seasonal K_c and K_L values used in the irrigation management of large gardens including lawns and mixed planting garden beds in South-Eastern Australia.

Landscape coefficient (K_L)

The water requirements of a group of ornamental plants or landscape planting are influenced by a range of factors, mainly the water use characteristics of the plant (species factor: K_s).

The water demand of the area is also influenced by other factors. Adjustments need to be made for the nature of the microclimate (microclimate factor: K_{mc}) and the number, coverage, shape and architecture of the plants in the group (density factor: K_d) (Figure 5.10).

The methodology for the development of the Landscape coefficient (K_L) is outlined in Costello and Jones (2000). The Landscape coefficient (K_L) is a single term that accounts for all factors that may influence the water requirements of a group of plants.

$$K_L = K_s \times K_d \times K_{mc}$$

where:
K_s – Species factor
K_d – Density factor
K_{mc} – Microclimate factor

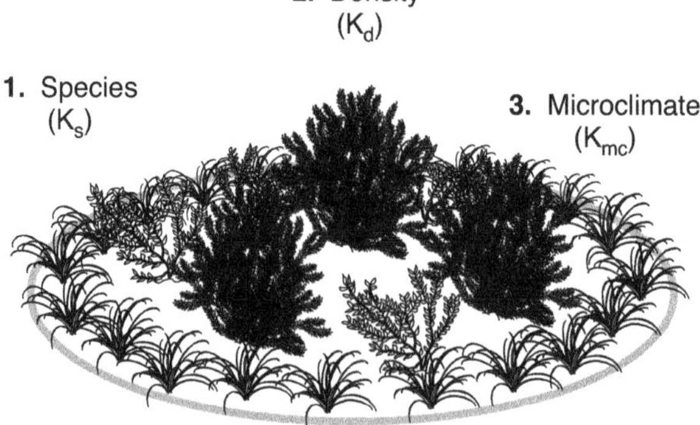

Figure 5.10. Components of the landscape coefficient (K_L) term

Table 5.9. Species (K_s) values for use in landscape coefficient (K_L)

Plant type	Maximum appearance (lush)	Acceptable appearance	Low maintenance (lean)
Trees	0.90–0.95	0.70–0.75	0.45–0.50
Shrubs	0.60–0.65	0.45–0.50	0.30–0.35
Desert plants	0.40–0.45	0.30–0.35	0.20–0.25
Ground cover	0.70– .80	0.50–0.60	0.30–0.40
Mixed (trees, shrubs, ground cover)	0.90–1.0	0.75–0.80	0.50–0.55

Source: Irrigation Association (2011)

The microclimate factor (K_{mc}) takes into account site-specific conditions such as shade, wind and thermal radiation (from buildings). The range of K_{mc} values is generally 0.5–1.2.

The density factor takes into account the foliage coverage of the area and degree of multi-tiered plantings, such as trees with understorey plantings. The range of K_d values is generally 0.3–1.2.

The Irrigation Association (IA) (USA) has adapted the Costello and Jones (2000) data to present them in a form that allows ready use within the various irrigation industry water management training programs including the irrigation auditing courses.

Table 5.9 outlines the range in species values (K_s) expected for the various factors that determine the landscape coefficient (K_L) value.

The values presented in Table 5.9 cover a broad range of species within, each plant type. Crop coefficients for the particular species should be used if they are available.

The role of the microclimate factor is to adjust the estimated water demand for site conditions where the demand is significantly increased or reduced due to siting and physical features and structures that alter the solar, thermal and atmospheric properties experienced by the plants (Table 5.10). Roof overhangs, which shade plants and garden beds situated on the southern side (Southern Hemisphere) of buildings, usually require the use of K_{mc} to decrease value of estimated demand. Increasing the estimated demand is usually limited to exposed sites, full sun, which are windy and so plant water demand would be higher than expected for a typical site. When using Table 5.10, the low and high values should be viewed as the extreme values and not typical values.

The role of the density factor is to adjust the estimated water demand for site conditions where the demand is significantly increased or reduced due to the density of the planting (Table 5.11, Figure 5.11). If, for example, the planted area, including foliage cover, is less than 50% of the total garden bed area, then the estimated demand will be reduced (Figure 5.11). If the foliage coverage is between 70% and 100%, then it is suggested that no reduction be made. In situations where there is a combination of high demand from trees and very dense plantings

Table 5.10. Microclimate factor (K_{mc}) for different plant types

Vegetation	High (exposed, windy)	Average (sunny)	Low (shaded)
Trees	1.2–1.4	1.0	0.5–0.8
Shrubs	1.2–1.4	1.0	0.5–0.8
Ground cover	1.2–1.4	1.0	0.5–0.8
Mixed planting	1.2–1.4	1.0	0.5–0.8
Turfgrass	1.2–1.4	1.0	0.5–0.8

Source: Irrigation Association (2011)

Table 5.11. Density factor (K_d) for different plant types

Plant type	¼ to ½ ground cover	½ to ⅔ ground cover	More than ¾ ground cover
Low growing plants <400 mm	0.35–0.45	0.60–0.75	0.80–0.95
Small shrubs approx. 1 m high	0.35–0.50	0.70–0.80	0.85–0.95
Large shrubs, trees >4 m	0.40–0.55	0.75–0.95	0.95–1.0
Turfgrass	n/a	n/a	1.0

Source: Irrigation Association (2011)

within the garden bed, then it is reasonable to adjust the estimated demand above the typical full coverage condition values.

Comments on K_d and K_{mc}

For both K_d and K_{mc}, the conditions need to be severe in order to adjust the values. For example, in the case of K_d, if the open space represented 30–40% of the total area, then an adjustment should be made, for example, K_d 0.6. K_{mc} is adjusted if the site is shaded, or very windy and exposed. Landscape plantings that would warrant higher K_d values are shown in Figures 5.12 and 5.13.

Example – Landscape coefficient

Site: garden bed
Planting: hydrangeas (high water use) (K_c or K_s = 0.8)
Planting density: sparse (K_d =0.8)
Microclimate: sunny/normal (K_{mc} =1.0)

Figure 5.11. Sparsely planted garden bed that requires a reduced K_d to estimate water demand

Figure 5.12. High density and multilayer plantings require the use of a K_d greater than 1.0

ET_o: 40 mm/week (from reference tables (e.g. Appendix 3.4) or weather station)

$$K_L = K_s \times K_d \times K_{mc}$$

$$K_L = 0.8 \times 0.8 \times 1.0 = 0.64$$

$$\text{Weekly water use } (ET_c) = 0.64 \times 40 = 25.6 \text{ mm/week}$$

Figure 5.13. Example of high density garden bed at the Royal Botanic Gardens, Melbourne

Figure 5.14. Surface performance standard and presentation quality need to be determined for each site and each activity

Vegetation performance and crop coefficients – turf

Irrigated public open space (IPOS) Adelaide turf quality parameters

The *Code of Practice for Irrigated Public Open Space,* published by SA Water (2008), uses a graduated turf quality approach to determine the water budgets for sports grounds and similar sites. The surface needs to be assessed in terms of activity type, intensity of use, surface quality and presentation standards required (Figure 5.14). The turf quality ranges from 'acceptable' to 'lush', and is represented by four levels or standards (Plate 5.2). These are referred to as the turf quality visual standards (TQVS). For each performance level or standard, an image of the turf condition is used to provide a reference.

The amount of water required increases as the turf standard increases. Table 5.12 presents the K_c and associated factors relevant to each standard. The various visual standards relate to nominated turf situations (Plate 5.3). These are:

- elite sports – state/national competition (AAMI stadium/Adelaide Oval)
- premier sports – district/regional level competition (district turf cricket grounds/athletic tracks)
- local sports turf – local level competition. Local sports grounds/community parks
- passive recreation turf – non-sports turf. neighbourhood parks/passive recreation.

In this method, the water requirement is determined through the use of a soil moisture stress factor (K_{ws}), which adjusts the overall coefficient value to reflect the lower water demand. In some methods a separate stress factor is not used and instead the reduction is applied directly to the crop factor/crop coefficient value.

The following expression is used to determine the turf water requirements:

$$ET_c = K_c \times K_{ws} \times ET_o$$

Table 5.12. IPOS crop coefficient values for turf quality visual standards (warm season grass)

TQVS classification number	Ground/field designation	Crop coefficient (K_c)	Water stress factor (K_{ws})	Combined coefficient (K_t)
1	Elite sports turf – state/national	0.7	1.0	0.7
2	Premier sports turf – district/regional	0.7	0.6	0.42
3	Local sports turf – community parks	0.7	0.5	0.35
4	Passive recreation turf – neighbourhood parks	0.7	0.4	0.28

Notes:
1. A warm season grass crop coefficient (K_c) is used to determine ET_o.
2. K_t is the product of K_c and K_{ws}.

Example – IPOS turf quality standard

Determine the daily ET_c rate for the following sports ground which is required to provide Premier Sports rating surface. Grass is warm season grass, e.g. kikuyu.
Premier sports ground: TQVS 2.
Crop coefficient: from Table 5.12 $K_c = 0.7$, K_{ws} 0.6
Daily evaporation: 7.2 mm/day (ET_o data for Adelaide, from Bureau of Meteorology)

$$ET_c = K_c \times K_{ws} \times ET_o$$
$$ET_c = 0.7 \times 0.6 \times 7.2$$
$$= 3.0 \text{ mm/day}$$

Adjusting crop factors and crop coefficients for soil moisture stress

The K_c or CF value can be modified or adjusted to accommodate varying vegetation 'quality' levels as part of water conservation strategies. The terms 'allowable stress' and 'deficit irrigation' are sometimes used to describe this approach.

Ornamental plants are often managed to achieve a quality performance level that is lower than what might be considered for optimum or lush growth.

The adjustment of plant water use demand can be made either through direct adjustment of the K_c value or in a separate step using the water stress factor (K_{ws}), as shown above. The landscape coefficient (K_L) is also used to provide for overall adjustments due to soil water stress.

Site specific K_c, CF values – adaptive management approach

The K_c or CF value that will provide the highest water efficiency at a site is determined through a process of using an initial value, usually a high value, and then, by monitoring the response of the turf or landscape plantings to this value, adjusting it until a value is obtained that matches the site conditions and performance required. This process is referred to as adaptive management.

To be able to do this effectively, the irrigation system should be:

- uniform in application
- precise in control
- monitored in terms of water use.

Crop coefficient value adjustments can be initially made by 10% increments (changes to K_c of 0.10). For example, initial value selected for medium use shrubs may be K_c 0.60. Following a period of watering, say 2 to 3 months, and if the plants are healthy and performing well, a lower value of K_c 0.50 may be used. This process may be continued to the 5% level (K_c changes of 0.05).

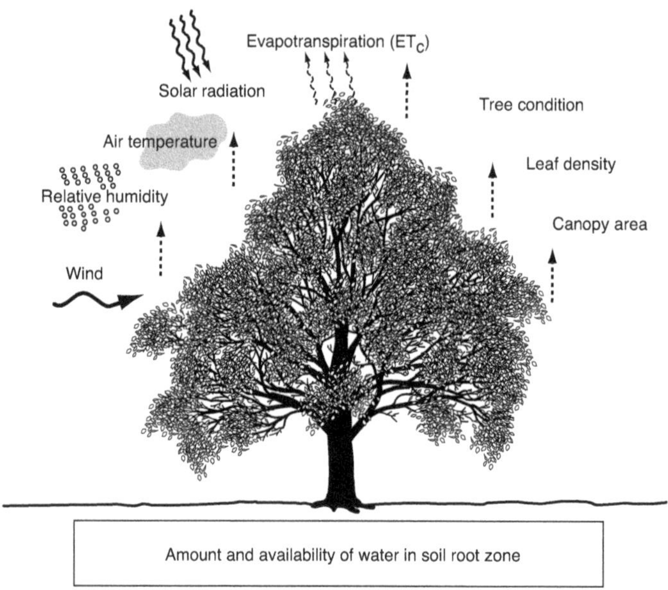

Figure 5.15. Many site and climate factors influence tree water use demand and behaviour

As an example, through a process of adaptive management, a reduced K_L or adjusted K_c value of 0.30 is used by the Royal Botanic Gardens Melbourne for some of the low water use garden bed areas. A K_c value of 0.5 was initially used at some sites.

Estimating tree water use

Estimation expression

Both the total amount of water used by the trees over a period (e.g. summer) and the water use rate are important in the overall water management of urban trees. The rate at which trees use water depends on numerous factors (Figure 5.15), including the:

- water use characteristic of the tree species (high, medium or low)
- size of the tree – the canopy area is the main factor
- density and area of leaves – as reflected by the leaf area index (LAI)
- site climate – the evaporative demand
- condition or health of the tree
- stage of development of the tree
- availability of water to the tree – trees, like all plants, use less water if it is not readily available in the root zone.

The key values that need to be considered when estimating the water use rate of a tree are:

- crop coefficient (K_c) or crop factor (CF). Note K_c is typically in the range from 0.2 to 0.9 for trees (Appendix 4.1).
- area (for transpiration) of tree canopy
- reference evaporation rate (ET_0).

The total foliage area influences the total potential transpiration rate of the tree. The leaf area index (LAI) provides a quantitative measure of the total leaf area available for transpiration. These data, however, are not readily available. LAI values are typically in the range of 2.0–5.0 for trees.

The relative water use rate and guide to K_c value for trees are presented in Appendix 4.

The tree water use estimation relationship is as follows:

$$\text{Daily water use} = K_c \times (C_A) \times ET_o \quad \text{L/day}$$

where:
K_c – Crop coefficient (includes species, LAI properties, competition with grass)
C_A – Canopy area of the tree (calculated assuming crown is a circle) (m^2)
ET_o – Evaporation rate (Bureau of Meteorology or weather station) (mm/day)

Example – tree water use estimation

Tree species:	*Eucalyptus* (various)
Overall height:	10.0m
Crown diameter:	5.0 m
Canopy plan/ground area (C_A):	19.6 m^2 (determined from area of a circle)
Crop coefficient (K_c):	0.6 (healthy growth, medium water use rate)
Evaporation (ET_o):	8 mm/day (summer day in south-east Australia)
Daily water use:	94.1 L/day (658.7 L/week)

The evaporation rate selected, in this example, is a typical daily value for Melbourne and on many occasions in summer this would be exceeded.

It should be noted that the volumes involved do not necessarily need to be supplied through irrigation. Except for trees growing with severely restricted root systems, such as trees growing in containers, water will be available from the soil, supplied by rainfall.

The water use rate calculated in this example represents a particular set of conditions. A single value of K_c has been used to reflect 'healthy growth'. The tree will survive and maintain condition on lower rates.

Figure 5.16 shows that the actual water use rate of a tree depends on the condition of the tree. If a tree needs to be maintained in a lush condition, with very strong growth, then large amounts of supplementary water may be required, if it is not supplied through rainfall. Trees can survive on relatively low amounts of water, with expected crop coefficient values in the range of 0.1–0.2.

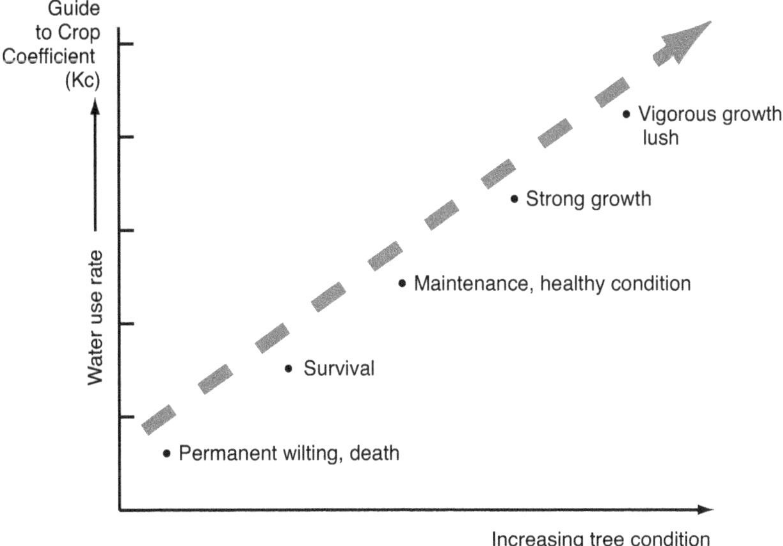

Figure 5.16. Tree water use rate is dependent on tree conditions

Table 5.13. Guide to turf establishment watering program (grass sprigging)

Time	Watering	Plant response
Week 1	3–6 times per day	–
Week 2	2–4 times per day	Root initiation
Week 3–5	4–6 times per week	Root extension
Week 6–8	2–3 times per week	Root extension and increased root density Rhizome spreading
Week 9–12	1–2 times per week	Root development Good grass coverage
Week 13 onwards	As required to keep soil moisture optimum	Healthy established plant, continued root development and growth

Water for turf establishment

A significant amount of water is initially required to successfully establish water-efficient grasses, such as kikuyu and couch. The role of the water is to ensure that healthy grass plants with good root systems are established, so that the ongoing reduced water requirements and water use efficiencies are achieved.

The establishment water performs several functions, including:

- keeping the newly planted vegetative matter in a moist condition
- providing a soil environment that is conducive to root initiation and development
- preventing overheating and burning of the grass when conditions are hot and dry.

The amount of water required to meet these requirements will depend upon the time of year of planting, soil, site conditions and weather. The amount of water applied is in excess of the water that is transpired by the grass. Water is used to provide an environment (surface and soil) in which the plant will grow and the roots will form, extend and the root system develop.

Table 5.13 illustrates a typical watering program for the sprigging or line planting of warm season grasses.

These watering regimes can require a lot of supplementary water. It is important that only light applications, say 3 to 5 mm, are made in the early stages. If rain occurs, then the need for irrigation will be significantly reduced. Although these applications apply excess water, they are needed to achieve successful establishment of grasses.

The volume required to establish a sports turf area may be in the order of 3 to 5 ML/ha if establishment occurs over the warmer summer months and significant rainfall is not received.

In the case of warm season grasses replacing cool season grasses, the water investment in the establishment phase, is generally recovered in the first 2 to 3 years.

Water budgets and annual irrigation volumes (AIV)

Water budget terms

Good water management requires knowledge of the amount of water being used and the water required. Efficient irrigation depends on knowing how much water should be applied at any point in time to meet the needs of the landscape. The amount water is required may be in the short term, days, and in the longer term, seasonally or annually.

The amount of water required should be estimated on a monthly and annual basis. The monthly estimate is used to guide and monitor irrigation management and the annual amount is used as part of planning and overall evaluation. Annual volumes are also sometimes used for requesting or justifying water allocations.

The following terms are relevant in the determination of water budgets:

- plant water requirement (PWR) – water used by plants
- effective rainfall (P_{eff}) – rainfall entering the plant root zone
- net water requirement (NWR) – depth of water (mm) or deficiency that needs to be made up through irrigation
- irrigation water requirement (IWR) – depth of water (mm) that needs to be applied including allowance for inefficiency of irrigation system
- (monthly) water budget – volume of irrigation water required for a particular month
- base irrigation requirement (BIR) – monthly irrigation volumes in IPOS Program
- annual water budget or annual irrigation volume (AIV) – total volume of water required for the year or irrigation season.

Irrigation volume

The total volume of water required for irrigation depends on a number of key factors including:

- total irrigated area
- crop/vegetation type and condition (crop coefficient or crop factor)
- climate – evaporation and rainfall
- irrigation efficiency.

Plant water requirements (PWR)

Initially the landscape or turf PWR or ET_c is determined (refer to the section 'Techniques to estimate rate of water use' earlier in this chapter).

$$\text{PWR or } ET_c = CF \times E_{pan} \text{ or } ET_c = K_c \times ET_o$$

Rainfall contribution

Rainfall will reduce the need for irrigation water. However, not all rainfall is effective: some is lost through runoff, evaporation, deep drainage below the root zone or is surplus to the amount that can be stored in the soil root zone. It is common to assume that 50% of rainfall is effective for plantings with relatively shallow root systems, such as turf. For landscapes with deep root systems, higher rainfall effectiveness factors can be used. However, caution is required for mulched areas, which usually absorb some rainfall and so reduce the effectiveness in terms of adding to soil moisture. Effective rainfall is covered in more detail in Chapter 9.

Net water requirement (NWR)

The irrigation system is required to supply the deficiency between the water needs of the turf and the rainfall contribution.

$$NWR = ET_c - P_{eff}$$

Irrigation system efficiency (IE)

Some water applied by the irrigation system does not reach the root zone and is effectively lost. This amount needs to be allowed for in estimating the irrigation needs of the site. A guide to irrigation efficiency values can be obtained from Table 4.16.

Irrigation water requirement (IWR)

The depth of water that needs to be supplied to the site is determined from the NWR for the appropriate period (e.g. month, year) and the Irrigation Efficiency (IE).

$$\mathrm{IWR} = \frac{\mathrm{NWR}}{\mathrm{IE}}$$

Conversion from irrigation water requirement depth (mm) to water volume

The IWR which is expressed in depth (mm) can be converted to water volume (kL or ML) using the following expression:

$$\text{Irrigation volume (V)} = \text{Irrigated area (m}^2) \times \text{Irrigation water requirement (mm) (L)}$$

Note: V/1000 = kL; V/1 000 000 = ML

Determination of monthly water budgets and annual irrigation volume

The monthly water budget is calculated using the IWR for the month and area to be irrigated.

$$\text{Monthly water budget} = \text{IWR (month)} \times \text{Irrigated area (m}^2)$$

The determination of the annual irrigation volume (AIV) or annual budget is determined by summing the monthly totals for either the whole year or the irrigation season. The monthly water budgets are valuable in determining irrigation schedules and monitoring the performance of the irrigation.

Example – water budgets and annual irrigation volume

Table 5.14 presents the calculation of annual irrigation requirements. The following data has been used in the example:
Site area: 1.4 ha (14 000 m^2)
Climate data: These are assumed values of evaporation and rainfall for this site. Actual site climate details can be obtained from Appendix 3 or through the Bureau of Meteorology, Climate Averages.
Plant: turf, warm season grass – couch
Crop factor (CF): 0.5 (strong growth)
Note: A single crop factor value has been used throughout the irrigation season.
Irrigation Efficiency (IE): 75% (assumed)
In this example, the annual irrigation water requirement/volume (AIV) is 6.572 ML.

The water budget can be prepared using crop coefficients (K_c) and reference evapotranspiration (ET_0) data.

Comments on water budgets and annual irrigation volume

The analysis shows that the expected total volume of water required to irrigate the sports field is 6.57 ML/year. The total depth of irrigation to be applied is 469.4 mm.

The monthly volumes can be used as target volumes during the actual irrigation of the site. It is important to recognise that the values predicted in a water budget analysis are determined using average historical climate data and the volumes actually required to be efficient need to be based on the climate data experienced during the irrigation season.

The impact of the plant water use rate on monthly water budgets is shown in Figure 5.17. In this case, it was assumed that the annual rainfall was 735 mm and the annual ET_0 was 1380 mm. Four K_C values (0.2, 0.4, 0.6 and 0.8) were used in the budget calculation. Effective rainfall was assumed to be 50% and the irrigation efficiency 75%. The water budget is expressed in mm. To calculate the volume, the monthly irrigation requirement is multiplied by area (m^2).

Climate scenarios for estimating future irrigation water volume

A range of rainfall scenarios can be used to estimate irrigation volumes. The following are some examples:

Table 5.14. Preparation of water budgets and annual irrigation volume

Parameter	Jan	Feb	Mar	Apr	May	Jun	Jul	Aug	Sep	Oct	Nov	Dec	Total
Evaporation (E_{pan}) (mm)	200	160	140	100	66	40	54	62	90	130	160	180	1382
Rainfall (mm)	48	40	50	60	70	70	64	70	72	70	60	56	730
Effective rainfall (P_{eff}) (mm)	24	20	25	30	35	35	32	35	36	35	30	28	365
Crop factor (CF)	0.5	0.5	0.5	0.5	0.5	0.5	0.5	0.5	0.5	0.5	0.5	0.5	
ET_c (mm)	100	80	70	50	33	20	27	31	45	65	80	90	691
Irrigation efficiency	0.75	0.75	0.75	0.75	0.75	0.75	0.75	0.75	0.75	0.75	0.75	0.75	
Irrigation water requirement (mm)	101.3	80.0	60.0	26.7	-	-	-	-	12.0	40.0	66.7	82.7	469.4
Monthly water budgets volume (kL)	1418	1120	840	374	-	-	-	-	168	560	934	1158	6572

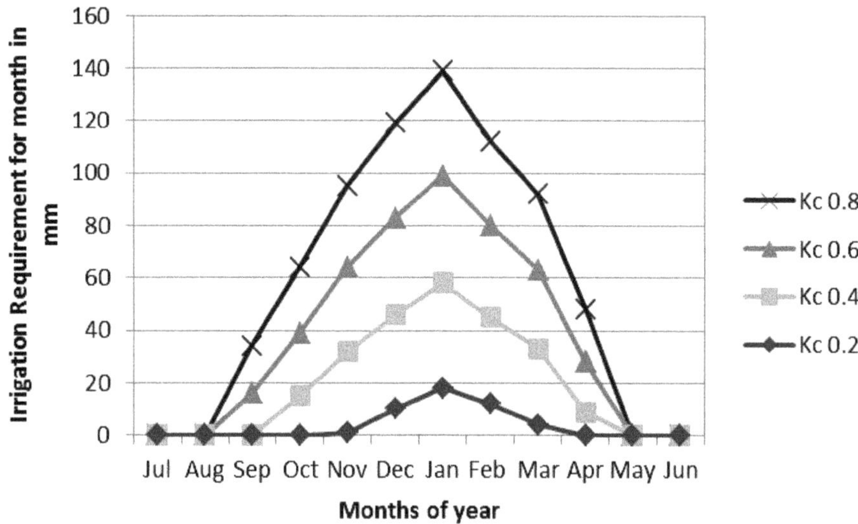

Figure 5.17. Monthly water budgets for range of K_c values

- long-term climate data – average or mean rainfall
- long-term climate data – decile 5 rainfall
- modified long-term climate data
- dry year (e.g. decile 1 rainfall, specific year in recent period)
- wet year (e.g. decile 9 rainfall, specific year in recent period)
- no rainfall year.

Planning for future water demand can be approached in many different ways. Although average data provides the basis for understanding the typical situation, the use of decile 1 rainfall provides a sound basis of what could be expected in a dry year; that is, 1 year in 10.

The irrigation water demands of a large parkland area were estimated using a number of scenarios in the example presented in Table 5.15. The scenarios were as follows.

1. The long-term data were for all years from 1855 to 2008 for rainfall and from 1955 to 2008 for evaporation. This indicates the extensiveness of data from the weather station being referenced.
2. The modified long-term data used the long-term evaporation data and reduced long-term rainfall. The amount of rainfall reduction, 17%, was determined from the analysis of the rainfall of the last 10 years at the Melbourne Regional Office location. A fixed percentage reduction was applied to each month of the long term average rainfall data.
3. The 'dry year' selected was 2003, which is the eighth driest year in the past 20 years. This year was used because there was a relatively dry summer (the third driest in the past 20 years. Summer rainfall was 101.8 mm in 2003, 72.8 mm in 2001 and 41 mm in 1997.

Table 5.15. Annual irrigation volume (AIV) for various climate scenarios

Scenario	Wet year	Long-term data (average)	Modified long-term data	Dry year	No rainfall
Water budget (ML)	11.45	15.00	17.70	25.05	36.55
Annual rainfall (mm)	793	650	542	493	0
Annual evaporation (mm)	1091	1215	1215	1171	1215

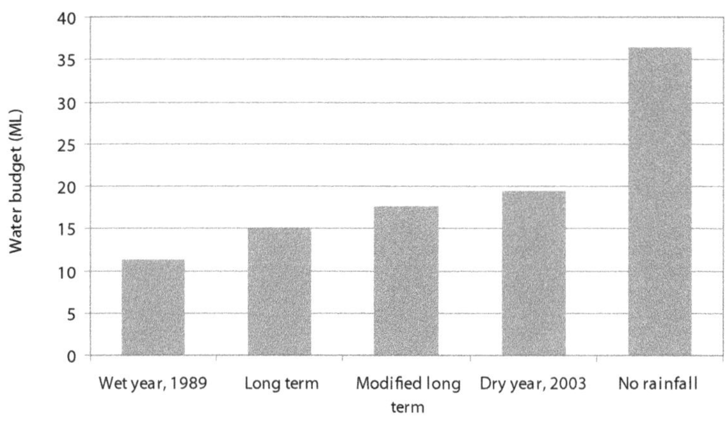

Figure 5.18. Graphical presentation of irrigation estimates for annual irrigation volume (AIV)

4. The 'wet year' was 1989, which is the fourth wettest year in the past 20 years, commencing in 1980. In 1989, there was 793 mm rainfall. There was 803 mm in 1995, 830 mm in 1992 and 844 mm in 1993 (wettest year).
5. In the 'no rainfall year', the long-term average evaporation was used in the water budget calculations.
   Table 5.15 is an example of water budgets prepared using the above scenarios.
   Figure 5.18 is a graphical presentation of the water budgets for nominated scenario years.

Plate 1.1. Roman water pressure regulation tower used for controlled delivery of water to the residents of Pompeii (100 AD)

Plate 1.2. Botanical landscape plantings provide high quality amenity and conservation (Royal Botanical Gardens, Melbourne)

Plate 1.3. Trees and grass create microclimates that provide relief during hot summer conditions

Plate 1.4. Salisbury wetland project achieving sustainable use of water through integrated stormwater and aquifer storage

Plate 2.1. Consequences of drought and water restrictions can be severe: bare grounds not able to deliver recreational services to the community

Plate 2.2. Green space provides many benefits including improved physical and mental health

Plate 2.3. Sports fields and surrounds provide for both active and passive recreation (Victor Trumper Oval, Paddington, NSW)

Plate 2.4. The grass-covered roof of Parliament House, Canberra, modifies the thermal environment of the building and reduces input energy requirements

Plate 2.5. Irrigated green space provides valuable habitat in urban areas

Plate 3.1. Stormwater runoff should be used rather than running to waste

Plate 3.2. Below-ground storage (300 kL) using plastic crate modules with plastic liner under construction

Plate 3.3. Protection of water quality in storages, including algae prevention, is a key water management issue

Plate 4.1. Good system design is the foundation of high efficiency of sprinkler systems

Plate 4.2. High-quality surfaces such as golf greens require excellent uniformity of application

Plate 4.3. Ring main supplying drip laterals for sports ground subsurface drip irrigation

Plate 5.1. Drought hardy and low water use plants are well suited to many urban landscapes (Royal Botanic Gardens Melbourne garden bed irrigated using landscape coefficient K_L 0.3)

Turf Quality Visual Standard

(Rhizomatous sp. Kikuyu/Couch)

	Classification No 1. **Elite Sports Turf** - State/National Competition AAMI Stadium/Adelaide Oval
	Classification No 2. **Premier Sports Turf** - District/Regional Competition Regional Cricket/Football/ Athletics
	Classification No 3. **Local Sports Turf** - Local Competition Local Sports Grounds/Community Parks
	Classification No 4. **Passive Recreation Reserve** - Non Sports Turf Neighbourhood Parks/Passive Reserves

Plate 5.2. IPOS turf visual quality rating photo reference (Source: J. Townsend, Micromet Pty Ltd, *pers. comm.*)

Plate 5.3. Elite turf surfaces such as the Melbourne Cricket Ground require the highest turf surface rating to be applied

Plate 7.1. Contact sports require turf surfaces to be uniform and provide a safe surface for impact

Rain Bird Corporation Agri-Products Div.
Uniformity Evaluation

Sprinkler Name	Rain Bird	Base Pressure (KPA)	310.3
Sprinkler Model	5000	Riser Height (CM)	0.0
Nozzle Size	3.0 RC	Set Screw Setting	NA
Flow Rate (LPM)	11.66	Degree of Arc	360
Date/Time of Test	04/22/02 13:57	Mins./Revolution	2.00
Testing Facility	User Created	Record Number	9353_15
Comment	3.0 Catalog data at 45 psi - rotor 2 - n		
Comment	Arc, mins/rev., and appl. rate modified to assume 360° arc		

Distr. Uniformity	77%	Min (mm/Hr)	2.1	Spacing
CU (Christiansen)	81%	Mean(mm/Hr)	3.4 3.6 (Theor.)	Rectangular
Sched Coeff (5%)	1.3	Max (mm/Hr)	34.7	14.0M x 14.0M

Plate 8.1. Sprinkler uniformity analysis software allows irrigation designers to select optimum combinations of products, nozzles and layouts (Source: Irrigation Design Department, Reece Pty. Ltd.)

Plate 9.1. Exposed fine water absorbing tree roots of elm tree close to trunk at Waite Institute, Adelaide

Plate 9.2. Rainfall collection rig used to measure rainfall passing through tree canopy (Royal Botanic Gardens Melbourne)

Plate 9.3. Narrow, irregular shaped areas are difficult to water efficiently with sprays

Plate 10.1. Rotating stream spray provides more uniform and lower application rates

Plate 11.1. Uneven grass cover is evidence of non-uniform irrigation application. Site: Arden Street, North Melbourne. (Source: NearMap Pty Ltd, www.nearmap.com/)

Plate 11.2. Typical grid layout pattern for sports ground system evaluation

Plate 11.3. Example of more intensive testing of lawn sprinklers using in excess of 40 catch cans

Plate 12.1. Planning that includes all water on the site is required to achieve efficient and sustainable use of water

Chapter 6

Managing soil water and irrigation scheduling

Soils and plant growth

Role of soil

Soil is often treated very simplistically in water management discussions. It is represented in terms of a water reservoir that is filled, emptied and refilled. Soil, in fact, is a very complex medium and influences plant growth and water behaviour in many ways. The foundation for a sustainable landscape and turf sites is a healthy soil.

In addition to being the source of water and nutrients for plant growth, the soil environment, including temperature, air content and organic matter, needs to provide the conditions that support healthy growth. The soil conditions also need to physically allow for root initiation and extension to allow the plant to exploit available resources.

One of the important roles of the soil is to provide structural support or anchorage of large plants, such as trees (Figure 6.1). Soil is much more than a simple water reservoir.

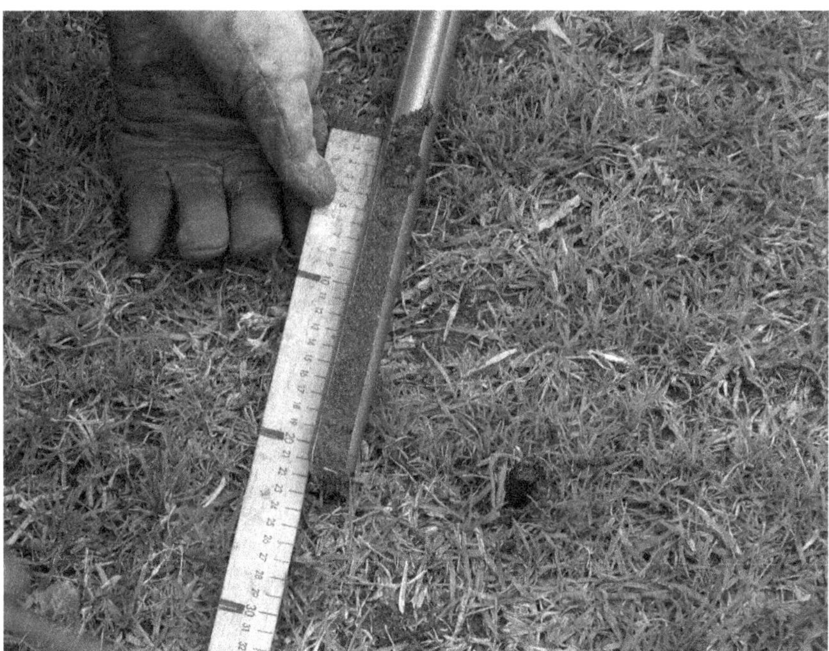

Figure 6.1. Examination of soil section provides information about the soil type, structure and root system

Table 6.1. Soil particle size ranges and objects to demonstrate relative sizes

Soil particles	Soil particle diameter (mm)	Objects with similar relative sizes
Gravel	>2	
Very coarse sand	1–2	Soccer ball
Coarse sand	0.5–1	Tennis ball
Medium sand	0.25–0.5	Golf ball
Fine sand	0.1–0.25	Play marble
Very fine sand	0.05–0.1	Match head
Silt	0.002–0.05	Sesame seeds
Clay	<0.002	Table salt

Source: McIntyre and Jakobsen (1998)

The properties that a soil exhibits and, hence the way it influences plant growth, depend on soil types, soil structures, soil layers, nutrient availability and organic content. Healthy plant growth can be achieved when all of the influencing soil factors are catered for and are in balance.

The soil is made up of the following parts:

- mineral particles
- water
- air
- organic content.

Soils are highly variable in composition and properties. The proportion of solid, air and water varies greatly. Two key terms are commonly used to describe soils: soil texture and soil structure. Soil texture is the relative amount or proportion of sand, silt and clay particles in the soil. The organic matter content is not included in the classification of soil texture.

Soil particles are defined in terms of diameter. Soil particles have different shapes and sizes. The largest soil particle classification is gravel, which is greater than 2 mm, and the smallest is clay, which is less than 0.002 mm (Table 6.1).

Soils are categorised on soil particle proportions. Although, for simplicity, soils are sometimes referred to as being sand, loam, sandy loam, silt and clay, there are several other classifications that are used. These include silt loam, sandy clay loam, clay loam, sandy clay and silty clay.

Soil particle analysis is used to classify a particular soil, according to a soil texture diagram or triangle. There are several soil texture triangles: the USDA and Australian triangles are commonly used in irrigation (Figure 6.2). The main variation between these two is in the definition of silt and sand particles. Australian soil texture defines silt as 2 to 20 microns and USDA defines silt as 2 to 50 microns. Irrigation auditing references, which mainly originate in the USA, tend to use the USDA soil texture triangle.

The important issue in water management is the determination of the water properties of the site soil, rather than the formal classification of the soil.

Soil structure is defined as the arrangement of particles and pores in soils. The aggregation or clumping together of particles that provides an openness or porosity of the soil is a key feature. The organic content is another important component of soil structure. Several factors, including the origin of the soil, organic content and recent soil management practices influence soil structure. Soil structure is also important in terms of influencing the water properties, such as water-holding capacity and infiltration rate.

Healthy soil environment

A soil that can support healthy plant growth is a core requirement of a sustainable landscape.

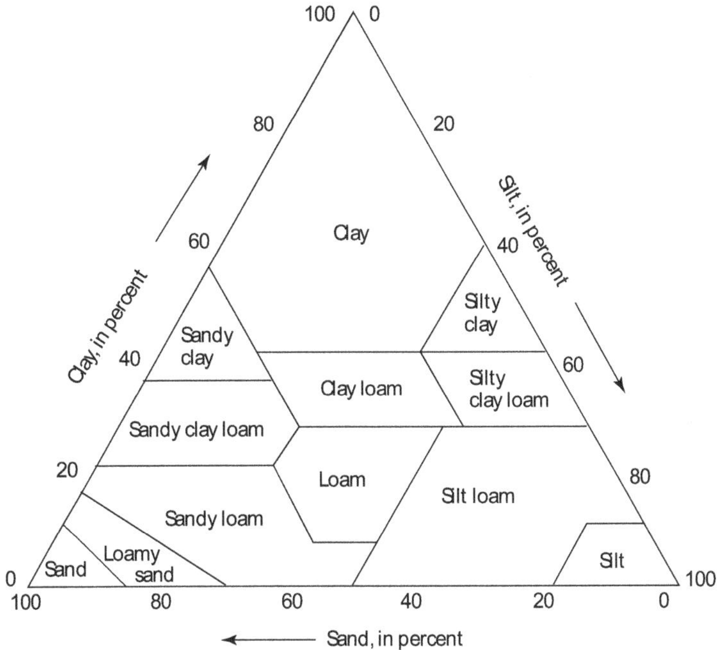

Figure 6.2. Soil classification triangle (Source: adapted from USDA 1967)

Soil properties can be evaluated in terms of the following:

- physical properties
- chemical properties
- biological properties.

The following series of lists show the soil characteristics that could be considered necessary for a soil environment that will support healthy landscape plantings.

The physical properties of good soil include:

- generous water-holding capacity
- good infiltration rate
- ready water movement within soil profile (Permeability)
- open structure to allow gaseous exchange
- well structured (not compacted)
- stable aggregates (resist breakdown)
- acceptable density (allows air and water movement and root expansion).

The chemical properties of good soil include:

- availability of nutrients in adequate and balanced amounts
- soil pH within acceptable ranges
- soil salts below nominated threshold
- low levels of toxins (excess element concentrations and specific toxins such as heavy metals)
- low levels of pesticides, insecticides (organic synthetic compounds).

The biological properties of good soil include:

- generous amounts of organic matter
- living organisms, decaying matter and humus
- soil habitat suitable for soil animals such as worms
- diversity of organisms in soil

- strong soil respiration activity (CO_2 production)
- physical and chemical conditions (thermal, moisture, gaseous) favouring organic processes.

Evidence of unhealthy soil conditions includes:

- crusting
- erosion
- water logging
- compaction
- runoff and ponding
- bare areas
- stale odour (lack of oxygen)
- excess drainage
- shallow root systems
- poor plant growth.

Plants and soil water

Varying soil moisture conditions

Most soils experience wetting and drying cycles over the seasons. The top layer or surface soil layers can rapidly change from saturated conditions to dry in just over a few days. The soil environment is dynamic in terms of the water, chemical, physical and organic conditions.

The rate of change in soil moisture is generally slower at greater depths in the soil profile. It is the behaviour of the upper soil layers– the top 300 mm – that has greatest impact on water management practices, such as irrigation.

The water properties of a soil can be enhanced through practices that improve soil structure. Incorporation of organic matter is often an effective technique in improving soil structure in sandy soils.

The addition of organic matter to sandy soils can encourage hydrophobicity (water repellency), in some situations.

The key properties of a soil that are important, in terms of irrigation (Figure 6.3), are the:

- water-holding capacity of the soil
- water infiltration rate into soil
- water movement through the soil (percolation)
- stability of soil e.g. erosion, breakdown.

Although the focus of irrigation is on overcoming deficits in soil moisture, the negative consequences of saturated and waterlogged conditions should also be recognised. Saturated conditions can change soil particle distribution and damage soil structure and also remove

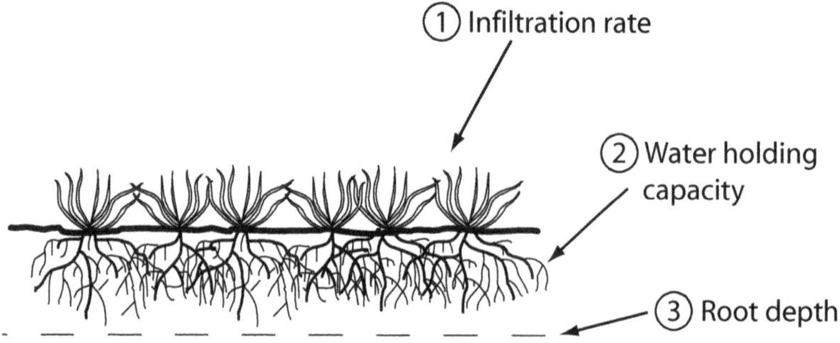

Figure 6.3. Key water-related properties of soil

beneficial nutrients from the soil through leaching. Saturated soils exclude oxygen and so can directly impede plant growth.

Construction and maintenance of irrigated sites so that excess water is removed, either on the surface, subsurface (drains) or both, should be part of the water management of the site. The design of effective drainage systems is important in achieving sustainable turf or landscape plantings. The role of soil properties in the design and management of horticultural sites is well described in McIntyre and Jakobsen (1998).

Water movement in soils

The water in the soil mass is subjected to a range of forces or pressures that determine both the direction of movement, the distance of spread of water and the rate of movement. The three main forces that dictate water movement are gravity, cohesion and adhesion.

Gravity forces naturally pull objects downwards. In the case of soil water, there is always a strong tendency for water that is not held more strongly by other forces to move directly downwards through the soil profile. This is referred to as gravitational water.

The force of cohesion between water molecules holds a water droplet together. The water molecules are attracted to each other.

For water to remain in contact with other materials or surfaces, there needs to be forces of attraction between the water and the other surface: this is the force of adhesion. Thin layers of water (films) on soil particles are held by the force of adhesion.

The net water movement in the soil is determined by the relative strengths of the forces acting on the water at any point in time. When the soil is saturated soil, water movement is normally downwards, dominated by gravity forces, and when water has drained from the soil, water movement, if any, is mainly determined by adhesion forces. The rate of movement of water laterally is generally much slower compared with the rate of movement downwards.

The movement of water within small soil pore spaces is determined by adhesive forces, once excess water has drained from these spaces by gravity. The direction of movement of water can be laterally or upwards, under the influence of adhesive forces. Water that moves under adhesive influences is capillary water and it is very important in determining the distribution of water within the soil. The upward movement of capillary water is used in perched watertable constructions to supply water to the overlying root zone. Capillary water is also important with subsurface drip systems, where lateral coverage is required to be achieved away from the delivery point.

When capillary water has been removed from soil pore spaces, the remaining water is held onto the soil particle surface in a very thin layer with large adhesive forces. This water is not available to the plant and is referred to as hygroscopic water.

Water extraction from soil

The removal or extraction of water from the soil by the plants depends on the density and health of the root system and the availability of water. The greatest extraction takes place in the top layers of soil (Figure 6.4). Water is generally more available in these upper layers and most plants develop root systems more densely in the upper layers. There are obviously exceptions; deeper-rooted plants, such as trees, may extract more water at greater depths.

In the management of an irrigated site, it is important to note that most water extraction takes place preferentially in the upper layers of soil. There are various estimates that provide a guide to the relative removal of water from different soil layers. Figure 6.4 has been prepared with reference to soil extraction rates reported by Pair (1969). It is estimated that 40% of the water is removed from the top 25% of the root zone.

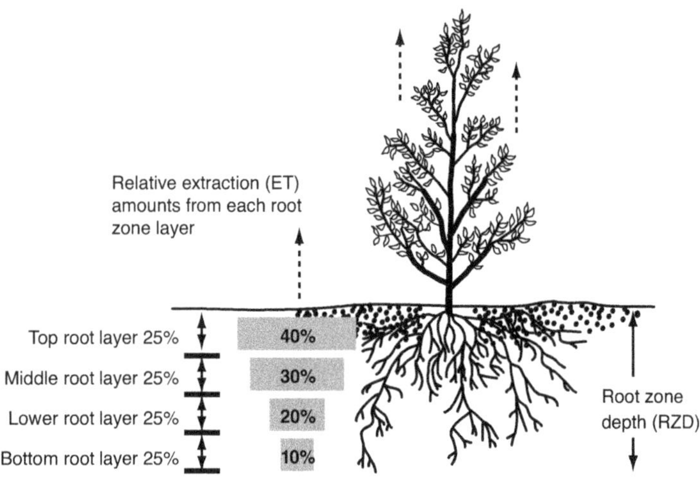

Figure 6.4. Relative water extraction from various soil layers (Source: Pair 1969)

Soil types and properties

Key soil water properties

A key property of a soil is the amount of water available for the plants. There are numerous factors or characteristics of the soil that influence the water stored and the amount available. These include:

- soil texture
- soil structure
- depth of soil layers and overall depth of soil.

Figure 6.5. Excavating a hole with a shovel can provide extremely valuable information about the soil and root system

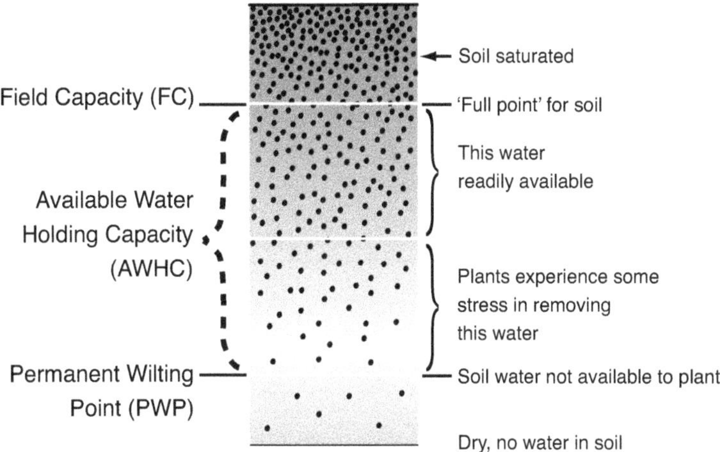

Figure 6.6. Soil water properties and associated terms

Gaining an understanding of the soil properties at a site is fundamental to achieving efficient use of irrigation water (Figure 6.5).

There are a number of terms used to describe the various water properties of soils in irrigation management (Figure 6.6). The available water-holding capacity (AWHC) of a soil refers to the amount of water that is stored and available to the plant. The value of AWHC is commonly expressed in millimetres per metre (mm/m) of soil. It is also expressed as centimetres per centimetre (cm/cm), millimetres per centimetre (mm/cm) and centimetres per metre (cm/m).

There is a limit to how much water can be stored in soil. The absolute limit is determined by the total void space in the soil. If the void space is saturated, then some water will drain away and so is not available to the plant. The maximum amount of water that can be held by the soil and not drain away, following saturation, is the field capacity (FC). The lower limit of available water is the amount of water that cannot be removed by the plant. It is limited by the capacity of the plant to extract the water and strength of the forces with which the water is held by the soil. This is the permanent wilting point (PWP). The water available to the plant in the soil is the amount stored between FC and PWP. This is the AWHC of the soil.

The volume of water that can be stored in soil and available to the plant typically ranges from around 6% to 20% of the total soil volume. This is commonly expressed in terms of millimetres of water per 1000 mm (or mm/m) of soil. A sandy loam soil, for example, has an AWHC of 120 mm per 1000 mm of soil.

Figure 6.7 illustrates the difference between the total amount of water that can be stored by the soil and the water that is available to the plant. An open soil, such as sand, which is composed of relatively large particles, cannot hold much water in total, compared with loams and clays, because water readily drains from the large pore spaces. However, the plant can readily extract moisture from the sandy soil, because it does not hold water with strong forces. The net result is that sands are low in terms of AWHC, mainly due to the inability to retain large amounts of water. Clay soils, which comprise many fine particles, can hold the largest amount of water. However, the extraction of moisture is much more difficult and significantly more water remains in a clay soil, than a sandy soil, when plants are wilting. The net result for clay soils is that they have a considerably higher AWHC.

The water-holding benefits of loam soils are illustrated by their high AWHC. This is achieved through a combination of high FC and reasonably high PWP. It is the difference between FC and PWP that determines the AWHC value.

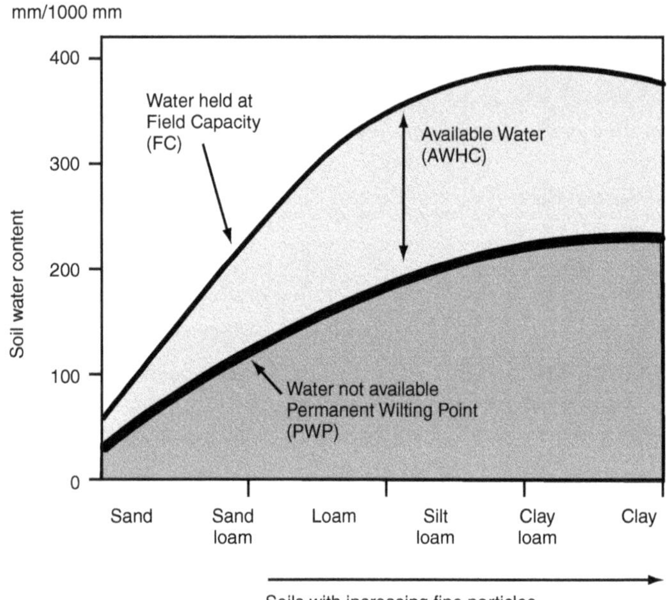

Figure 6.7. Relationship between available soil water and soil type

The benefits of well-structured soils and the incorporation of organic matter are evident in Table 6.2. Sandy loam increases from 140 mm/m to 200 to 300 mm/m with the inclusion of organic matter. A well-structured clay soil exhibits an AWHC of 200 mm/m, compared with 140 mm/m for a clay soil.

Infiltration rate

The rate at which water enters the soil, the infiltration rate (IR), is another important property of a soil. Open soils, such as sands, with high infiltration rates readily absorb rainfall, whereas

Table 6.2. Guide to soil water properties for soils of different textures

Soil texture	Water in soil at field capacity mm/m	Water remaining at permanent wilting point (PWP) mm/m	Available water (AWHC) mm/m
Sand	90	20	70
Loamy sand	140	40	100
Sandy loam	230	90	140
Sandy loam with organic matter	290	100	200–300
Loam	340	100	220
Clay loam	300	160	140
Clay	380	240	140
Well-structured clay	500	300	200

Note: These values are for specific soils within each of the classifications. Values will vary from site to site and between soils with the same texture classification.

Source: Handreck (2008)

Table 6.3. Guide to key soil water properties (AWHC and IR)

Soil type	Available water-holding capacity (AWHC) Range (mm/m)	Available water-holding capacity (AWHC) Average value (mm/m)	Infiltration rate (IR) mm/h
Sand	60–80	70	>30
Fine sand	80–100	90	20–30
Sandy loam	100–140	120	15–20
Loam	180–220	200	10–15
Silt loam	160–180	170	8–12
Clay loam	120–180	150	5–10
Clay	120–160	140	1–5

Note: These values are a guide only and site specific information is recommended for detailed irrigation analysis.

in the finer soils, much water can be lost due to runoff, because the precipitation rate often greatly exceeds the infiltration rate. On the other hand, some of the rainfall on sands can be lost, because the rainfall water may drain below the root zone of the plant. It is not necessarily lost to the environment, because this water may be refilling or recharging the groundwater.

A low infiltration rate soil often exhibits low rates of movement of water through the soil profile. The movement of water through the soil is referred to as percolation. The rate of water movement through the soil (percolation rate) is measured as the hydraulic conductivity of the soil. In most cases, it is expressed in terms of the rate of movement under saturated conditions (K_{sat}). The units of mm/h and m/day are commonly used.

The following is a guide to K_{sat} values and description of relative permeability rates:

- low – less than 0.5 m/day
- medium – 0.5 to 5.0 m/day
- high – greater than 5 m/day.

The movement of water through heavy clay soils may be very slow: less than 0.05 m/day. On the other hand, free draining soils may be in excess of 20 m/day. The potential range in duration for water movement within the soil mass is therefore great.

Table 6.3 provides a guide to key soil water properties including the AWHC and IR.

The infiltration rate in Table 6.3 is presented as a relatively narrow range and provides a general guide only. The actual infiltration rate varies over time depending on a range of factors, including soil moisture level and percolation rates. Infiltration rates can be measured in the field. The techniques include using a section of pipe, minimum of 150 mm diameter, 300 mm long, and observing the decrease in ponding water depth over time. This provides an indicative value only. More comprehensive testing can be done using dual infiltration rings and calibrated water reservoirs to measure the water intake rate more accurately (Figure 6.8).

The infiltration rate is influenced by soil texture, soil structure, initial soil moisture levels, characteristics of the precipitation event and time. Generalised curves indicating changes in infiltration rate over time are presented in Figure 6.9.

The variation in soil infiltration rate over time is important in terms of irrigation management. The graphs in Figure 6.9 show that initially infiltration rates are relatively high and then, in the case of the heavier clay and loam soils, the infiltration rate decreases rapidly. The residual or steady infiltration rate is referred to as the base infiltration rate. The implication of this characteristic is that if the application of water is continued at rates higher than it can infiltrate, then it will run off. In the case of clay soils, this may take only 5 to 10 minutes. In lighter loam soils, it may be 10 to 20 minutes.

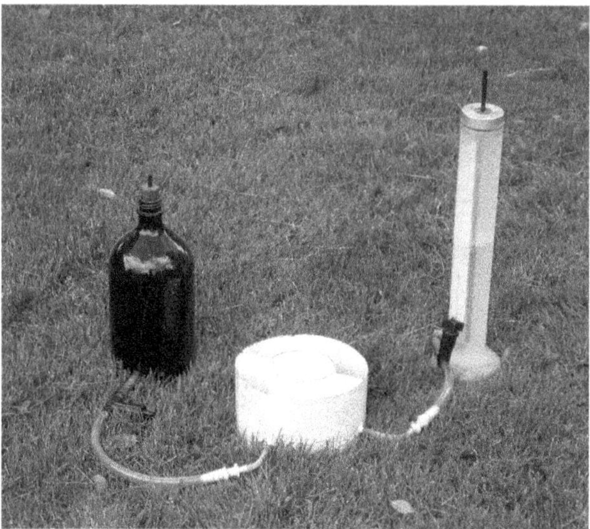

Figure 6.8. Rate of water depletion in a flooded ring provides a measure of soil infiltration rate (using device called an infiltrometer)

Because achieving effective infiltration is a key aspect of good irrigation, strategies need to be put in place to prevent waste from runoff. The approach that is used is the 'cycle and soak' technique, where irrigation amount is delivered in a number of smaller applications, over an extended period of time.

Although the expected period for runoff to occur can be theoretically determined for a site, observation of the operation of the system is recommended.

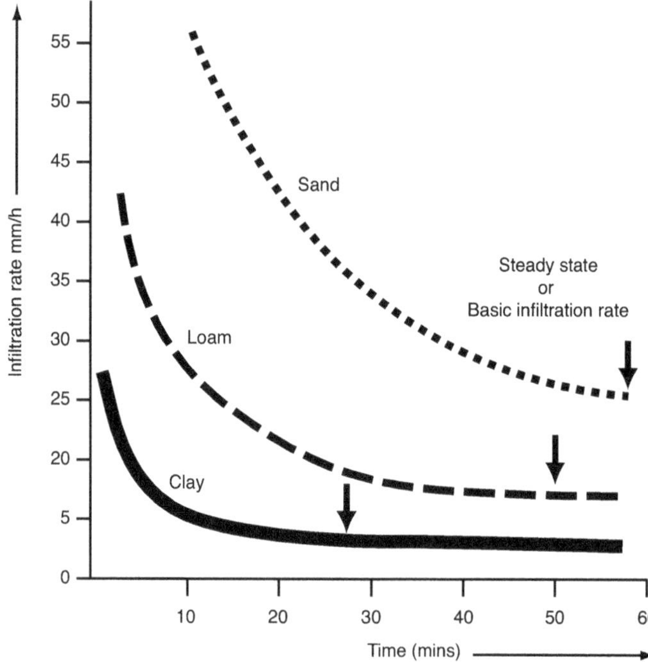

Figure 6.9. Infiltration rate curves for selected soil types

The soils of many urban landscapes are often highly variable. The spatial variability in soil properties across the site can be significant. In addition to changes in texture, there can be much variability in organic content, compaction and the presence of inorganic material, such as bottles, iron, bricks or stones. These soils are sometimes referred to as 'urbic' soils.

In many sites, there are discrete soil layers. These may be naturally formed layers or the result of construction and management practices. Sporting grounds that are regularly top dressed with sandy loams soils will often have discrete soil layers, with a sandy loam layer overlying another usually heavier or finer soil.

The water properties of each layer need to be considered when evaluating the water available to the whole plant. For example, should there be an underlying layer of low permeability soil, this will influence the water content of the soil above, and will also directly influence the soil water conditions and suitability for root growth at that level. Encouraging root development through the various layers is important.

Another property or characteristic of urban landscapes is soil compaction. The forcing of soil particles together so that pore spaces are reduced results in reduced water-holding capacity, lower infiltration rates and poorer gaseous conditions (less oxygen) for root growth, as well as inhibiting the initiation and development of roots. The bulk density (BD) of the soil, the mass per unit volume (in kg/m^3 or g/cm^3), provides an indication of the openness (porosity) of the soil and the degree of compaction. Bulk densities typically range from around 1.1 g/cm^3 for fine clays to 1.6 g/cm^3 for sands in soils that are considered suitable for root growth. Compacted soils may have BD approaching 1.8 g/cm^3 for sands and 1.5 g/cm^3 for clays. Bulk density values tend to increase with soil depth. Devices, such as penetrometers, can also provide an indication of surface compaction.

Water content and soil water tension

The amount or volume of water and the force with which it is held are both important when considering the water used by plants and irrigation management. The force (tension) required to be exerted by plants depends on the soil type and the amount of water remaining in the pore space. As water is extracted from a soil, the size of the force, or tension, required to remove the water increases. In the case of fine soils, such as clay, the small pore spaces can hold water very tightly.

The term soil matrix or water potential, which describes the water tension, expresses the strength of water forces, in terms of energy levels. Water potential is measured in kilopascals (kPa). When measuring the force with which water is held within the soil it has a negative value. A water tension or suction of 20 kPa (expressed as a positive value) is equivalent to matrix or water potential of –20 kPa.

The actual amount of water available to a plant depends on the ability of the plant to remove the water, as well as the amount of water stored in the soil. The water available in the soil is therefore not only a property of the soil, but also reflects the capacity of the plant to extract the water.

The management of the soil moisture of a landscape requires knowledge of the soil moisture status. The status may be described in terms of either an amount (mm or percentage volume), tension (–kPa) or suction (kPa).

The relationship between the water tension and a designated soil water condition or property is specific to a particular soil and site. However, it is possible to provide a guide to the tension value associated with the different soil water conditions. Table 6.4 provides a guide to the range of tension values associated with the various soil moisture conditions that are relevant to the management of irrigated sites. It should be noted that these values are provided as a guide only and values will vary for particular sites, plants and soil types.

Table 6.4. Guide to soil condition and water tension

Plant response	Soil water condition	Guide to water tension values (kPa)	
		Coarse texture soil e.g. sand	Fine texture soil e.g. clay
Stress – lack of oxygen	Saturated soil	0	0
→	Field capacity	–10	–30
↓	Limit of 'readily available water' type 1 Plants	–20	–40
	Limit of 'readily available water' type 2 plants	–40	–80
Increasing plant stress	Available soil water 50% depleted	–60	–100
→	Permanent wilting point	–1500	–1500
	Oven dry	–	–

Soil texture and water tension

The force with which water is held by the soil is important for several reasons. These include:

- It determines the amount of water actually available to the plant.
- It determines how the plant responds to the increasing difficulty of extracting moisture from the soil.
- It is used in the measurement of the soil moisture status.

Readings from soil moisture devices, which measure water tension, need to be converted to soil water content values in some cases. These are then expressed as a percentage (volume/volume %) or depth (mm).

Each soil has its own characteristic relationship between the water tension with which the water is held and the amount (volume) or content of water that is held. This is called the soil characteristic curve or the soil moisture release curve.

A generic soil release curve is presented in Figure 6.10.

Plants, soil water tension and readily available water

The range of values of water tension that can exist in the soil is large, ranging from 0 kPa to more than -1500 kPa (wilting point-limit of plant extraction). When the soil is saturated, the

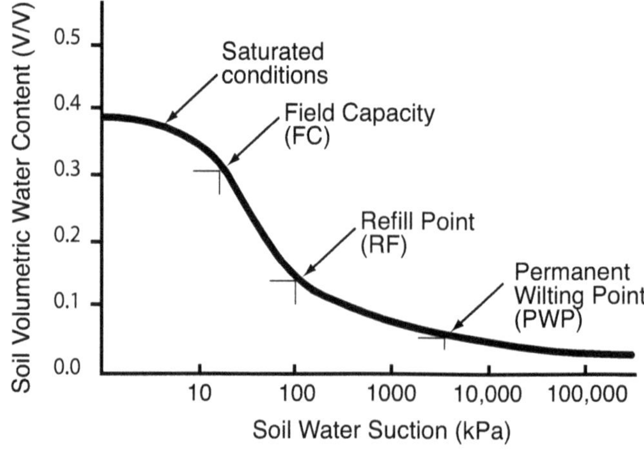

Figure 6.10. Generic soil water release curve

soil tension is close to 0 kPa and when it is very dry, and the plant is capable of exerting very large extraction forces, it can be in excess of –1000 kPa.

Soil moisture is often managed in horticultural crop situations so that water is continuously and readily available. Under these conditions crop yield is maximised. Only small degrees of stress, in terms of extracting water from the soil, are tolerated. In the case of ornamental plants, some water stress can be tolerated without any negative impact on the quality of the vegetation.

The term readily available water (RAW) is used to describe the amount of water that can be removed prior to plant stress or a negative impact on plant performance. RAW is expressed in terms of depth of water (mm), for a particular root zone layer, and also in terms of water tension (–kPa). At times, RAW is expressed in terms of a property of the soil (mm/m). In the case of some horticultural crops, such as berries and vegetables, low tension needs to be maintained to achieve good productivity. In Table 6.5 this is represented by Type 1 plants. Strawberries, for example, are managed in the –20 to –30 kPa range and tomatoes in the –60 to –80 kPa range.

Irrigated sites can be managed according to the soil water tension values. In broad terms, when soil moisture tension reaches approximately –60 kPa, further extraction requires much increased force to extract water and some plant stress would be expected. In the case of ornamental plants, greater levels of tension can usually be tolerated. This is represented by Type 2 plants in Table 6.5, but it should be noted, that only one value is quoted for ornamental plants. Ornamental plants can be managed to allow significantly greater stress. For example, the Royal Botanic Gardens Melbourne program irrigation for some landscapes at a tension of –300 kPa with no long-term effects on the landscape quality.

Soil water reservoir

In addition to the soil properties, the plant water use characteristics also directly influence the total amount of water available.

The key characteristics influencing available water include the:

- distribution of roots
- depth of roots
- density of roots in each soil layer
- plant response to increasing water tension.

The total amount of water stored in the root zone and potentially available to the plants is determined using the AWHC of the soil and the depth of the root zone. Where the soil profile is made up of soil layers, the available water in each layer is determined.

Indicators for a healthy irrigated soil

Table 6.5 identifies the range of indicators that may be used to evaluate soil condition as part of achieving a sustainable landscape.

Irrigation scheduling

Scheduling principles

The operation of the irrigation system, so that soil moisture is maintained, within defined levels or limits, is the underlying basis of irrigation scheduling. Good irrigation scheduling has two components: the application of the correct amount or depth of water and the timing of water application.

The actual soil moisture levels to be maintained depend upon the required performance of the plants and the properties of the soil. In some circumstances, the soil is maintained close to

Table 6.5. Soil health indicators

Category of indicator	Soil health/quality indicators	Indicator measure/unit
Physical	Water storage	AWHC (mm/m)
	Infiltration rate	IR (mm/h)
	Percolation, permeability	m/day
	Compaction – penetrometer	kPa
	Bulk density	kg/m^3, g/cm^3
Chemical properties	Soil acidity	pH
	Salinity	EC, TDS
	Nutrients	N (mg/L)
	Concentration of sodium to calcium – risk of soil particle dispersal	SAR
	toxins	mg/L
Biological properties	organic content	mg/m^3
	Respiration rate	Rate of absorption of CO_2

optimum soil moisture levels (close to field capacity) so that the plants always have ready access to water and growth is not limited in any way by water. Strong lush growth is then produced. In other circumstances, the soil moisture level is deliberately maintained at a partially full level, so that there is some stress on the plant. This strategy, which is referred to as deficit irrigation, is sometimes used as a means of saving water and also in encouraging deeper root systems.

The plant performance, soil water properties and irrigation performance all need to be taken into account in irrigation scheduling. The scheduling techniques that are available include:

- observation of foliage condition – signs of stress, wilt, leaf curl or leaf drop
- soil appearance and feel
- soil moisture measurement – using sensors
- canopy or foliage temperature – determine water stress index (elevated leaf temperatures)
- water balance – programmed based using pre-set times
- water balance – using daily ET_c.

Requirements for precision irrigation scheduling

Precision scheduling involves having a thorough knowledge of the soil properties, plant responses to soil moisture levels and daily ET rates. The following information is required in order to make sound irrigation scheduling decisions:

- existing soil moisture level
- water-holding capacity of the soil
- plant evapotranspiration rate ET_c (daily)
- irrigation precipitation or application rate
- irrigation system efficiency
- rainfall contribution
- plant response to available soil moisture.

Irrigation scheduling is based on systematically tracking the soil moisture level and adding the required amount of irrigation water in a timely and efficient manner. The timing of the irrigation event and the run times for irrigation delivery are the two core components of irrigation scheduling.

Weather stations generating reference evapotranspiration data (ET_0) are increasingly popular in the scheduling of irrigation. This approach is well suited to turf, because the performance of the crop or turf over a defined area can readily be predicted. Turf is reasonably uniform in terms of water use and the root system is also reasonably well defined and uniform for each soil type.

The irrigation scheduling of landscape plantings or turf requires knowledge of the actual soil moisture conditions, so that the timing of irrigation decision making is correct. Although good efficiency can be achieved through close monitoring of plants and site conditions some type of soil moisture sensing provides actual, real-time information on the soil moisture status. The soil moisture sensor provides feedback to the irrigation controller or manager, to aid in irrigation scheduling decisions.

Shallow and deep root systems

The depth of the root system influences the scheduling approach. When scheduling irrigation for urban trees, for example, the following need to be taken into account.

- Soil moisture levels may be highly variable throughout the tree root system.
- The irrigation application may be needed to replenish the soil water deep within the soil profile, rather than to replace daily water needs with shallow applications.
- The selection of the refill point is influenced by the multiple soil layers and variability in soil moisture in the different layers.
- Movement of water through the soil profile to replenish lower soil layers can take considerable time, perhaps weeks or months.

Determining irrigation depth

Soil water storage

Precision irrigation is based on the application of the correct amount of water. Although the term 'amount' is commonly used, a more appropriate term is the 'depth' of water. Rainfall is measured and described in millimetres (mm). Irrigation should also be described in mm. In both cases, it is the depth of water that is important.

A primary requirement of good irrigation is that the depth of water applied by the irrigation system should be appropriate to the actual water storage capacity in the plant root zone. If it is greater than the capacity of the soil storage, then some water will be wasted. The first step is to determine the soil water storage capacity of the soil.

Total available water (TAW) or stored water (SW)

The depth of water that is in the root zone and is available for use by the plants is referred to as the total available water (TAW). The TAW is dependent on the depth of the active water-absorbing roots in the soil (called the root zone depth – RZD) and the AWHC of the soil.

Root zone depth (RZD)

The depth to which plants extend their root system is a key property, in terms of irrigation management. Plants with deep root systems have access to greater water reservoirs. The range in rooting depths is extensive and depends not only on the plant species but also soil type, degree of compaction, gaseous exchange capacity and oxygen diffusion, and watering regime. Root systems tend to develop to greater depths in soils that are more open, such as sands. In heavier and compacted soils, root depths are generally shallower. In some situations, root development may be limited by an underlying high resistance and low permeability soil layer.

The root zone depth selected for irrigation management is critical, because it determines the storage depth on which other key irrigation decisions are made.

Some plant species are more opportunistic in establishing root systems than others. Within the turfgrass family, there are marked differences in both the extensiveness of the root systems and the depth of the roots. In a study, Short and Colmer (1999) ranked turfgrasses in

Table 6.6. Guide to effective root zone depths

Plant	Effective root zone depth (guide only) (mm)
Annual flowers	100–150
Fruit trees	300–600
Ground covers	100–150
Ornamental shrubs	150–300
Ornamental trees	300–600
Turf – cool season	100–200
Turf – warm season	150–300

Note: These values are a guide only. Site investigation is recommended.

terms of root exploration. Couch, kikuyu and buffalo were excellent; tall fescue and zoysia were good and ryegrass was poor. Root exploration capacity is a strong feature of the warm season grasses, which have greater access to soil water reserves.

Most trees have strong exploratory root development characteristics, both laterally and vertically. Root spread is typically in the range of two to four times the crown diameter if there is no impediment to root expansion. This characteristic provides very good access to soil water reserves. However, in the urban environment, there often is a restriction to root development.

Table 6.6 is provided as a guide only. The surest way of establishing the active or effective root zone depth is to take a soil sample or dig up a section of the landscape planting or turf and observe the roots in the soil. In addition to the species and soil structure influencing root zone depth, the mowing practices on turf and the watering regimes all influence the root zone depth.

The terms effective and active are used in describing the root zone depth. Although some plants – trees in particular – can develop very deep roots (e.g. tap roots), it is the depth of roots, which are actively absorbing soil moisture, that determine the effective root zone depth. Many grasses, for example, are reported as having roots to 500 mm–1000 mm, however, the actual depth of active roots may be only 150 mm–300 mm. In some cases, it is much less.

Determining total available water (TAW) or stored water (SW)

Determining the depth of water that can be stored in the soil and potentially available to the plants is the foundation of good irrigation management. It is determined in the following way:

$$\text{TAW or SW} = \text{RZD} \times \text{AWHC}$$

Both TAW and SW refer to the water storage capacity of the root zone.

Example – Total available water (TAW) or stored water (SW)

Determine the TAW for the following site.
Kikuyu grass growing in sandy loam soil
RZD: 150 mm
AWHC: 120 mm per 1000 mm (from Table 6.3)

$$\text{TAW or SW} = 150\text{ mm} \times (120/1000)$$
$$= 18\text{ mm}$$

TAW is used to determine the depth of irrigation that should be applied. In this case 18.0 mm is the maximum depth of water that can be held within the root zone. This would take the soil from permanent wilting point to field capacity.

Table 6.7. Water available in multiple soil layers of a tree

Soil layer range (mm)	Soil type	AHC (mm/m)	Available water per layer (mm)
0–200	Sandy loam – high organic content	200	40
200–300	Sandy loam	120	12
300–500	Clay loam	150	30
		Total available water (TAW)	82

Note: Soil water-holding properties determined from test of site sample.

Example – Total available water in multiple soil layers

Table 6.7 is an example of the calculation of the water available to a plant, such as a tree, that may have several different soil layers. The water stored in each layer is determined separately, and then summed.

In order to calculate the desired irrigation depth, there are two key irrigation management factors that need to be considered, the refill point (RP) and the system application efficiency (E_a).

The refill point (RP)

The conventional approach to irrigation scheduling involves determining a set soil moisture level (a nominated degree of stress) at which irrigation water is applied. This is referred to as the refill point (RP). Enough irrigation water is applied to raise the soil moisture level to the 'full' point, which is the field capacity of the soil (Figure 6.11).

The TAW represents the total water storage capacity available to the plant, when the soil is at field capacity. Management of the irrigation system commonly requires soil moisture level be maintained between 'full' and 'empty'. At the 'full' level, the soil is at field capacity (FC). If it is 'empty' it is at the permanent wilting point (PWP). If irrigation is applied so that the 'full' level is exceeded, then water will be wasted, either through deep drainage or runoff. If the soil moisture is allowed to be depleted to the PWP, then plant stress and permanent damage will occur.

Good irrigation is about managing the soil moisture level so that the plant is maintained in a healthy condition. In many situations a degree of water stress may be acceptable. The point at which the soil storage is replenished is the refill point (RP).

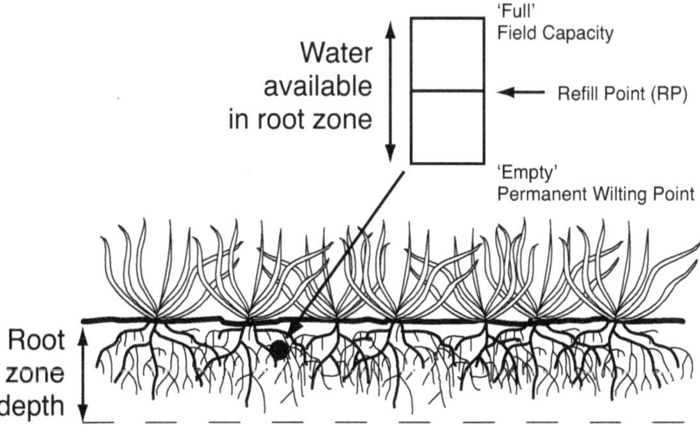

Figure 6.11. Water storage capacity of soil and the refill point

How much water should be depleted before initiating a new irrigation event? In some cases, the soil storage is maintained close to 'full'. This is often the case with drip irrigated crops, for example strawberries, that require readily available on a continuous basis. Allowable depletion under these situations is low. It may only be in the vicinity of 10% to 20%, which means that soil water is close to field capacity and the plant continually has access to readily available water. On the other hand, there are many situations, such as parkland trees, in which soil water storage is allowed to be depleted so that the majority (60% to 70%) of the soil storage is removed before irrigation.

Management allowable depletion (MAD)

The proportion of the TAW that is allowed to be removed, before commencing irrigation, is referred to as the management allowable depletion (MAD). This is an irrigation management decision and not a property of the soil.

The value of MAD will depend both on the availability of the water in the soil and the plant's capacity to extract water. The water tension or forces with which the water is held will influence MAD. Reference to the soil water release curve, which describes the relationship between soil water volume and water tension, and knowledge of the behaviour of the species to soil moisture stress will inform the selection of an appropriate MAD value. As a guide coarse, sandy soil will allow higher MAD values than fine soils, which hold the water more tightly. For example a MAD value may be 60% for sand and 40% for clay.

As a general guide, it is common practice for turf and landscape irrigation systems to be managed with 50% depletion. The optimum allowable depletion level depends on the soil type and the plant species.

The depth of water to take the soil from the RP to FC is referred to as the refill depth (RD) or, in the case of irrigation auditing programs, as the working storage (WS).

The irrigation manager decides how much of the TAW is removed before irrigation is initiated (at the RP). The WS is determined from the value of MAD and the TAW in the root zone.

A MAD value of 50% means that 50% of the TAW is removed and then irrigation is initiated. The way in which MAD is used is as follows:

$$\text{WS or RD} = \text{TAW} \times \text{MAD}$$

Example – Water storage

Site: sports ground
Turf: kikuyu grass
Assume MAD: 50%
TAW = 18.0 mm (from previous example)

$$\text{WS or RD} = 18.0 \times (50/100) = 9.0 \text{ mm}$$

Calculating irrigation depth (ID)

The delivery of water to the plant root zone, by the irrigation system, is not 100% efficient. Water can be lost due to runoff, drainage below the root zone, poor uniformity, wind drift and evaporation. The irrigation application efficiency (E_a) takes these into account. A guide to the application efficiencies for various irrigation methods is presented in Chapter 4.

The irrigation system needs to apply a depth of water to both refill the soil storage and allow for the inefficiency of the irrigation method.

Irrigation depth

$$ID = \frac{\text{WS or RD}}{\text{Application efficiency } (E_a)} \text{ (mm)}$$

Example – Irrigation depth

From above.
Sports ground: kikuyu grass
Sprinkler system: assume E_a 75% (0.75)
Allowable depletion (WS or RD): 9mm

$$ID = \frac{9}{0.75} = 12 \text{ mm}$$

In this situation, the irrigation system will need to apply 12 mm to ensure that 9.0 mm is effectively delivered into the turf root zone. It is assumed that 3 mm will be lost, due to inefficiencies, in the process of delivering the water into the root zone.

Determining the volume of water to be applied

Although irrigation management can be discussed in terms of the depth of water, it is then necessary to determine the volume of water for applications such as system design, flow management, analysis of water consumption, water resource availability and costing.

$$\text{Volume (V)} = \text{Area } (m^2) \times \text{Depth (mm)} \quad \text{(L)}$$

Note that when m^2 are multiplied by mm, the resulting volume is in litres (L).

Example – Volume of water

Determine the volume of water required to be delivered by the irrigation system to apply a depth of 11 mm to a sports oval of 1.2 ha.
Site: sports oval
Turf: kikuyu
Area: 12 000 m^2 (1.2 ha)
Soil: sandy loam
ID: 12 mm
Irrigation volume (V): 144 000 litres (= 12 × 12 000)
In this situation the irrigation system needs to deliver 144 000 litres (144 kL) per irrigation event.

Timing or frequency of irrigation

The timing of irrigation needs to meet the changing water demands of the plant and the moisture level of the soil. The key factors to consider are the:

- water use characteristics of the plant (K_c or CF)
- climate conditions (ET_0) – evaporative demand
- soil moisture level and soil water storage.

Irrigation Interval (T_i)

The time between irrigation events represents the frequency of irrigation. If the storage is large or plant water use is low, then there will be longer irrigation intervals (T_i). The water extraction rate (due to evapotranspiration) is extremely variable and is strongly dependent on the weather conditions.

$$T_i = \frac{WS \text{ or } RD}{ET_c}$$

Example – Irrigation interval

Determine the interval between irrigation events for an area of turf, assuming that the area experiences an evaporation rate (ET_o) of 8 mm/day for several days?

Turf: kikuyu
K_c: 0.5
ET_c: 4 mm/day (refer previous example, where $ET_c = 0.5 \times 8 = 4.0$ mm/day)
WS or RD: 9.0 mm (refer previous example)

$$T_i = \frac{9.0}{4} = 2.25 \text{ days}$$

Irrigation is generally commenced at a similar time each day. The irrigation interval needs to fit in with this format rather than irrigate at different times of the day. In the example above, the irrigation interval would be every 2 days rather than attempting to accommodate a 2.25 day interval.

The following factors can restrict the timing of irrigation events:

- sporting events and training
- maintenance practices
- water supply limitations including flow rate, pressure and access
- environmental conditions such as wind
- regulatory requirements such as operation times for recycled water application.

Some conditions or factors tend to necessitate more frequent irrigation:

- shallow root systems
- plants with a high water demand
- soil with low water-holding capacity
- soils with low infiltration capacity.

Programming of irrigation controller

The programming of an irrigation controller requires an irrigation schedule to be installed. This schedule is based on the soil properties and the expected plant water demand that would occur for a designated period, say a month. The plant water demand is determined using the historical long-term climate data for the locality. The irrigation schedule includes an irrigation frequency that will deliver an amount of water to meet the expected typical demand of the plant, for that particular period.

The determination of the historical plant water demand is usually based on E_{pan} data from the Bureau of Meteorology or historical ET_o data, if it is available. This irrigation schedule is referred to as the Base Schedule in the Irrigation Australia Limited (IAL) auditing program. Calculation of the Base Schedule is outlined in Section 'Preparing a base schedule' later in this chapter.

An individual irrigation schedule should be developed for each month of the irrigation season and the irrigation control program adjusted accordingly, to allow for variation throughout the season. This approach will provide a coarse correction to the irrigation program. It will prevent significant over- and under-watering.

The optimum ET scheduling approach is to use daily ET_0, to determine the amount of water used by the landscape, and then adjust watering to replenish soil water reservoir at the right time. Rainfall is also monitored on a daily basis and taken into account in the scheduling.

An alternative (not ET) approach is to use soil moisture sensors to inform the control program of the actual status of the soil moisture and initiate irrigation at a pre-set level. A combination of ET and soil moisture sensing can be used to achieve very precise and timely irrigation.

ET scheduling

The daily calculation of plant water demand (ET_c) provides precise information for irrigation management. The irrigation frequency will reflect the actual weather conditions experienced by the plants. The amount or depth of water is generally not altered, but the frequency of irrigation or interval will change. If plant water demand is low, then irrigation may only occur once per week, rather than twice per week.

This technique is based on assuming that there is an initial point when the soil water storage is full (at FC) and then daily adjustments are made to the inflows and outflows of water. When the RP is reached, then the decision is made to irrigate. It is referred to as a water balance or cheque book approach.

The potential inflows are rainfall and irrigation. The outflow is extraction of water by the plant. Figure 6.12 illustrates the range of processes involved when rainfall or irrigation, occurs.

Table 6.8 demonstrates the cheque book technique. It is based on a site where the total water that can be stored in the root zone is 24 mm and that irrigation is initiated when 50% of the water is depleted. The RP is therefore 12 mm. It is assumed that K_c is 0.5.

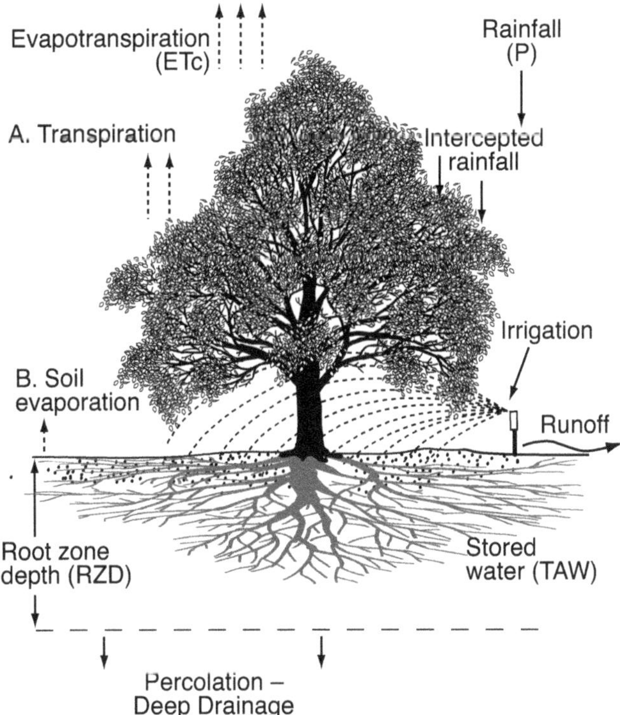

Figure 6.12. Processes involved in water balance of an irrigated landscape (tree)

Table 6.8. ET_0 schedule using cheque book (water balance) technique

Date	10 Jan	11 Jan	12 Jan	13 Jan	14 Jan	15 Jan
ET_0 (mm)	8	9	8	2	7	10
ET_c (mm)	4	4.5	4	1	3.5	5
Rainfall (P) (mm)	0	2(1)	0	12(2)	0	0
Effective rainfall (P_{eff}) (mm)	0	0	0	6	0	0
Daily gain/loss (mm)		–4.5	–4	5	–3.5	–5
Soil water level (mm)	20	15.5	11.5	16.5	13	8 + 12 = 20
Irrigate	No	No	No	No	No	Irrigate

Notes:
(1) Rainfall of 2 mm or less is ignored.
(2) Effective rainfall is assumed to be 50% in this case.
The soil refill point (RP) is 12 mm.

The treatment of rainfall in a daily cheque book approach varies. Although general rules, such as assuming 50% of rainfall is effective, are often used when extended periods, such as a month, are being assessed, daily analysis allows a more precise assessment.

If the soil reservoir level is such that the rainfall can be used and not overfill the storage, then the majority of rainfall can be included. There will, however, be some losses, due to water sitting on the ground or re-evaporating from the top surface layer of soil or from thatch. If only small rainfall events occur (e.g. <2 mm), these are often ignored. If a large rainfall event occurs (e.g. 30 mm), and there is only capacity for 10 mm of storage, then the balance of the rainfall will need to be disregarded.

In the example here, the net rainfall that enters the soil root zone is assumed to be 50% of the rainfall that occurs.

Scheduling using soil moisture sensors

Much irrigation scheduling is carried out without direct reference to the actual soil moisture conditions. Soil moisture sensors provide the feedback to assist in the irrigation decision-making process. The basic principles that are applied are that irrigation is initiated when a threshold soil moisture value is reached. These will normally be the RP and FC. The threshold value may be expressed in either tension readings or volumetric readings (broadly the two types of sensors that are used in irrigation management).

It is necessary to assess the site soil conditions when determining the appropriate threshold points. Soil type, soil water properties, root zone depths and sensor calibration requirements all need to be taken into account.

Tension type sensors, such as gypsum blocks and granular matrix sensors (e.g. Watermark) can provide economical soil moisture sensing. A tension type sensor has the benefit of providing a direct reading of the level of tension forces that the plant is exerting on the soil water. Further details on soil moisture sensors are covered in Chapter 10.

The positioning of the soil moisture sensor for scheduling is important. The sensor must provide a reading that is representative of the plant and soil conditions. There is often considerable spatial variability in urban landscape sites, including areas of turf. Lateral variability can be due to changes in plants, soils, microclimate and irrigation distribution.

Figure 6.13 shows the suggested positions for the vertical positioning of sensors. If only one sensor is available, then installing the sensor in the mid point of the active root zone is recommended. Ideally multiple sensors can be used to obtain a more comprehensive reading of the soil moisture conditions and water movement through the soil.

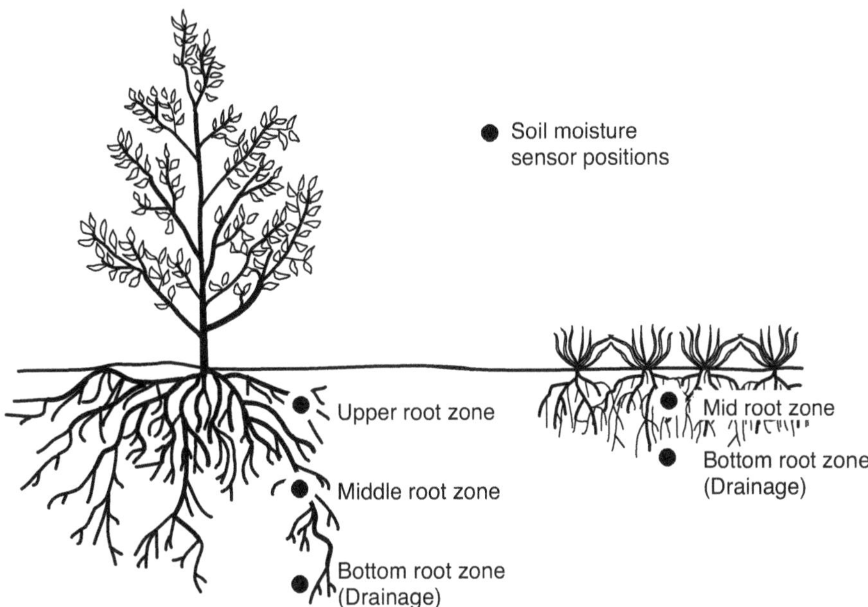

Figure 6.13. Soil moisture sensor positions for scheduling and monitoring

Preparing a base schedule

Steps involved

An important outcome from the evaluation of an irrigation system is information that allows an optimum irrigation schedule to be developed for each part of the irrigation season. Both the correct depth of application and typical irrigation frequency are determined. This is referred to as the base schedule. The schedule should include not only the absolute amount of irrigation that should be applied, but also how it should be applied, in terms of irrigation run times.

In the IAL Irrigation Efficiency Course (IEC) and the superseded Certified Landscape Irrigation Auditor (CLIA) Course, the preparation of a base schedule is a core component.

The steps involved in preparation of a base schedule are:

1. Determination of the plant water requirement (PWR)
2. Determination of the irrigation water requirement (IWR)
3. Determination of the irrigation depth and frequency
4. Determination of the optimum run times and total run times
5. Determination of the water volumes and costs.

The determination of each of these key areas is outlined here. The terms and abbreviations used here are similar to those used in the IEC programs delivered by Irrigation Australia Ltd (IAL) (website: www.irrigation.org.au).

There are, however, some variation in acronyms and terminology.

Step 1: Plant water requirement

The following information is required:

- vegetation identification (e.g. warm season or cool season grass)
- required plant performance level e.g. lush, elite, acceptable
- reference evaporation (e.g. E_{pan} or ET_o)
- CF, K_c, K_L, K_{mc} and K_d).

The PWR or ET_c is calculated using the following expression:

$$\text{PWR or } ET_c = K_L \times ET_o$$

or

$$\text{PWR or } ET_c = CF \times E_{pan}$$

Example

The following values have been used to illustrate the various steps in determining the base schedule.
Turf: kikuyu
Turf crop coefficient (K_c): 0.35
Evaporation ET_0: 200 mm for January
Soil: Sandy loam
AWHC : 120 mm per m
RZD: 180 mm
Irrigation system: Pop-up sprinklers
Irrigation Precipitation Rate (PR): 12 mm per hour (mm/h)
Irrigation Distribution Uniformity (DU): 75% (this value is determined from a catch can test)

$$\text{PWR or } ET_c = 0.35 \times 200 = 70 \text{ mm (for month of January)}$$

This represents a daily PWR or ET_c of 2.3 mm (70 mm divided by 31 days in January)

Note that rainfall is often not taken into account when determining an irrigation schedule. It is assumed that there is no rain and should rain occur then the irrigation is cancelled and recommenced when the soil moisture level has depleted to the RP.

Step 2: Irrigation water requirement (IWR)

The IWR is the amount of water that needs to be applied, over a designated period of time such as a month, to ensure that the PWR is satisfied. It takes into account losses associated with the inefficiencies of the irrigation system. It should be noted that this technique uses DU, which is a uniformity measure, as the efficiency of application of irrigation. A detailed description and determination of DU are presented in Chapter 11.

$$IWR = \frac{PWR}{DU}$$

Example (above): DU = 75% (determined during audit catch can test)

$$IWR = \frac{70}{0.75} = 93 \text{ mm}$$

This means that 93 mm of irrigation will need to be applied during January. However, if it rains, then this will reduce the amount of water that needs to be applied.

Step 3: Scheduling – depth and frequency

The following information is required:

- root zone soil type (obtained during site inspection)
- soil AWHC
- active RZD (obtained from site inspection)
- MAD.

The depth of water to be applied depends on the water that can be stored in the root zone. This is calculated using the AWHC of the soil and the RZD. Not all of this amount can be removed or extracted before the next irrigation, because this would result in permanent wilting of the plants.

To determine the RD or WS the following expression is used:

$$\text{RD or WS} = \text{MAD} \times \text{AWHC} \times \text{RZD}$$

Example

Soil: sandy loam, AWHC = 120 mm per 1000 mm.

RZD: 180 mm.

MAD: 50%.

$$\text{RD} = \frac{0.5 \times 120 \times 180}{1000} = 10.8 \text{ mm or } 11 \text{ mm}$$

The frequency of irrigation is determined using the RD and the average daily water use rate for that month.

In the above example, the daily water use rate is 2.3 mm/day and the RD is 11 mm. This means that it would be expected, under average conditions, that irrigation program would occur every 5 days (11 divided by 2.3 = 4.8 days)

This frequency value provides a guide only. It is a basis on which to plan irrigation for that month. The actual irrigation will need to be adjusted to take into account the actual weather conditions experienced.

Step 4: Optimum run times and total run times

The following information is required:

- system net precipitation rate (obtained from catch can test) (PR_N)
- time to runoff (TRO).

The net precipitation rate is the rate determined using the results of a catch can test, which measures the actual irrigation water hitting the ground.

The run time (RT) for the irrigation is calculated using the ID and the PR_N.

$$\text{RT} - \frac{\text{ID (mm)}}{\text{PR}_N \text{ (mm)}}$$

Example: ID is 11 mm. $PR(_N)$ is 12 mm/h.

$$\text{RT} = \frac{11 \times 60}{12} = 55 \text{ mins}$$

Note: run time is expressed in minutes, so the term is multiplied by 60 to convert to hours.

The operation of the irrigation system continuously for 55 minutes may cause losses as a result of runoff. The soil may not be able to cope with this application. It is recommended that the water be applied in a number separate applications if there is a risk of runoff and wastage. This is referred to as 'cycle and soak'. The key value is the amount of time that the system can be operated without runoff occurring. Ideally the irrigation system has been designed to have a precipitation (application) rate that is less than the soil infiltration rate. This is not always the case.

The situations in which runoff is a risk is with relatively high precipitation rate (sprays in particular) systems, heavy or fine soils, compacted soils and sloping terrain. When conducting an irrigation audit with catch cans, it is valuable to observe soil conditions, during the test, and note if or when runoff occurs.

If runoff occurs in a short period of time (e.g. 10 minutes) then the controller will need to be programmed to apply the required depth in multiple stages or cycles. If the soil has a high infiltration rate (e.g. >15 mm per hour) then the system may be operated continuously to apply the required depth.

$$\text{Number of cycles (NC)} = \frac{\text{RT}}{\text{TRO}}$$

Example

Assume TRO: 15 minutes
RT: 55 minutes.

$$\text{NC} = \frac{55}{15} = 3.7 \text{ cycles}$$

In practice, the system should be operated for four cycles each of 12.5 minutes. An adequate time needs to be left between cycles (e.g. 1 hour). This timing requirement may be difficult to achieve within the time windows for irrigation and also the irrigation controller needs to have the programming capability to handle this. The total run time (TRT) for the irrigation system is sometimes required. It allows the total volume of water to be determined. The total run time for the month is calculated using the required IWR for the month and the PR of the system.

$$\text{TRT for month} = \frac{\text{IWR}}{\text{PR}}$$

Example

IWR for January: 93 mm
PR:12 mm/h.

$$\text{TRT} = \frac{93 \text{ mm}}{12 \text{ mm/h}} = 7.75 \text{ hours or } 465 \text{ minutes for January.}$$

This means that the irrigation system would need to be operated for 7.75 hours in the month of January to be able to apply the required depth of 93 mm. However, if rain occurred, then the run time would be reduced. It is important to note that these calculations are based on historical climate data, and the weather conditions that actually exist need to be monitored and taken into account.

Step 5: Water volumes and costs

The following information is required:

- flow rate for the station (QS)
- water cost.

The term 'station' refers to the group of outlets, such as sprinklers, operated simultaneously from a remote control valve (solenoid valve). The flow rate of individual outlets can be determined through collection of the flow or through reference to the performance charts. In the case of sprinklers, if the sprinkler make, model and nozzle size or designation is known, then flows can be determined. This requires measurement of the pressure at the sprinkler. In the case of a sprinkler system, the flow rate of an individual sprinkler multiplied by the number of sprinklers in the station provides the station flow rate.

Example

Assume sprinkler flow rate: 45 L/min

Assume number of sprinklers in the station : 5

$$QS = 5 \times 45 = 225 \text{ L/min}$$

The Total Volume of water for that station (TVS) is:

$$TVS = QS \times TRT$$

Example

QS: 225 L/min
TRT: 465 minutes (7.75 hours)

$$TVS = 225 \times 465 = 104\,625 \text{ L} = 104.63 \text{ kL}$$

The cost of irrigating an area can be calculated using the volume required per station, the number of stations and the cost of water. It may be necessary to calculate each month separately, because there may be different numbers of sprinklers in the various stations.

$$\text{Total cost of water (Month)} = \text{Number of stations (NS)} \times \text{TVS} \times \text{Water cost}$$

Example

Assume NS: 12
TVS: 104.63 kL
Cost of water: $1.50 per kL.

$$\text{Total cost of water} = 12 \times 104.63 \times 1.50 = \$1883 \text{ (January)}$$

Scheduling during dry periods

The core information required for the efficient management is the depth of water required to refill the root zone (WS or RD) and the rate of water use by the plants (ET_c or PWR). Although there is the potential to adjust both the depth and the timing, the main adjustment for efficiency is timing. As a general principle, the same irrigation depth should be applied and the irrigation interval adjusted to match the actual water needs of the plant.

The planned schedule is based on long-term, average climate data. Actual conditions, including high and low evaporative demand and rainfall, require adjustment of the timing.

It is generally not recommended that smaller depths are applied and the same frequency or interval maintained. This approach leads to increased losses and shallow root systems.

Monitoring the soil moisture level, on a daily basis, allows precise timing decisions to be made on irrigation. Although this is the recommended approach, there are often local constraints that make changes to the schedule difficult. These include watering only being allowed on specific days and also the absolute number of irrigation days may be limited. The use of the landscape can also have an impact on these decisions.

At the end of the month, a comparison can be made between the water actually applied and the amount that was expected to be applied. Variations in these two amounts need to be explained. They may be due to variations in weather or due to poor irrigation management decisions, if excess water has been applied. The efficiency of delivery/management is measured using the water applied compared with the water required. This is referred to as the management efficiency or scheduling efficiency and is discussed in more detail in Chapter 11.

Deficit irrigation

The application of irrigation water during periods of limited supply requires rationing of the water to achieve the optimum outcomes. The underlying requirement is that the vegetation be maintained in a condition that, although it may be stressed, it will be able to return to healthy condition and full growth. The application of irrigation amounts less than the recommended refill amount, depth and frequency, is referred to as deficit irrigation.

Deficit irrigation is a technique to improve water use efficiency. The aim is to achieve acceptable level of outcomes or productivity and to use less water. The ratio of outcomes to water used is increased. An understanding of plant response to soil moisture stress and behaviour of soil water is required.

The techniques available to apply water, at less than the optimum amount, include:

- lower crop coefficient values (K_c and CF)
- higher depletion before refill (higher MAD)
- increased water stress factor (lower K_{ws})
- extended periods between waterings.

The use of the above will result in lower estimated plant water requirements. In most circumstances, it is preferable to apply volumes to significantly refill the root zone and then allow this water to be depleted to an acceptable level, rather than to refill the root zone at frequent intervals with small amounts of water. A technique used to improve the water use efficiency of grape vines is the alternative watering of sections of the root zones of the vines. This is referred to as partial root zone drying.

Leaching

A risk for saline soils is that under-watering will result in an increase in salt levels. The application of high saline water will also result in an accumulation of salts.

The application of excess irrigation water, above the plant needs, for the purpose of moving soluble salts below the root zone is referred to as leaching. The amount of irrigation water to be applied, in addition to plant needs, is referred to as the 'leaching fraction' or leaching requirement. This amount is dependent on the nominated threshold salinity level of the soil and the salinity of the irrigation water.

Higher levels of soil salinity are acceptable for salt-tolerant plant species. This will require less leaching. Reference lists, detailing soil salt tolerant levels for ornamental plants is presented in Stevens *et al.* (2008). The expression to determine the leaching fraction is:

$$\text{Percentage leaching requirement} = \frac{EC_{iw\ \text{(irrigation water)}}}{EC_{dw\ \text{(drainage water)}}}$$

Soils are also leached when over-irrigation occurs and following significant rainfall events. The application of water needs to be sufficient to cause saturated conditions in the soil profile so that water, including soluble salts, is moved below the root zone.

In a situation where the soil EC_{dw} is 5.0 dS/m and the EC_{iw} of the irrigation water is 1.0 dS/m, the leaching fraction is 0.20. This means that the irrigation depth to be applied would need to be increased by 0.20 or 20%.

When irrigating with saline water, and leaching may be required, strategies should be adopted to reduce the risk of short-term build up of salts. These strategies include: watering at night; using low salinity water (rainwater, freshwater or potable) to flush the soil; not allowing the soil to dry out excessively; and maintaining soil in a healthy condition so that soil infiltration and percolation are maintained at adequate levels. Monitoring of soil quality, in particular soil chemistry, is essential.

Quick reference to terms and acronyms

Table 6.9 lists the terms and associated acronyms that are relevant to soil water management.

Table 6.9. Terms used in base schedule – summary/quick reference

Acronym	Term	Brief explanation
AWHC	Available water-holding capacity	Depth of water stored per metre
CF	Crop factor	Used to determine ET_C from E_{pan}
DU	Distribution uniformity	Measure of application evenness
E_{pan}	Pan evaporation	Evaporation reference from Bureau of Meteorology
ET_c	Evapotranspiration rate of plants/crop	Water use rate of plants
ET_o	Crop reference evapotranspiration rate	Reference crop water use rate
ID	Irrigation depth	Gross water depth applied
IWR	Irrigation water requirement	Irrigation depth including plant water needs, rainfall and efficiency allowance
K_c	Crop coefficient	Used to determine ET_C from ET_o
K_d	Plant density factor	Adjustment for density of planting
K_{mc}	Microclimate factor	Adjustment for variation in microclimate conditions
MAD	Managed allowable depletion	Proportion of water that is removed prior to irrigation
NC	Number of irrigation cycles	Separate number of operation of irrigation outlets per irrigation event
PR	Precipitation rate	Depth of water applied or incident on ground per hour
PR_G	Gross precipitation rate	Calculated using system flow rate and irrigated area
PR_N	Net precipitation rate	Application rate determined using can test readings
PWR	Plant water requirement	Water demand of plants
QS	Flow rate of controller station	Flow rate for one group of laterals that are operated simultaneously
RD	Refill depth	Depth of water between refill point and field capacity
RTM	Run time multiplier	Time adjustment for irrigation event to allow for non-uniformity
RZD	Root zone depth	Depth of active roots
RT	Run time	Total time of operation of station per irrigation event
T_i	Time interval	Period (days) between irrigation events
TRO	Time for run off	Operating time until saturated soil causes surface runoff
TRT	Total run time for system	Duration from initiation of irrigation to finish for a program
TVS	Total volume for system	Amount of water used for whole system over designated period

Chapter 7

Best management practice (water management and irrigation)

Introduction

The goals for an irrigated turf and landscape site could be described as:

> 'Achievement of high water use efficiency, conservation of water resources and protection of the environment'.

The ultimate goal is a sustainable site that will deliver the required outcomes to the community.

It is important to have a framework that guides the progress towards the identified goals and provides specific advice on the approach that should be adopted for each of the key areas. A water management plan provides the broad framework for this approach and includes the goals, targets and strategies. It is the starting point in the pursuit of sustainable water use.

Best management practices (BMP) encourage good water management through provision of advice on each aspect of water use and through creating an environment in which continuous improvement in performance is the accepted norm. BMPs cover all areas of water management relevant to the site. These include land management, waterways, wetlands, groundwater, water storages, chemical use and nutrient management.

The BMPs covered in this chapter relate to the site water management issues that cover irrigation and specific irrigation BMPs. Irrigation BMPs cover the system design, performance standards, installation, scheduling of irrigation and maintenance.

Best management practice approach to water management

Adopting best practice in the water management of a site first requires considering some fundamental aspects of the site. A thorough review of the landscape and the water cycle of the site will provide the information necessary to make sound judgments about water issues. An overall assessment of the land use practices and water use should be the starting point for implementation of best practice in water management. Assessment of the whole landscape area may result in a re-evaluation of the use of the area, the management practices and the extent of the irrigation.

The first step in the water management process is the determination of the outcomes and services expected of the site. Some turf areas for example may be able to be maintained at lower performance/quality levels, and hence require less water, than areas that may have been previously maintained in a lush condition.

The next step is to determine the amount of water required to deliver the landscape outcomes. This stage involves preparation of a water budget or an annual irrigation volume (AIV). Water budgets provide a target value against which irrigation can be managed and evaluated. This can be done on weekly or monthly basis. This value established using historical rainfall and evaporation data. This value needs to be adjusted to take into account the actual weather conditions that existed during the irrigation period, when comparing actual water consumption with the target irrigation volume.

Once the outcomes have been defined and the target water volume determined, then the irrigation water needs to be applied to the highest standards. This involves applying water efficiently and scheduling irrigation to ensure that there is no over-watering and the application matches the weather conditions and the site.

Summary of best practice water management

The following provides some guiding steps in the approach to best practice water management:

1. A site assessment is performed and the landscape outcome determined.
2. Water budgets are prepared and the target annual irrigation volume is determined.
3. Efficient irrigation application techniques are employed.
4. Weather-based scheduling is carried out.
5. Water consumption is measured and reviewed.

BMP for turf and landscape site design and development

In the design and development of a turf or landscape site, the following principles should be adopted:

1. The site design should incorporate and respect existing natural assets, plants, soil and water, and features. Native vegetation should be used wherever possible.
2. Plants that are suited to the local climate (rainfall and temperature) should be selected. Low water use and drought-tolerant plants should be a priority.
3. Areas that require supplementary watering though irrigation should be minimised. Irrigated areas should be restricted to areas with high value uses.
4. Irrigation water should be applied effectively and efficiently.
5. The site should be shaped and contoured to facilitate stormwater collection and use on the site.
6. Permeable materials should be incorporated to facilitate rainfall infiltration into subsoil or diversion of stormwater to target areas.
7. Effective drainage should be installed to provide a healthy soil environment and to capture and reuse drainage water.
8. Provision for flood management should be included on the site so that the vegetation and soil is protected.
9. Natural treatment systems, such as wetlands and biofiltration systems should be incorporated into the design.
10. The ground should be treated (including use of vegetation) to minimise soil erosion and transport of sediment to connected waterways. Slopes are a particular risk.
11. High water use plants should be located in areas that are naturally moister, such as at the bottom of slopes or on drainage lines.
12. Storages should be designed to minimise evaporation losses.
13. Storages should be constructed to minimise leakage to subsoil. Synthetic liners may be required, if compacted soil (e.g. clay) layers are not appropriate.

Preparation of water budgets and annual irrigation volume

Knowing the target amounts of water required and the volume used are essential to good water planning, site water management and irrigation scheduling. The volume of water required depends on a number of key factors, including:

- total irrigated area
- landscape/vegetation type and condition
- climate – evaporation and rainfall
- irrigation efficiency.

The method of determination of irrigation requirements are outlined in Chapter 5.

Principles of efficient irrigation

There are the four key principles that need to be implemented to ensure that the irrigation of turf and landscape areas is efficient (Figure 7.1):

1. The amount of water applied should be appropriate to the plants and soil.
2. The timing of water application should be matched to vegetation and weather conditions.
3. Water should be applied uniformly and effectively.
4. Water should be applied to the plant root zone without wastage through runoff, deep drainage, ineffective coverage or other water loss sources.

In summary, good irrigation is the efficient application of the right amount of water at the right time in the right place.

Irrigation BMP

Landscape BMP principles

The 'IA Turf and Landscape Best management practices, December 2010' (Irrigation Association 2010), have been largely adopted in the United States, Canada and Australia and are available from Irrigation Association (USA) website (http://www.irrigation.org).

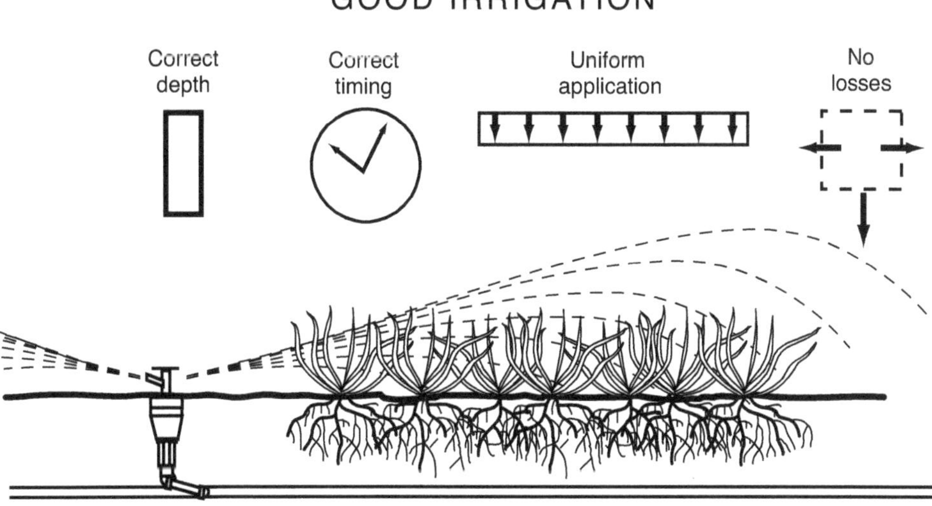

Figure 7.1. Good irrigation practice involves effective water application, minimum losses, correct depth and correct timing

The principles described for landscape irrigation BMPs are:

1. Assure the overall quality of the irrigation system.
2. Design the irrigation system for efficient and uniform distribution of water.
3. Install irrigation to meet the design criteria.
4. Maintain the irrigation system for optimum performance.
5. Manage the irrigation system to respond to the changing need for water.

These five principles provide a framework within which irrigation managers can identify the key areas to be considered and the standards that should be achieved.

The BMPs are supported with 'practice guidelines' that detail specific requirements for each of the key areas. The practice guidelines include recommendations that help achieve the BMPs such as the desired distribution uniformity or the use of technology to manage water resources better.

Complying with these principles/requirements involves achieving the following:

BMP 1. Assure overall quality of the irrigation system
- effectively distributes irrigation water
- designed for efficiency and uniformity
- installed according to specifications
- regularly maintained
- managed properly.

BMP 2. Design the irrigation system for efficient and uniform distribution of water
- designed for efficiency and uniformity
- specific criteria for site established
- relevant site details considered
- appropriate components selected
- compliance with local codes and regulations.

BMP 3. Install irrigation to meet the design criteria
- install according to design specifications
- install components to meet performance specifications
- installed system to be efficient and distribute water uniformly
- installation contractor/installer shall be licensed, experienced and reputable.

BMP 4. Maintain the irrigation system for optimum performance
- system regularly maintained
- serviced components to meet irrigation design specifications
- maintenance to result in sustaining efficient and uniform distribution of water
- maintenance contractor shall be licensed, experienced and reputable.

BMP 5. Manage the irrigation system to respond to the changing need for water
- conserve and protect water resource
- schedule adjusted to respond to changing weather
- schedule meets needs of healthy plants without wastage.

Best practice guidelines

The following practice guidelines, which support the implementation of the principles, are largely based on the document from the IA (USA). The idea of practice guidelines is to provide principles and strategies that can be applied locally to improve the performance and efficiency of irrigation systems.

Practice guidelines

PG 1. Quality of irrigation systems

- use qualified irrigation designer
- use experienced, licensed, appropriately trained, reputable contractor
- comply with codes and regulations
- maintain system maintained for efficient operation and performance
- match application of water to plant needs and site
- audit performance regularly.

PG 2. Design

- comply with codes and regulations
- protect environment including water source
- system capacity to meet expected peak demand
- system application (sprinklers) DU_{LQ} 75%
- irrigation schedule prepared for all zones
- use hydraulic design with minimum pressure variation and do not exceed maximum design velocities
- use zone layout to allow all areas of varying requirements to be watered independently
- allow for constraints, time and physical, associated with the use of the area (e.g. withholding periods, sporting events)
- equipment selected to meet performance specifications, minimum maintenance and reliable operation
- provide for expansion and technology upgrades.

Design for turf and landscape water management efficiency

- use dedicated irrigation water meter
- reduce dependence on potable supplies – seek alternative water sources
- design low flow (e.g. drip) systems for narrow, irregular areas
- select precipitation and application rates to suit site soil conditions and avoid runoff
- identify watering window to accommodate use of the site
- allow for wind in selecting sprinklers and sprays
- hydraulic design to incorporate pressure regulation and flow control
- incorporate environmental sensors (wind and rain)
- irrigation scheduling to include ET
- use controller that allows individual zone watering, cyclic watering and sensor input.

PG 3. Installation

- verify hydraulic design and identify site constraints
- install according to design specifications
- variations to design to be approved and recorded
- prepare as-built plans using surveyed/GPS information
- test system to verify equipment performance specifications
- audit system to determine DU% and zone precipitation rates
- use monthly program irrigation schedule for each zone
- provide training in system use and maintenance to site owner/manager
- provide product warranties and guarantees for installed equipment.

PG 4. Maintenance

- establish a maintenance program and report outcomes to site owner/manager
- verify system supply flow and pressure

Figure 7.2. Obstruction of water stream from sprinkler nozzles interferes with distribution and contributes to non-uniformity

- verify backflow prevention equipment is functioning correctly according to local regulations
- check functioning of all equipment in particular sprinklers, sprays, emitters and solenoid valves
- ensure plants, including grass, do not obstruct distribution from sprinklers and sprays (Figure 7.2)
- repair and replace faulty equipment in accordance with original design specifications
- review system design and performance with regard to growing and changing vegetation.

Figure 7.3. Regular checking of system operation to maintain good performance is essential

PG 5. Guidelines for managing supplementary irrigation

- understand system capacities, capabilities and performance
- establish a water budget based on plant water requirements and system performance
- measure water use and evaluate according to plant needs
- conduct audits and rectify performance deficiencies (Figure 7.3)
- use ET_0 in scheduling decisions
- adjust schedule to meet changing water needs
- check operation and calibration of sensors
- verify that landscape is healthy and in acceptable condition
- prepare written report on the system performance, assessment of water supply and strategies for changing water supply conditions, including drought. A site water management plan (see chapter 12) should incorporate these items because they are fundamental to achieving sustainable use of water.

Australian urban irrigation best management practices

System design

The BMP principles and practice guidelines provide a sound framework for the design, installation and management of urban irrigation systems.

Further details on each of the BMP principles are outlined in *Urban Irrigation – Best Management Practices* (Irrigation Association of Australia 2006). Website: www.irrigation.org.au

These guidelines include the following key performance values to be used in the design of urban irrigation systems:

1. System design and component selection should achieve a minimum *operational or field* distribution uniformity (DU) of 75% and emission uniformity (EU) of 75%.
2. Water source capacity selected to meet irrigation needs in no more than 10 hours in a day. Note that, in some circumstances – for example, to comply with withholding periods for recycled water – irrigation will need to be completed in shorter periods.
3. Metered supplies – design flow not to be greater than the lowest flow determined for each of the following criteria:
   - flow that results in 10% pressure loss of static pressure through the meter
   - flow that does not exceed 75% of the maximum flow rating of the meter
   - flow that results in maximum of 1.5 m/s in the service line.
4. Allowance of at least 70 kPa in static pressure in mains supplies for future supply network expansion and increased demand
5. Mainline and lateral pipe sizes selected so that the maximum velocity does not exceed 1.5 m/s.

Sports grounds – best practice

Design guidelines for safe irrigated sports grounds

The requirements for a sports ground to be safe are that it should be level, even and the surface conditions such that the area can be used without injury or harm (Plate 7.1). The potential injury risks associated with an irrigated sports ground include:

- presence of holes, depressions and sudden changes in levels
- abrupt changes in surface conditions and traction properties
- raised and protruding equipment (e.g. pop up sprinklers that may cause tripping)
- exposed or sharp surfaces or edges that may cause injury

- unstable surface conditions that do not support player impact loads.

Irrigation design considerations that support the development and maintenance of a safe and functional surface include:

- sprinkler heads with small exposed tops
- sprinkler top and other exposed surfaces are safe for users (e.g. rubber tops)
- anti-drain valves installed (avoid soggy patches)
- compacted trenches to minimise subsidence
- remote-operated master valve to reduce risk of flooding
- flow monitoring to detect leaks
- precision control, including sensors
- regular maintenance
- regular testing of uniformity of application.

Sprinkler irrigation design guidelines for sports grounds

The development of sustainable sports grounds is a high priority for the managers of open space turf sports facilities. The following are the broad requirements of irrigated sports grounds:

- healthy turf surfaces
- efficient in the use of water
- safe for users.

The practice guidelines outlined here are to assist in the achievement of high-quality irrigation systems.

Overall system performance

- The irrigation system has the capacity to meet the daily peak water demands of the landscape within water supply and functional constraints.
- The design of the system, including selection of equipment and installation, allows water to be applied uniformly.
- The application of water by the system is effective in delivering water into the turf root zone.
- The system uniformity of application performance standards are: design DU 85% minimum and scheduling coefficient ($SC_{5\%}$) less than 1.2 and field DU_{LQ} 75% minimum.

These uniformity coefficients are described in more detail in Chapter 11.

Player safety

- Valve boxes should be located outside the active playing area.
- Valve boxes should be of robust construction to withstand expected impact loads and fitted with non-slip lids or covers.
- Sprinklers should have small (not greater than 80 mm diameter), exposed tops and fitted with rubber top covers.
- Sprinklers should not to be located in high-use areas (e.g. goal square).
- Sprinklers should be of robust design and construction to reduce risk of injury from damaged heads or malfunction, such as sprinklers being stuck in the raised position.

System design, layout and zoning

- The pipeline route should be a ring main design, and positioned outside the active play area, to deliver water efficiently to sprinkler laterals (Figure 7.4).
- Sprinklers should be grouped into zones that have similar watering requirements. The selection of zones should reflect potential variations due to soil, microclimate and use patterns (e.g. goal squares and training areas).

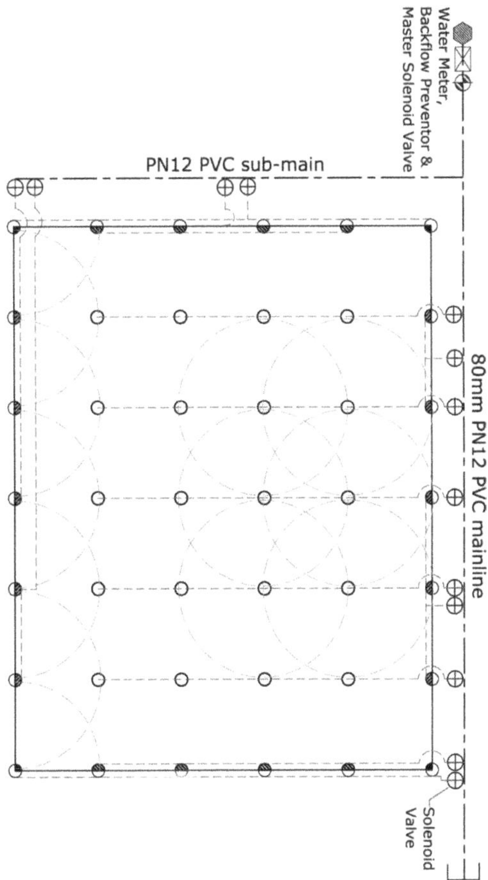

Figure 7.4. Soccer field irrigation design to best practice guidelines including part-circle boundary sprinklers (Source: Rainlink Australia Pty Ltd)

- Coverage by the irrigation system is not to extend beyond the designated playing area of the ground. Part-circle sprinklers should be used on boundaries to avoid or minimise overthrow into non-target areas.

Effective application

- System design precipitation rate should be selected to be less than typical soil infiltration rate (Figure 7.5).
- Only sprinklers of similar precipitation rate to be operated from the one solenoid valve. Part and full circle sprinklers not to operate on the same circuit.
- Sprinklers should be fitted with an anti-drain facility.
- Pop-up sprinklers should have adequate lift height to clear surrounding grass.

Hydraulic design

- Hydraulic design should allow sprinklers to operate at optimum flow and pressure conditions.
- Pipework should be designed to keep pressure loss, within a sprinkler group, within acceptable limits. Lateral pressure variation should not be greater than 10% overall within a sprinkler zone.

Figure 7.5. Compacted soils and high application rates can lead to significant runoff

- Pipework should be designed so that the maximum flow velocity, within the pipe network, does not exceed 1.5 m/s.
- Pressure regulated solenoid and other supply valves should be used to allow optimum operating pressure at each sprinkler head in a lateral, and throughout the system.

Water supply – safety

- Backflow prevention devices should be selected to meet system hydraulic and regulatory requirements of the local water authority and be effective in providing ongoing safe operating conditions.

Water supply – general

- A remote controlled master valve should be installed on the supply side of the irrigation system to ensure that the distribution pipelines and laterals are not pressurised unless the irrigation system is programmed to operate. This reduces the risk of pipe bursts and leakage within the irrigation distribution pipe network.
- Isolation valves should be installed upstream of the solenoid valves and at strategic points within the network.
- System flow rate should be within specified limits of the rated capacity of supply pumps and water meters.

Control system

- The control system selected should allow each station to be operated to meet irrigation requirements in an effective and timely manner. Ease of programming and flexibility in programming are key requirements.

- The flexibility of the control program should accommodate any time and date constraints imposed by water authorities.
- The control system should be capable of accepting inputs from environmental sensors including rain gauges, soil moisture sensors and system monitoring devices.
- The system should be designed and equipment selected to provide for future compatibility with central control.
- Provision for local weather data should be used in irrigation scheduling decisions.
- Controller features should include water budgeting, cycle and soak, global and individual station changes, percentage adjust options, multiple cycles and day skip.

Monitoring and measurement

- Dedicated measurement of irrigation water consumption is required, as is provision for automatic monitoring.
- A digital flow meter should allow continuous monitoring of system flow performance, including breaks and valve malfunctions, and alarm generation facility.

Plans and documents

- Accurate records of irrigation designs, using GPS, and incorporated into the organisation's GIS system should be kept. Irrigation infrastructure positions should be referenced to the site engineering and landscape features.
- Final 'as-built' drawings (AutoCAD format, GPS-generated) should be prepared. Notes outlining changes to specified works should be included on the as-built drawing.

Best practice irrigation using recycled water

The following is a guide to assist in irrigating with recycled water. Reference should also be made to the requirements of local authorities including environmental, health, water and local government. These comments should be read conjunction with the IAL *Urban Irrigation Best management Practices* Guidelines (Irrigation Association of Australia 2006).

In the case of recycled water, there are aspects of irrigation management that require particular attention or have specific requirements, in order to minimise risks. These are:

1. **Method of delivery or application** – Consider the suitability of alternative irrigation methods. Drip is generally better suited for the application of recycled water.
2. **Precise scheduling** – Irrigate to meet plant needs using precise applications and timed to meet water demand for each separate area of water demand, based on weather conditions, e.g. daily ET_0 data.
3. **Hydrozoning** – Design the irrigation delivery system so it applies water according to the particular requirements of an area, including water demand and microclimates.
4. **Uniform application** – Design and maintain the system to achieve high uniformity of application in the field (field DU greater than 75%; field EU greater than 85%). It should be operated and maintained to be able to meet these performance requirements.
5. **Sprayed droplet size** – Select sprinklers and sprays to minimise small droplets, which are a wind drift risk. Nozzle size, nozzle trajectory and operating pressure are influencing factors.
6. **Monitor atmospheric conditions and rainfall** – Use a weather station that monitors wind speed and wind direction to inform about conditions suitable for use. Rain gauges inform about the risk of runoff. Operate spray equipment to avoid wind conditions that will result in significant drift. Daytime watering with recycled water is increased risk, due to higher wind speeds (Figure 7.6). Avoid watering close to rainfall events, to reduce risk of runoff of nutrient rich solutions.

Figure 7.6. Operating sprinklers in windy conditions is wasteful and drift is a potential hazard with recycled water

7. **Monitor soil conditions** – Use soil moisture sensing equipment to help irrigate efficiently, to monitor soil conditions and schedule irrigation. This equipment will also provide information about possible risks associated with salinity increases.
8. **System hydraulic regulation and control** – Regulate pressure and flow to ensure hydraulic design performance standards are achieved so that delivery occurs under optimum flow and pressure conditions. Incorporation of isolation valves, throughout the pipe network to allow control and management of flow so that risk of damage is reduced should a break occur, and also to allow for maintenance activities.
9. **No overthrow** – It is important to avoid overthrow with recycled water. One way of doing this is to use part-circle sprinklers along boundary areas.
10. **Soil health** – Maintain soil in a healthy condition for healthy plant growth and optimum soil water properties. Effective infiltration, permeable structure and good water-holding capacity are ideal.
11. **Water balance** – Prepare a water balance for the site, indicating on a monthly basis, the volume of water that will be used by the vegetation of the site. This requires the use of monthly evaporation and rainfall data. Management of the water source, when plant water demand is low (e.g. winter), needs to be factored in to the overall management of the water on the site. A nutrient balance is also recommended.
12. **Monitoring water use and environment** –Monitor the water source, soil, water consumption and potential environmental impacts. This assists with watering efficiently, protects the asset and provides the data necessary to demonstrate regulatory compliance.

Irrigated Public Open Space (IPOS) Program (Adelaide)

A code of practice, which facilitates the achievement of high irrigation efficiency, has been developed in Adelaide. The code is known as Code of Practice – IPOS was developed by a group of turf managers, irrigation industry, parks and leisure personnel, representing local government, schools and sporting bodies, who are responsible for significant areas of irrigated space in Adelaide.

Gerry Charlton, who was with Salisbury City Council, provided leadership in coordinating the development of a best practice approach to open space turf irrigation management. The program has been published by SA Water (2008) and provides the basis for a partnership between the water authority and the open space water consumers. Water is supplied under a range of conditions, which encourage efficient use of water.

The benefits of the program are described as:

- fit-for-purpose turf
- significant water savings translating into reduced watering costs
- increased flexibility for the irrigator in the case of water restrictions
- recognition of the level of irrigation efficiency in assessing exemption applications in the case of water restrictions
- efficient irrigators are not penalised for efficiencies already demonstrated
- community recognition/accountability/demonstration of efficient watering practices
- potential aversion of public health and safety issues – particularly in the case of school ovals and sporting grounds
- increased amenity levels associated with appropriately irrigated open space.

The program is a six step process.

Step 1: Policy and commitment

Implement a strategic approach to irrigation underpinned by a policy statement and commitment by an organisation to appropriately resource and manage irrigation sites.

Step 2: Irrigation system performance

Ensure that systems are functioning to the appropriate performance standard with periodic system audits and ongoing regular maintenance.

Step 3: Horticultural maintenance

Ensure that an appropriate horticultural maintenance program is in place to maintain soil structure and turf nutrient requirements.

Step 4: Determine the baseline irrigation requirement

Determine the baseline irrigation requirement using long-term average climate data to set the monthly irrigation schedule and consumption.

Step 5: Management of irrigation schedule

Amend the determined irrigation schedule on a regular basis to account for climatic variance in any given season to the long-term average. This will ensure that turf is receiving the required water to maintain it at the required predetermined quality level.

Step 6: Monitoring and reporting performance

Monitor irrigation consumption against irrigation requirement and report on irrigation efficiency and turf quality.

Copies of the IPOS Code can be downloaded from SA Water website (www.sawater.com.au).

Irrigation best practice summary

The principles of best management practice irrigation can be summarised as follows:

1. Assure overall quality of the irrigation system.
2. Design the irrigation system for efficient and uniform distribution of water.
3. Install irrigation to meet the design criteria.

Table 7.1. Best practice checklist for irrigated sites

Category	Performance criteria	Compliance
1. Site	1.1. Accurate, detailed plan of site, boundaries, features and water infrastructure	
	1.2. Vegetation and soil survey	
2. Water source	2.1. Supply (pressure and flow) verified	
	2.2. Water quality tested	
3. Irrigation design	3.1. Design by qualified irrigation designer	
4. Irrigation application efficiency	4.1. Uniform application and delivery of water (field distribution uniformity >75%) (emission uniformity >85%)	
	4.2. Precipitation rate matched to terrain and soil properties	
	4.3. Irrigation operated only during optimum environmental conditions	
	4.4. System designed and operated to provide optimum hydraulic conditions	
	4.5. All irrigation equipment maintained in effective working order	
5. Water management efficiency	5.1. Irrigation depth matched to plant root zone and soil	
	5.2. Irrigation timing based on ET_c needs and soil moisture conditions	
	5.3. Sensors to prevent irrigation following rainfall – rain shut down, soil moisture sensor	
6. System maintenance	6.1. Documented maintenance plan and checklist	
	6.2. Records of maintenance and repair	
7. Water use	7.1. Water use records kept (e.g. monthly)	
8. Soil condition	8.1. Garden bed and tree areas mulched	
	8.2. Turf aeration, dethatching program and soil amelioration	
	8.3 Soil water storage optimised through deep root zone management	
Outcome	Irrigation system compliance with best practice	Overall rating

4. Maintain the irrigation system for optimum performance.
5. Manage the irrigation system to respond to the changing need for water.

Irrigation best practice checklist

Table 7.1 provides a checklist of each of the aspects of irrigation that need to be implemented in order to comply with best management practice and achieve high irrigation efficiency.

Chapter 8

Designing irrigation systems

Irrigation system design approach

An irrigation system design is an equipment solution to the challenge of efficiently delivering water into the root zone of plants. The process involves collecting and analysing much detailed information about the site, the soil, the plants, the water supply and water equipment (Figure 8.1). It requires a skilled designer to develop the right solution for the site. It should be carried out by appropriately qualified irrigation designers.

The overall design approach consists of a number of stages.

1. **Planning:** aims, policy, regulatory issues, information and design brief
2. **Design:** performance standards, system requirements, hydraulic design and equipment selection
3. **Quotation:** tendering, contracts and warranties
4. **Installation and commissioning:** testing, operating procedures and instruction.

It then becomes the responsibility of the owner or manager of the system to ensure that the system is operated to achieve the performance standards identified in the planning stage of the process.

Figure 8.1. Precision surveying using instruments with electronic measurement of distance, elevation and angles, and with GPS capability, provides a sound basis for good irrigation design

Table 8.1. Steps in the irrigation design process

Step	Description
1	Information collected
2	Water supply identified
3	Irrigation method selected
4	System capacity determined
5	Applicator (sprinkler or emitter) selected and layout determined
6	Zoning and pipe layout selected
7	Hydraulic design selected
8	Pump selected
9	Control, communications, monitoring and measurement
10	Plans and specifications prepared

Design process

Design tasks

Designing an irrigation system requires many different, and sometimes conflicting, factors to be taken into account. A well-designed irrigation system allows the appropriate amount of water to be applied effectively and efficiently, at the correct time. The integration of the many equipment components, into an efficient irrigation system is the underlying task.

Steps in the design process

The basic steps a designer may follow, when developing a design for a new irrigation system are shown in Table 8.1.

Irrigation designer

The design of an irrigation system requires the designer to have knowledge of crops, soils, system hydraulics and irrigation equipment performance. It is the designer's responsibility, in consultation with the site manager, to identify factors that may have a significant influence on the system and to recommend a design solution that optimises all of the competing needs of the irrigation system.

Turf and landscape managers should request that those undertaking the design of their system should be appropriately qualified in irrigation, hydraulics and engineering. The Certified Irrigation Design (CID) program, which is available in the USA and Australia, is recommended by Irrigation Australia Ltd (IAL) as a minimum standard. Lists of Certified Irrigation Designers (CID) are available on the IAL website (www.irrigation.org.au). The American Society of Irrigation Consultants (ASIC) includes irrigation designer members from Australia. Lists of independent irrigation designers are on the ASIC website (www.asic.org).

Information collection

Site plans and documentation

An irrigation system design should reflect the particular needs of the site, including vegetation, soils, climate and water resource management. Accurate plans are required of the property, vegetation on site, infrastructure, services and precise details of the areas to be irrigated. The availability of geographic information systems (GIS) plans, which have all relevant features accurately located, using precision global positioning system (GPS) technology, are a valuable asset both for the designer and the overall ongoing management of the asset. The accuracy of site plans is essential for the location and positioning of pipelines and key infrastructure features, such as meters and valves. It is important that all changes are regularly recorded.

The topography of the site influences the hydraulic design, the system control and the application of water. A contour plan is essential for sites on sloping or undulating terrain. The accuracy of these plans needs to match the performance requirements of the system. Contour plans, with elevation intervals at 1 m or less are required for drip irrigation systems.

Surveying developments

The surveying requirements for irrigation design vary greatly, depending on nature of site and works to be completed. Accurate coordinates of ground positions (latitudes and longitudes) and elevations are required in most situations. Referencing to property boundaries (cadastral surveying) may also be needed. Although the elevation (reduced level, RL) is required in some form, referencing to the Australian height datum (AHD) may also be required.

Spatial data can be captured rapidly and accurately using total station survey instruments, which are an advanced form of theodolite. The incorporation of electronic distance measurement and GPS means that precision measurements, including elevation, can be made of all landscape features of a site. Survey data management systems that store, process and export files to other platforms, such as computer aided design (CAD) programs, are invaluable in building an accurate base for design.

Local climate

An understanding of the climate at the site is vital in achieving optimum irrigation. The water use rate of the landscape is driven by the evaporative demand of the atmosphere. Obtaining evaporation data (Class A evaporation pan) from the Bureau of Meteorology or the use of local site weather data to determine the reference evapotranspiration (ET_0) are essential. Analysis of rainfall data is vital for the estimation of total irrigation requirements. Rainfall data is also necessary for the determination of the stormwater harvesting potential for water storages.

An appreciation of the strength and direction of wind is important in the design and operation of the sprinkler or spray irrigation systems. This influences the layout of sprinklers and sprays. The closer spacing of sprinkler heads is an important technique used to achieve acceptable uniformity under windy conditions. Wind speeds will also influence the selection of heads and nozzle trajectory angles. Low trajectories are sometimes used for windy sites.

Vegetation survey

A vegetation survey should be carried out to identify plants, cultural management requirements and water use characteristics. Assistance of a horticulturist may be required to identify particular plants and learn details of plant water use characteristics and possible risks, such as sensitivity to foliage wetting, that may be relevant in the design process.

The key vegetation information that needs to be obtained includes crop factors/crop coefficients, root zone depth, growth cycle characteristics, sensitivities to irrigation techniques, water quality and plant response to soil moisture stress. The location and density of roots for large plants (e.g. trees) also has an impact on the system design, including in determining water delivery method and installation of pipelines and irrigation laterals.

Soil survey

Soils may vary greatly across sites and this will directly influence the design of the irrigation system. The infiltration rate of the soil influences selection of the applicators and this, in turn, determines the run times for the operation of the irrigation equipment.

The water-holding capacity of the soil (AWHC), together with the root zone depth, are needed to determine the total available water (TAW) and hence are used in determining both irrigation application depth and timing of irrigation. Soil types and hydraulic properties are critically important for drip irrigation system design, as the spacing of emitters is directly dependent on soil properties. Soil properties are also important in terms of deciding on the construction techniques to be employed, including installation of pipelines and placement of drip emitters.

Other factors

Other important factors to consider include:

1. Site management practices
   - machinery used: time of use, type of use (e.g. risk of soil compaction)
   - access: times when irrigation can take place
   - sports grounds maintenance practices (e.g. aeration, coring – depth of buried pipes, risk to buried drip systems).
2. Labour
   - availability, skill and training requirements.
3. Costs
   - The funds available for the project need to be clearly defined. Limited funds may favour a staged project. The irrigation design will need to take this into account. Allocation of funds for future expansion, including allowing for increases in flow capacity, needs to be considered.
4. Electricity
   - Power capacity will sometimes determine the pump capacity and will directly influence the hydraulic design of the system and the operating cost.
   - Off peak tariffs will often determine the time of irrigation and the time window available.

Water supply

The matching of the available water supply to the site water requirements is a core task in the design process.

Water supply characteristics

The total irrigated area and the irrigation method will both be strongly influenced by the characteristics of the water source. The key issues to consider are:

- quantity available – this will determine the total area that can be irrigated
- flow rate – this will determine area that can be simultaneously watered and also the size of components (e.g. pumps and pipes)
- quality – this will determine suitability for specific plants, irrigation method, equipment materials and irrigation management practices.

The characteristics of the various potential water sources are outlined in Chapter 3.

Water quality considerations

The quality of water has an impact on the method of application, operation of the system, the ongoing sustainability of the irrigated site, the selection of water filtration and water treatment equipment and the selection of materials used in the system.

Water quality needs to be considered under the following categories: chemical, physical and biological. Water quality, including the impact on irrigation, is discussed in detail in Chapter 3.

Irrigated areas and hydrozones

In the first instance, the area to be irrigated is defined according to the overall requirements of the site or facility. The next step is to assess the total irrigated area in terms of how it is to be watered to achieve high efficiency. This requires breaking the area down to hydrozones. A hydrozone is an area of the landscape that has its own unique water requirements.

There are numerous factors that may cause a section of the site to have water demands different to adjoining areas. These include:

1. The evaporation demand of the area
   - exposure to full sun or shaded
   - windy or protected sites
   - influence of thermal loading from nearby structures, buildings and carparks.
2. The plant type(s)
   - need to group plants of similar water requirements.
3. The soil water properties
   - the water-holding capacity of the soil
   - the infiltration rate of the soil
   - the amount of stored water in the root zone
   - the root zone depth
   - the slope of the area – risk of runoff and distribution of moisture in the soil with subsurface drip irrigation
   - deep drainage properties
   - water repellency of soil.
4. The natural drainage characteristics
   - natural soil moisture conditions, wet or dry, due to movement of surface water.
5. The irrigation technique
   - To achieve precision application, irrigation system performance needs to be consistent across the zone.
6. Hydraulic considerations
   - Flow and control constraints may cause areas to be irrigated separately.

System capacity

Peak plant water demand

The irrigation system is required to deliver to the plant or landscape water in amounts that can maintain the plants in a healthy condition and a performance standard appropriate to the site to deliver the required outcomes. Initially the peak water requirement should be determined and then the system capacity designed to meet this demand. If the peak demand cannot be met, then plant stress will occur and there could be permanent damage. The irrigation system does not directly supply water to the plant. The soil root zone volume acts as storage. If the soil water storage is large, then the peak requirement can be reduced, because water will be taken from the soil and may be replenished, at a lower rate, by the irrigation system. If the storage is small, such as in shallow rooted turf in sandy soil, then deficiencies in peak demand will show up as stress in the turf, because the soil does not have the capacity to act as a significant storage buffer. This may be acceptable at some sites.

An irrigation system designed to meet the long-term average water demand is likely to result in some plant stress at some stage, because the demand can be significantly exceeded for short periods, even for several days.

Table 8.2. Daily ET_0 for various design scenarios

Scenario number	Scenario	ET_0 (mm/month)	Peak demand daily ET_0 (mm/day)
1	Average daily ET_0 based on month of highest evaporation (Jan)	165	5.3
2	Average daily of median (decile 5) of month of highest evaporation (Jan)	180	5.8
3	Average daily of decile 9 month of highest evaporation (Jan)	230	7.1
4	Decile 9 of all evaporation (ET_0)	–	14
5	Highest daily ET_0	–	18

It is the responsibility of the irrigation designer, in consultation with the site manager, to determine the appropriate plant water demand that needs to be satisfied. It is a case of balancing risk (the consequences, if demand is not met) with the increased cost of systems with higher capacities.

It is common practice in urban irrigation to use the average daily evaporation rate in the month of highest evaporation, as the basis for determining the required irrigation system capacity. This rate does not meet the peak requirements of the plant. Rates higher than the average, such as the highest demand in the highest evaporative month, can be used, but there are consequences in terms of the system design, larger flow capacities and higher costs.

There are a range of options in determining a peak demand value. The approaches range from average demand in the month of highest evaporation to the maximum daily experienced at the site. The consequences of not meeting the selected or design peak demand need to be considered. In some circumstances, if extreme conditions exist, then assumed constraints may be able to be relaxed. For example, longer watering windows may be used if there are extremely hot and demanding conditions.

Table 8.2 illustrates the range in water demand values that could be expected for various evaporation scenarios.

The implications in selecting ET_0 values, for plant water demand estimation, based on actually daily ET_0 values is severe in terms of the design of the capacity of the system.

The design flow rate will increase by 9% if decile 5 is used rather than the average and it will increase by 34% if the average daily evaporation for the month that represents 9 out 10 highest (decile 9) evaporation rates. It will obviously be much higher if the design is based on evaporation rates for single days.

While average and median (decile 5) monthly evaporation and rainfall data can be used, based on month of highest evaporative demand, it is important to recognise the consequences when the daily demand will not be met. It is a case of considering the risks.

The key system capacity consideration is the amount of water that can be delivered to the site in a specified period (flow rate). The following factors affect this value:

- plant water demand
- irrigated area
- time window for irrigation (available delivery period)
- irrigation system efficiency.

If the daily peak water demand is increased from 6 mm to 9 mm, for example, and the delivery period is fixed, the required flow rate will increase by 50%. This will have a very significant effect on the cost of the system, because larger pumps, pipes, valves and fittings will be required. This requires a cost–benefit analysis to determine the optimum solution.

If the system has a capacity that does not allow the peak demand to be satisfied for a period of several days, it may be feasible to include an extra irrigation event to meet this demand. This is in effect extending the assumed time window for irrigation. If not, it will mean that the plants will be stressed, because water will be extracted from the soil reservoir beyond the designated refill point (RP). Rather than refilling at an assumed managed allowable depletion (MAD) of 50%, it may mean that the soil moisture level is closer to a MAD of 70–80%.

The situations that provide the greatest risk, in terms of not meeting peak demand are sites in which the soil reservoir is small compared with the expected water demand. Plants with very shallow root systems growing in low AWHC soils and plants growing in containers are at risk of running out of water and experiencing permanent wilting point (PWP). Plants with deep root systems, and generous water storages, provide good resilience against the risks of the system not meeting peak demand.

Calculating plant water demand

The method of determining the plant water requirement is outlined in Chapter 5. The basic approach used is:

$$ET_c = K_c \times ET_o$$
$$\text{or } ET_c = CF \times E_{pan}$$

The period of time over which the plant water requirement is determined will vary according to the purpose. If an estimate is required of the volume of water to irrigate the landscape over the irrigation season, then monthly data can be used. If scheduling of irrigation is the task, then daily data is required.

Weekly irrigation requirement

The reference management period, for many organisations responsible for open space is a week. This is a convenient period in terms of general resource management, such as labour and facility use, and also is commonly used for irrigation management. Determining the weekly irrigation requirements is therefore convenient in many cases. Weekly ET_o and irrigation requirements for Australian capital cities are presented in Appendix 3.5.

The following is an example of calculation of the weekly irrigation water requirements for a site.

Example – Weekly irrigation requirement

Site: sports ground
Turf: kikuyu
Turf crop coefficient (K_c): 0.4
Irrigation efficiency (E_a): 80%
ET_o Weekly: 45 mm
Weekly ET_c: 18.0 mm (= 0.4 × 45)
Weekly irrigation water requirement (IWR): 22.5 mm (= 18.0/0.80)

Note: irrigation efficiency is expressed as a decimal in this calculation.

Determining system capacity or flow rate

The system flow rate can be determined, taking into account the time available for delivery, and the total area to be irrigated.

$$\text{Flow rate} = \frac{\text{Volume (litres)}}{\text{Time (hours, minutes, seconds)}}$$

Volume to be delivered is calculated using

$$\text{Volume} = \text{Area (m}^2) \times \text{Depth (mm)}$$

Example

System design flow rate to meet the following requirements.
Site area: 1.5 ha (15 000 m^2)
ID: 22.5 mm/week
Time window: 8 hours per night, 2 nights per week. Total 16 hours

$$\text{Volume of water: Area} \times \text{Depth} = 15\,000 \times 22.5 = 337\,500 \text{ L}$$

$$\text{Design flow rate:} \left(\frac{337\,500}{16.0}\right) = 21\,094 \text{ L/h}$$

$$\text{Converting to L/s: Flow rate } \frac{21094}{60 \times 60} = 5.86 \text{ L/s}$$

The time available for delivery directly influences the system design. Longer delivery times reduce flow rates and hence reduce the size of hydraulic components. Lower flow rates reduce the cost of the system.

A consequence of water restrictions regimes is that often the available delivery time is reduced. This means that high flow rates are required to achieve the same depth of application. Higher flow rates provide greater flexibility in the management of irrigation systems. There is greater opportunity to apply the required depth of water to designated areas (Table 8.3). Also, there is greater flexibility when irrigation is carried out. This also provides greater opportunity to capture any rainfall that may occur.

Higher flow rates allow shallow applications to be avoided. Shallow applications (e.g. 5 mm or less) result in shallow root systems and are a source of significant water loss and inefficiency.

The design flow rate is generally a compromise between meeting the water needs of the plant and the costs associated with higher flow rates, which provide higher capacities and greater flexibility in irrigation management. In many urban situations, the system capacity is determined by the size of the mains water meter and the hydraulic conditions in the supply mains. Under these circumstances, it is a matter of developing an irrigation design that satisfies the available site flow rate. If this flow rate is not adequate, one strategy is to investigate the possibility of installing a larger meter. This requires approval by the local water authority.

Table 8.3. Irrigation design capacity – flow rate and area covered

Delivery time	Required flow rate – L/s								
	5 mm gross application			10 mm gross application			20 mm gross application		
	4 h	6 h	8 h	4 h	6 h	8 h	4 h	6 h	8 h
Area									
0.5 ha	1.74	1.16	0.87	3.48	2.32	1.74	6.96	4.64	3.48
1.0 ha	3.47	2.31	1.74	6.94	4.62	3.47	13.88	9.24	6.94
1.5 ha	5.20	3.47	2.60	10.40	7.94	5.20	20.80	15.88	10.40
2.0 ha	6.94	4.63	3.47	13.88	9.26	6.94	27.76	19.52	13.88
3.0 ha	10.42	6.94	5.20	20.84	13.88	10.42	41.68	27.77	20.84

Note: The application depth (mm) or irrigation depth is a gross application. The net amount entering the root system will be dependent on the efficiency of application.

Table 8.4. Guide to water meter size and flow rate (based on service line velocity of 2.3 m/s)

Water meter size (mm)	Flow rate (L/s)
20	0.6
25	1.0
32	1.5
40	2.75
50	4.5
65	6.25
80	9.0
100	14.0

Note: The flow rate shown is based solely on maximum service line velocity of 2.3 m/s. Other hydraulic constraints may prevent this flow rate being achieved. Physical conditions at each site determine actual flow rate.

Metered mains supply

The available design flow rate from metered mains supply is sometimes determined using a number of constraints. The flow rate that meets the most limiting constraint, that is the lowest flow rate, is used as the design flow rate. These constraints include:

- maximum service line velocity is 2.3 m/s
- design flow rate not to exceed 75% of the maximum rated flow of the meter (this information available from the manufacturer)
- pressure loss through the meter not to exceed 10% of the static pressure of the mains supply.

These criteria limit the flow rate, based on excessive friction and velocities, which may cause water hammer, when remote valves close (shut down) suddenly. Table 8.4 is a general guide to the flow rates available through various size service lines and meters. The values do not take into account the changes to the flow rate as a result of the reduced pressure and the friction loss characteristics of the particular meter in use.

Selection of irrigation applicators and layouts

Principles

The device that delivers the water to the soil is critical to the success of an irrigation system. The requirement of effective delivery, with minimum of losses, is demanding. The key aspects to consider are the:

- coverage or wetted area
- precipitation or application rate (should be less than soil infiltration rate)
- effectiveness of application – losses due to wind, evaporation, drainage
- mechanical properties – design, materials, operation, functioning and reliability.

Sprinkler selection and layout

The process of identifying a particular sprinkler product and layout is an iterative one. It is a case of choosing options, checking and then considering alternatives. The layout pattern is sometimes determined by the shape of the area to be irrigated and the planting layout and sometimes by the need to optimise the area to be covered.

The layout patterns may be square, rectangular or triangular (Figure 8.2).

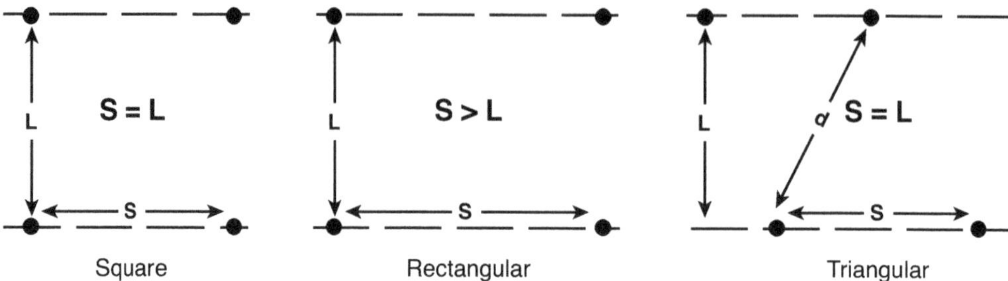

Figure 8.2. Layout patterns for sprinklers and sprays

Square and rectangular patterns are often used to cover small landscape areas and rectangular or square-shaped grass or lawn areas. Triangular patterns generally allow wider spacing, and hence lower costs, when large areas, such as turf, are to be watered. The equilateral (all sides equal in length) triangular pattern is often used in sprinkler design. As the name suggests, the distances between each sprinkler head are equal. The perpendicular distance between rows of sprinklers, in this arrangement, is calculated by the following:

L (perpendicular spacing between rows) = 0.866 S (spacing between sprinklers)

The infiltration rate of the soil plays an important role in the selection of the correct outlet and spacing. The precipitation rate of the proposed design solution is calculated and then checked against the required system performance standards.

There are many possible combinations of products, nozzle sizes, operating pressures and spacing that directly affect both the precipitation rate and the uniformity of application. It is for this reason that access to quality irrigation performance test data is essential to achieve an optimum solution.

Selecting a sprinkler or spray

The key requirement for sprinklers is that they apply water uniformly, that there is no runoff and that the required depth is delivered.

The following are the major performance factors to be considered for sprinklers and sprays:

- operating pressure (kPa, m)
- coverage
  - wetted diameter (m)
  - wetted radius (m).
- discharge rate (L/min, L/s)
- precipitation rate – system (mm/h)
- uniformity of application – distribution uniformity (DU).

The key selection factor for sprinklers and sprays is the coverage, the wetted diameter and/or the wetted radius. In many situations, the sprinkler or spray selected will be determined directly by the dimensions of the area to be covered. If the width of a lawn is 5.0 m, then a spray, that has an effective wetted radius of 5.0 m, will generally be selected.

In the case of covering large areas, such as sports grounds, the sprinkler head is often selected according to the maximum effective coverage that can be achieved. This is, however, dependent on the available pressure. Sprinkler distance of coverage increases with operating pressure. Sprinkler layouts with larger spacings require fewer sprinkler heads and generally this means lower costs. The uniformity of application needs to be taken into account, as part of the sprinkler layout design phase.

Software programs, such as SpacePro (Center for Irrigation Technology, Fresno), provide designers with comprehensive data on sprinkler performance (Figure 8.3) All key performance

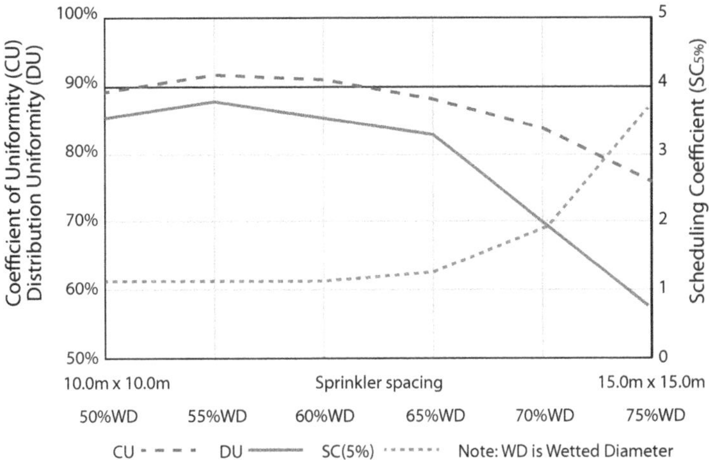

Figure 8.3. Sprinkler profile test analysis showing effect of spacing on uniformity

parameters can be considered simultaneously and a sprinkler solution that meets the system performance standards obtained.

The performance data presented in Figure 8.3 show that the key indicators of performance CU, DU and SC all deteriorate as the spacing progresses from 10 m × 10 m to 15 m × 15 m, for this sprinkler. As the spacing apart of sprinklers reaches around 65% of WD, there is a marked decline in overall uniformity of application (CU and DU) and the measure of ratio of extreme depths applied (SC) increases. These data represent the performance on a particular sprinkler model, one nozzle combination and one operating pressure.

Good design takes into account all of the factors that can potentially have an impact on the performance of an installed system (Figure 8.4).

Figure 8.4. Significant overlap of sprinkler distributions is necessary to achieve uniformity of application

Table 8.5. Guide to sprinkler spacing in windy conditions

Pattern	Wind speed (km/h)	Maximum spacing
Square	0–5	55% of WD
	6–12	50% of WD
	13–20	45% of WD
Triangular	0–5	60% of WD
	6–12	55% of WD
	13–20	50% of WD

Note: WD is the wetted diameter of the sprinkler. Achieving high uniformity may require closer spacing than indicated here.
Source: These data are based on recommended spacings outlined in Pair (1969).

Sprinkler spacing for windy sites

For wind-prone sites, manufacturers' data should be consulted, but as a guideline, where irrigation cannot be scheduled to avoid periods of moderate wind greater than 15km/h, then sprinkler spacing should not exceed 50% of wetted diameter. It is a general sprinkler design recommendation that sprinkler spacing be reduced to maintain acceptable performance under windy conditions.

Table 8.5 is a guide to reduction in recommended sprinkler spacing, according to various wind speeds.

Assessment of wind speeds and direction are critical when designing for the use of recycled water in sprinkler and spray systems. Operating conditions and buffer distances are determined by wind speed and direction.

Selecting a drip emitter

The selection and layout of drip emitters is strongly influenced by the soil properties of the site. The initial requirement is the determination of the type of drip system to be used: point source dripper or line source dripper (e.g. dripline).

If the site is to be watered to achieve continuous wetting along a line or the whole area is to be watered, such as turf or densely planted garden beds, then dripline products will be appropriate. For watering of individual plants, such as shrubs spaced at intervals of several metres, then individual point source drippers are usually appropriate.

The selection of pressure-compensating (PC) or non-PC emitters is another important selection. If the terrain is undulating, then PC emitters will be required to achieve constant delivery rates.

Factors to consider in the selection of a particular emitter, such as emitter construction, flow regulation pattern, filtering technique, anti-siphon and non-drain, are outlined in Chapter 4.

Key drip system design considerations include the emitter delivery rate and emitter spacing. Both the spacing between consecutive emitters along a drip supply line, referred to as the emitter interval (S_{ei}), and the spacing between driplines or drip laterals (S_{ls}), need to be selected. The choice of emitter flow rate is influenced by soil type, soil moisture distribution required and the overall system hydraulics and design.

The effective and reliable operation of a drip system requires hydraulic regulation and control equipment including filtering, pressure regulation, flow control, air venting and flushing (Figure 8.5).

Drip irrigation system flow rate needs to ensure a minimum flushing velocity of 0.5 m/s can be achieved in all drip supply lines.

Figure 8.5. Automatic flushing valves are essential for reliable operation of drip systems

Precipitation rate and application rate

Precipitation rate of sprinkler systems

Although the sprinkler profile provides performance details for individual heads, it is the overall performance of combinations of sprinklers that is important. The two key performance characteristics of sprinkler systems are the uniformity of application and the precipitation rate. The uniformity of sprinkler systems is covered in Chapter 9 and 11.

The parameters required to calculate the theoretical precipitation rates, referred to as the Gross precipitation rate (PR_G), are the dimensions of the sprinkler layout and the discharge rate of the sprinklers or spray heads.

$$\text{Gross precipitation rate } (PR_G)\ (\text{mm/h}) = \frac{\text{Outlet flow rate}}{\text{Wetted area}}$$

$$\text{Gross precipiation rate } PR_G = \frac{Q \times 60}{S \times L} \qquad \text{mm/h}$$

where:

Q – Flow rate from one outlet (sprinkler or spray) (L/min)

S – Spacing or interval between outlets along lateral pipe (m)

L – Perpendicular spacing between lateral pipes (m)

Example

Sprinkler layout pattern: equilateral triangle

Sprinkler interval spacing along lateral (S): 15.0 m

Sprinkler spacing between laterals (L): 13.0

Note: The perpendicular spacing between lateral pipes, in an equilateral triangular pattern, can be calculated by multiplying the spacing (S) by 0.866.

$$L = 0.866 \times S = 13.0 \text{ m}$$

Sprinkler flow rate (Q): 40 L/min

$$PR_G = \frac{40 \times 60}{15 \times 13} = 12.3 \text{ mm/h}$$

Precipitation rate considerations

Each of the following precipitation rate characteristics needs to be taken into account, as part of the design process.

1. The operation of mixed full and part-circle sprinklers, on the same pipe circuit, is likely to result in uneven application, due to varying precipitation rates. Half circle sprinklers, fitted with similar nozzles, will apply twice as much water as full circle sprinklers, if operated from the same pipe circuit. These two types should not be connected to the same circuit. However, sprays are now often designed to deliver amounts of water, in direct proportion to the area or sector covered. This means that they will have similar precipitation rates for each of the various sector types (quarter, half or full circle). This type of spray head is referred to as matched precipitation rate (MPR) head and they can be connected to the same pipe circuit and solenoid valve. These are recommended for all landscape sites.
2. Adjusting the arc of coverage of a part-circle sprinkler system changes the precipitation rate. If half circle sprinklers are reduced to quarter circle, the precipitation rate will double.
3. Reducing the radius (e.g. by adjustment) may significantly alter uniformity.

Drip system mean application rate (MAR)

The application rate, referred to as the mean application rate (MAR), of drip systems, is an important system performance parameter. This rate determines the appropriate run time for each irrigation event. The MAR is determined from the emitter flow rate, emitter interval (i.e. spacing between emitters along the dripline) and the spacing between rows of drippers (Table 8.6). The MAR for drip systems is calculated using the following expression:

$$MAR = \frac{\text{Emitter flow rate (L/h)}}{\text{Emitter interval} \times \text{Dripper spacing}} \quad \text{(mm/h)}$$

It should be noted that drip system application rates can be very high, despite the fact that the emitters have low flow or discharge rates. During operation of a drip emitter, the initial area made wet may be as small as 100 mm diameter. A 2.0 L/h dripper wetting this small area has a high application rate (theoretically it may be 200 to 300 mm per hour). In some soils this may cause tunnelling. However, the water moves out of this saturated zone and is effectively then wetting a greater area. The actual application rate is then much lower.

The method of calculation of the MAR here is based on assuming that all the discharged water is spread evenly over the area between the surrounding drippers. The delivered water, however, is not spread evenly over the area between the emitters. This MAR approach is useful for knowing the rate of application of water to an area and for use in the management of the drip system.

The application rate can be calculated for individual drippers (point source). The assumed wetted area that is watered at any time changes as the water spreads out, and difficult to gauge. The flow rate of the emitter (L/h) provides a more convenient measure in the management of point source drip systems. The run time multiplied by the emitter delivery rate produces the volume of water delivered. Both the distance between emitters along a supply line and the spacing between adjacent lines provide the reference area to determine a theoretical precipitation rate (Figure 8.6).

Table 8.6. Drip system application rate for combinations of emitter interval and dripline spacing.

Emitter interval (S_{ei})	Dripline spacing (S_{ls})	Mean application rate (MAR) (mm/h)			
		Emitter flow rate 1.0 L/h	Emitter flow rate 2.0 L/h	Emitter flow rate 3.0 L/h	Emitter flow rate 4.0 L/h
0.3 m	0.3 m	11.2	22.4	33.6	44.8
	0.4 m	8.3	16.6	24.9	33.2
	0.5 m	6.7	13.4	18.1	26.9
	0.6 m	5.6	11.2	16.8	22.4
0.4 m	0.3 m	8.3	16.6	24.9	33.3
	0.4 m	6.3	12.6	18.9	25.0
	0.5 m	5.0	10.0	15.0	20.0
	0.6 m	4.2	8.4	12.6	16.7
0.5 m	0.3 m	6.7	13.4	20.0	26.8
	0.4 m	5.0	10.0	15.0	20.0
	0.5 m	4.0	8.0	12.0	16.0
	0.6 m	3.3	6.6	9.0	13.3

Hydraulic design

Optimum hydraulic operating conditions

Sound hydraulic design is critical to achieving effective and efficient, functioning of the system. Pressure and flow rate are closely interrelated. Friction losses in pipes are a major influencing factor on the selection of pipes and associated irrigation equipment. The key hydraulic properties that influence the amount of friction that occurs are the velocity of the water and the turbulence of the flowing water. The pipe diameter, which determines velocity

Figure 8.6. Application rate of a drip system is dependent on spacing between driplines and the emitter interval along the dripline

Table 8.7. Pipe (PVC) flow rates for friction rate and velocity limits

Pipe inside diameter (mm)	Flow rate for friction of 1 m/100 m L/s	Flow rate for friction of 1 m/100 m L/min	Flow rate (L/s) for velocity of 1.5 m/s	Flow rate (L/min) for velocity of 1.5 m/s
25	0.3	18	0.8	48
32	0.6	36	1.2	72
40	0.8	48	2.0	120
50	1.7	102	3.5	210
65	2.9	174	4.3	258
80	4.5	270	7.5	450
100	9.0	540	15	900
125	16.0	960	20	1200
150	21.0	1260	25	1500
200	54.0	3240	50	3000

Note: These values are approximate only. Reference should be made to manufacturers' performance data on specific pipe products.

for a given flow rate, and pipe material, smooth or rough, are both important in influencing the friction (loss) rate. The pipe length and associated fittings determine the total amount of friction in that part of the system.

Friction or pressure loss is commonly expressed in m per 100 m or 1000 m length of pipe. The unit of hydraulic pressure (m) represents the pressure resulting from a column of water (referred to as head) 1 m high.

Pipe sizing and selection

The provision of the correct pressure and flow rate at each outlet depends on the selection of the correct water transfer and distribution components (pipes, valves and fittings). A thorough understanding of hydraulics is essential in making these important decisions, because small variations in size can have large influences on the hydraulic performance of the system.

A problem, in some turf irrigation systems, is the use of pipes that are too small in diameter, with the result that excessive pressure variations occur, because of high friction losses. High water velocity is another problem. In addition to high friction losses, high velocities produce water hammer (pressure surges), when automatic (solenoid) valves are opened and closed quickly. Maximum velocities, usually 1.5 m/s, are specified. Minimum velocities, in the range of 0.3 to 0.5 m/s, are needed if line flushing is required.

Pipe flow rates corresponding to a maximum friction rate of 1 m/100 m and maximum velocity of 1.5 m/s are presented in Table 8.7. As a general guide, the friction rate selected will be in the range of 0.4 to 1.4 m/100 m of pipeline.

The design performance specifications stipulate the maximum amount of pressure loss or pressure variation that is acceptable in each part of the system.

Although changes in pipe sizes may seem small, the consequences can be profound in the overall performance of a system. For example, for a flow rate of 240 L/min (4.0 L/s), the difference in friction rate between 50 mm and 65 mm is nearly three times more for the smaller pipe. Over a mainline length of 200 m, this would represent a difference of 90 kPa between the two pipe sizes. Typical pipe friction charts for PVC and poly pipes are presented in Appendix 8.

Pipe selection should be carried out by personnel skilled in system hydraulics, such as qualified irrigation designers and water management engineers (Figures 8.7 and 8.8).

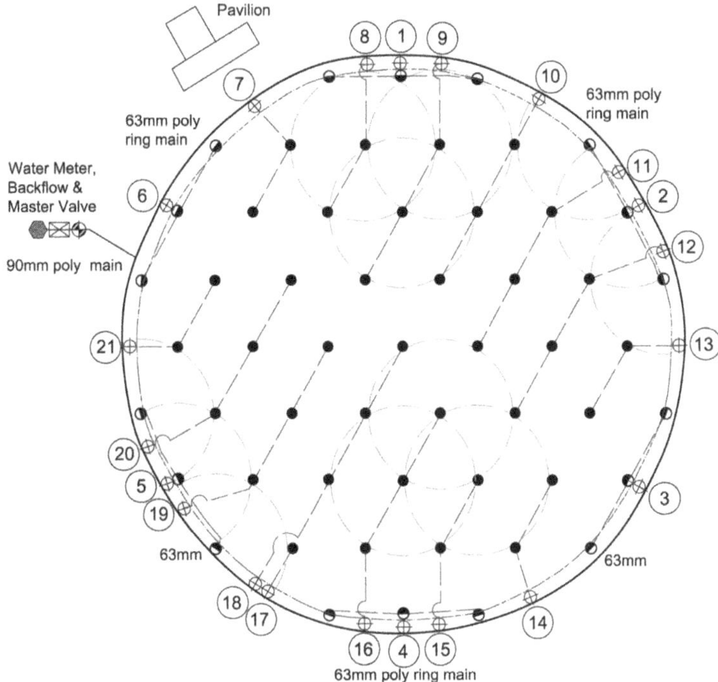

Figure 8.7. Sports oval irrigation design showing sprinkler and pipe layout (Source: Rainlink Australia Pty Ltd)

Figure 8.8. PVC pipe for turf sprinkler system being installed on a sports ground

Table 8.8. Pressure ratings for plastic (PVC) pipes

Rating designation	Working pressure rating (MPa)	Working pressure rating (kPA)	Working pressure rating (m)
PN 4.5	0.45	450	45
PN 6	0.60	600	60
PN 9	0.90	900	90
PN 12	1.20	1200	120
PN 15	1.50	1500	150
PN 18	1.80	1800	180

Note: These ratings apply to water at temperature 20°C. Higher temperatures require rating to be reduced.

Pipe materials and pressure rating

The pipe materials available for general urban irrigation include PVC, ABS and polythene. Copper and other pipe materials are sometimes used as well. The soil environment for pipes can be quite harsh (Figure 8.9). PVC pipes are available as either solvent weld joint (SWJ) or as rubber ring jointed (RRJ). It is often recommended that for diameters 80 mm and larger that RRJ be used because it is considered more effective and reliable for the larger pipe sizes.

A range of polyethylene (PE) pipes are used. These include low density (LD), medium density (MD) and high density (HD). LDPE pipes are used for microirrigation emitter laterals, whereas MDPE and HDPE are used for pipes that have to operate at higher pressures.

The pressure rating (PN) of pipes is a key consideration in selection and design (Table 8.8). The PN number refers to the maximum working pressure for the pipe. A PN 9 represents a pressure rating of 900 kPa (0.9 MPa) or 90 m (pressure head of water in metres).

Pipes and fittings are covered by a number of Australian Standards, which are listed in Appendix 10.

The design and installation of pipelines is critical to the long-term success of an irrigation system. In addition to friction pressure losses, it is necessary to consider the potential stresses that the pipe may experience. Rapid closing valves, directional changes in water flow, negative pressures and top compression loads can all contribute to the weakening of the pipe system.

Figure 8.9. Buried pipe hazards include flow constriction caused by tree roots

Table 8.9. Maximum length of dripline for selected emitter flows

	Emitter flow rate: 1.6 L/h			Emitter flow rate: 3.0 L/h		
	Emitter interval (S_{ei})			Emitter interval (S_{ei})		
Inlet pressure	0.3 m	0.4 m	0.5 m	0.3 m	0.4 m	0.5 m
100 kPa	68 m	87 m	105 m	45 m	57 m	70 m
150 kPa	86 m	110 m	133 m	57 m	73 m	88 m
200 kPa	99 m	127 m	154 m	65 m	84 m	107 m
250 kPa	109 m	140 m	170 m	72 m	93 m	112 m

Note: Maximum length is based on flat ground.
Source: Netafim (2007).

Installation of thrust blocks, air release valves, vacuum breakers, non-return valves and bedding of pipes on sand, to provide firm and stable support, are all techniques used to protect the integrity of pipelines. Design and installation requirements are outlined in standard AS/NZS 2566 *Buried Flexible Pipelines.*

Sprinkler lateral design

The variation in operating pressure between outlets, such as sprinklers, along a lateral should be maintained within acceptable tolerances. Pressure variation can arise from friction in the pipes and fittings and also from variations in elevation.

To achieve the desired aim of all sprinklers operating at the optimum hydraulic conditions, it is necessary to select pipes that produce acceptable levels of friction (pressure) loss. This will usually mean a range of pipe diameters are used along the lateral because the flow rate along the lateral will vary. This is both an economical and hydraulically acceptable solution. The most common technique is to establish the amount of pressure and flow variation that is acceptable along the lateral between the first and the last sprinkler. A maximum pressure variation of 20% of sprinkler operating pressure has sometimes been used. For systems that require higher levels of performance, a total pressure variation of 10% is appropriate. As an approximate guide, the discharge rate of sprinklers and sprays will vary by approximately 50% of the pressure variation.

Microirrigation (drip) lateral design

The discharge of water from drip emitters in each planting situation should be even, and within the specified limits of variation. Pressure variation along a dripline will cause flow variation, particularly if non-pressure-compensating (PC) emitters are being used. The source of pressure variation, other than elevation changes, is friction in the pipes.

Table 8.9 is a guide to the maximum lengths of dripline for combinations of emitter flow rate and emitter interval. Lower flow rate emitters, at wider emitter intervals, can be installed on longer driplines.

Dripline spacing and total dripline length

The spacing between driplines has two important influences on the irrigation system design: the wetted coverage and soil moisture distribution and the cost of the system.

If driplines in SDI systems are spaced beyond recommended distances, uneven growth will occur. Sometimes this is referred to as 'striping' on turf areas. Once the system is installed, there is no opportunity to change the spacing. It is therefore wise to be conservative in the spacing of driplines if good uniformity is essential or required.

Table 8.10. Total dripline lengths for various dripline spacings

Spacing between driplines (mm)	Number of dripline runs	Total dripline length (m)
400	250	25 000 (25 km)
500	200	20 000 (20 km)
600	165	16 500 (16.5 km)
700	140	14 000 (14 km)
800	125	12 500 (12.5 km)

Note: The layout is based on an irrigated area, 100 m long and 50 m wide (0.5 ha).

The cost implications of dripline spacing are, however, significant. It is not only the cost of materials, but also the cost of the installation. Soils with low lateral spread of water, and hence closer spacing, will be more expensive to irrigate than soils that allow wider spacing.

Table 8.10 illustrates the impact of differences in spacing and total dripline required.

The underlying design requirement is that the drip design be based on the properties of the soil and the site performance standards required.

Microirrigation system reliability

A drip system has special requirements in terms of ensuring reliability of operation and effectiveness of delivery. The many small orifices and pathways are a potential risk in terms of system failure. Drip systems need to be designed and installed to minimise these risks (ASAE 2003). The risk of emitter blockage is the main consideration.

Some of the causes are:

- poor quality water with suspended particles
- suck back of soil into the emitter from soil surrounds
- root intrusion
- poor installation practices, contaminated lines.

The following items, in addition to pipes, fittings and a controller, are required for a drip system:

- control valves – master valve and zone control valve
- backflow prevention (if potable supply)
- flow meter
- filter
- pressure regulation equipment
- air release and vacuum breaker valves
- flushing devices/valves
- remote control valves (solenoid valves).

Flushing

In addition to protecting the irrigation system through filtration, it is necessary to make provision for the removal of accumulated deposits and debris from the driplines by forcing water through the pipe network at high velocity. The system design will need to include piping and valving to facilitate flushing of all driplines.

Back siphonage

There is a risk of soil particles and other debris being sucked back into the emitters when the water supply is shut down. The installation of vacuum relief valves on supply lines and the installation of flushing valves on the drip laterals is required.

Figure 8.10. Roots present a potential hazard for drip emitters and valves

Root intrusion

There is a risk of failure of the SDI systems and buried equipment in general due to root invasion (Figure 8.10).

Steps that can be taken to reduce the likelihood of root intrusion into drip emitters include:

- chemical (pre-emergent herbicide) application/injection (e.g. Trifluralin – commonly referred to as Treflan) applied at low dosages
- copper (metal) inside emitter to protect against root intrusion (Copper Shield)
- chemical (Treflan) embedded in emitter (Geoflow)
- selection of drip emitter designed to physically resist intrusion of roots
- management of the system to minimise root growth in the vicinity of the dripper.

Under-watering at times of high growth rates is a risk. Over-watering to reduce the risk of root intrusion is also not a sound strategy for efficient irrigation.

Valves – hydraulic regulation and control

Pressure is lost in a water distribution system where the flow is restricted and there is turbulence. Both manual and automatic valves create turbulence and are restrictive, so these devices are potential sources of significant pressure loss. The design of irrigation systems need to ensure that losses in fittings and valves are within acceptable limits. Both the velocity of water flow and the pathway through the device influence the amount of pressure loss. The resistance coefficients provide an indication of the relative pressure losses that will occur when water flows through the fittings and devices (Table 8.11).

The actual loss in pressure is determined using the following expression:

$$\text{Pressure loss (m)} = K \times \frac{v^2}{2g}$$

where:

K – resistance coefficient value

v – velocity of water flowing through the fitting or device

g – gravitational constant (9.81 m/s^2)

Table 8.11. Resistance coefficient values for fittings and valves (examples)

Fitting type	Resistance coefficient (K)
Pipe fittings	
Tee – Flow in line	0.3
Tee – Flow to branch	1.0
Sudden contraction (50%)	0.4
Sudden enlargement (100%)	0.6
Elbow (90°)	1.1
Valves	
Ball valve	0.1
Butterfly valve	0.2
Gate valve (fully open)	0.2
Non return valve	2.5
Angle valve (Manual)	5.0
Solenoid valve (50 mm, velocity 1.5 m/s)	6.0
Water meter (50 mm, velocity 1.5 m/s)	8.0
Pump suction foot valve with strainer	15.0
Backflow device – RPZ (80 mm, velocity 1.0 m/s)	60.0 (approx.)

Note: These K values are a general guide only. Reference should be made to manufacturer's performance specifications for each item and operating conditions.

All underground valves should be installed in robust enclosures (Figure 8.11).

Table 8.11 illustrates the way in which pressure loss increases with flow and also is dependent on the type of device.

Although the pressure losses through individual fittings or open valves may not be high, due to the total number of flow restrictions in the network, the total losses can be significant. Abrupt changes in flow direction and constrictions, with associated turbulence, are the main concerns. Backflow prevention devices can be the source of significant pressure loss, (for example, 3 m (30 kPa) or more).

The selection of filtration and water treatment equipment is a crucial component of hydraulic design. These items have requirements for flow rate ranges and also obviously incur

Figure 8.11. Installation of valve boxes requires sound foundations and provision for drainage

pressure losses. Manufacturer guidelines include recommended flow velocities and pressure loss curves.

Pumping systems

Role of the pump

Specialist expertise should be sought to ensure the selection of the correct equipment, including pumps, power units, drive arrangements and controls and to ensure that the full range of operating requirements will be satisfied, at highest energy efficiency and reliability.

A correctly selected pump, together with good irrigation system design, provides each outlet with the right hydraulic operating conditions. The pump provides the combination of flow and pressure. Irrigation systems have a requirement for a flow rate to satisfy the site needs. The pressure required is a function of the required flow rate, the topography and the irrigation system components selected.

In addition to consideration of flow and pressure issues, the system designer needs to consider the efficiency of the pump, the selection of the power unit and the installation arrangements of the pump unit so that it will operate efficiently and reliably.

Pump performance

A pump needs to be able to develop the pressure necessary at the required flow rate. The total pressure that needs to be developed by a pump to produce the desired hydraulic operating conditions is referred to as the total dynamic head (TDH). The unit of pressure commonly used in pumping and pipe friction analysis is m, which is equivalent to a 1 m high column of water.

The conversion between kPa and m is:

$$\text{m} \times 9.8 = \ldots \text{kPA}$$

$$\frac{\text{kPa} \times 1}{9.8} = \ldots \text{m}$$

Hydraulic pressure units and conversions are included in Appendix 11.

Total dynamic head (TDH)

The components of TDH are as follows (Figure 8.12):

- static discharge head (H_d) – total elevation – suction and delivery of pump
- static suction head (H_s) – height difference between water level and pump inlet
- pipe and fitting friction (H_f) – lateral, sub-main, mainline, suction line
- velocity head (H_v) – pressure or energy required in moving water
- outlet operating pressure (H_o) – required optimum pressure of irrigation outlet.

$$\text{TDH} = H_s + H_d + H_f + H_o + H_v$$

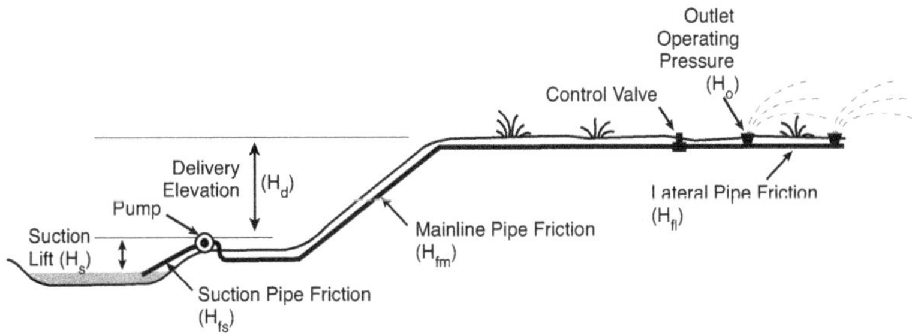

Figure 8.12. Components of the total dynamic head (TDH)

Table 8.12. Calculation of TDH example

Category	Designation/symbol	Head (m)
1. Static discharge lift	H_d	15.0
2. Static suction lift	H_s	5.0
3. Pipe friction and fittings loss– delivery and suction	H_f	8.0
4. Velocity head	H_v	2.0
5. Operating pressure	H_o	50.0
	Total	80.0

Note: $H_v = \frac{v^2}{2g}$ (m)

where:
v – velocity of water in the pipeline (m/s)
g – Gravitational acceleration = 9.81 m/s^2

Example

The calculation of the TDH is shown in Table 8.12.

Pump duty

The term 'pump duty' refers to the operating condition of the pump. The selection of a pump involves determination of the pump duty to satisfy the particular irrigation design. A pump that can produce this duty at high efficiency should be selected (Figure 8.13). The pump duty is expressed as 'flow rate at a specified pressure'. For example, the pump duty is 20 L/s at 70 m. The stated pump pressure of 70 m is required, when there is a flow rate of 20 L/s. A pump that

Figure 8.13. Pump set of multiple units capable of operating an irrigation system efficiently under varying duty conditions (Source: Australian Golf Course Superintendents Association (AGCSA), Clayton, Vic)

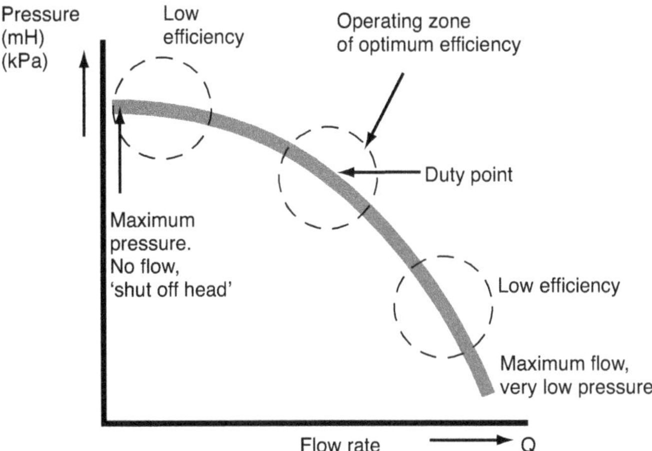

Figure 8.14. Centrifugal pump performance curve

can produce 20 L/s at 50 m is not a solution. The combination of flow and pressure are required simultaneously.

Centrifugal pump performance

There are several categories of pumps that are used for general water supply and irrigation. The most common category is the centrifugal pump, which is well suited to the delivery of water at medium to high flows, and low to high pressures. The centrifugal pump develops flow and pressure through the rotation of an impeller. The performance of the centrifugal pump can be regulated through the following:

- diameter of impeller
- speed of impeller
- number of impellers
- design of impeller.

Figure 8.14 provides a generalised representation of the performance of centrifugal pumps.

A typical manufacturer's pump curve is presented in Appendix 9. It is recommended that some allowance be made for wear and performance variation be included in the determination of a pump duty. The recommended safety factors for pump duties are an additional 5–10% discharge rate and an additional 5% pressure head (Irrigation NZ Inc. 2007).

Pump efficiency and energy consumption

The energy consumed in pumping is important for several reasons. They include:

- to determine the capacity of the power unit to drive the pump
- to estimate the cost of pumping
- to evaluate the environmental impact of the energy used.

The power required by a pump depends on the:

- flow rate (Q) (L/s)
- total dynamic head (TDH)
- pump efficiency (EP)
- properties of water.

$$\text{Pump input or shaft power (Ps)} = \frac{\rho \times g \times \text{TDH} \times Q}{\text{EP} \times 10^6} \quad \text{(kW)}$$

where:
ρ – Density of water 998 kg/m^3
g – Gravitational constant 9.81 ms^{-2}
Q – pump delivery flow rate m^3/h
EP – Efficiency (hydraulic) of pump
TDH – Total dynamic head (m)

Example
Pump duty: 5.0 L/s at 80 m
Pump efficiency (EP): 65%

$$Ps = \frac{998 \times 9.81 \times 80 \times 5}{0.65 \times 10^6} = 6.02 \text{ kW}$$

When the pump is producing the required duty, it will be consuming approximately 6 kW. A power unit with higher, say 9 kW, capacity than the 6 kW that is required to meet the specified duty, because there will be operating conditions – for example, full or maximum flow – that will require more power. This decision is the responsibility of a pump or irrigation specialist.

Pump energy consumption
The energy consumed by equipment is determined in the way:

$$\text{Energy consumption (kWH)} = \text{Power (kW)} \times \text{Time (h)}$$

Example
In the example above, if the irrigation system operated for 800 h/year, the energy consumption is:

$$\text{Energy} = 6.02 \text{ kW} \times 800 = 4816 \text{ kWH} = 4.816 \text{ MWh}$$

If the cost of the energy is $0.15 per kWh, then the total energy cost for the year is

$$\text{Pumping energy cost} = 4816 \text{ kWH} \times \$0.15 = \$772.40 \text{ per year}$$

Greenhouse gas production
The amount of greenhouse gas (kg CO_2) produced through the generation of this energy, assuming it is an electric motor, can be calculated using the appropriate emission factor. Values for the various states and territories are presented in Chapter 2 (Table 2.6).

Further information on emissions is presented in *National Greenhouse Accounts (NGA) Factors* (Commonwealth of Australia 2009).

In the above example, if the irrigation system was located in Queensland, the greenhouse gas production would be 4286 kg CO_2 or 4.286 tonne CO_2/year. This is based on an emission factor of 0.89 kg CO_2/kWh.

Types of centrifugal pump arrangement

There are many different impeller assemblies of centrifugal pumps used in irrigation. The pump arrangements are either horizontal or vertical.

Examples of horizontal arrangements include:

- single stage, close coupled to an electric motor
- single stage, belt drive to engine or electric motor.

Single stage means that the pump only has one impeller. The impeller is relatively large in diameter (e.g. 150 mm to 300 mm).

Examples of vertical arrangement include:

- multistage close coupled (direct drive) to electric motor
- multistage borehole pump close coupled to electric motor.

Multistage means that the pump has several impellers, perhaps 10 or more. Each impeller is relatively small in diameter, in many cases not bigger than 125 mm diameter. The pressure produced depends on the number of impellers. Each impeller progressively adds pressure to the water.

Factors in selection of centrifugal pumps

The following steps are involved in selecting a centrifugal pump (APMA 1987):

1. The irrigation system design duty is within the performance range of the pump under consideration.
2. The pump that allows maximum efficiency to be achieved is selected. This process requires the review of more than one pump, because the operating efficiency of pumps varies with the duty point and there is a zone of maximum efficiency for each pump (Figure 8.14). Selection of maximum efficiency ensures the minimum capacity power unit is required and also keeps operating costs to a minimum.
3. Determine the suction lift capability of pump to ensure it will work in proposed position. The term net positive suction head (NPSH) is used describe the suction head that needs to be available at the inlet of the pump to lift the water to the pump. If the elevation difference is too high or there are excessive restrictions, air may be expelled from the water and implode against the surface of the impeller. The phenomenon is called cavitation. In addition to making noise, it also causes pitting of the impeller. In time, this can cause a serious reduction in performance.
4. Determine the actual power required to operate the pump. This is required to select power unit capacity (e.g. kW of electric motor).
5. Review the cost of the pump. Cost is influenced by such factors as design, peak efficiency, materials, mounting details, power unit, control systems and accessories.

Importance of pumping efficiency

The efficiency of the pumping plant also needs to be considered. Maximising pumping efficiency has the benefit of lower ongoing operating costs and also minimising the negative impact of greenhouse gas production. The selection of the pump model that provides the highest efficiency is essential. This requires an understanding of both the operation and hydraulics of the irrigation system.

A 10% improvement in efficiency for a 20 kW electric pump that operates 800 h/year and has a life of 15 years represents a saving of 24 000 kWh. If the cost of electricity is 15 cents per kWh, then this represents a saving of $3600 over the life of the pump. Purchasing the right pump in the first instance is a sound decision in terms of energy efficiency and economics.

Pump protection

Efficiency requires the reliable operation of pump units. Monitoring and sensors are required for protection of the pump unit and the irrigation system. The main protection strategy is to shut down the pump. Some of the faults that may give rise to shutdown are:

- low water pressure
- high water pressure
- motor overload
- excessive current/amp draw
- pump failure
- excess flow rate.

Faults should be automatically recorded on the pump controller or irrigation controller database.

Irrigation control and monitoring

Controller features

The selection of a controller will be influenced by the scale of the project (number of sites and zones) and the sophistication required at the site. The widespread adoption of central control (multiple sites, controlled remotely from a computer host) provides a reference standard for all controllers. Increasingly, the features that are available on central control systems will be available on conventional controllers.

The controller features required to achieve efficient management of water include:

- multiple, independent programs – to allow plant need based watering
- multiple cycles
- extended watering time to accommodate drip systems
- water budgeting with global adjustment
- secure back up program memory
- electrical stability and uninterruptible power supply (UPS)
- flexibility in program days including cancellation e.g. rain, event
- station run times, 1 minute to 12 hours
- rain shut down input
- fault identification, alarm and reporting facility
- soil moisture sensor input
- weather data input.

The success of the irrigation system depends on all the key components, including the controller, functioning together as an integrated system to achieve the desired objective.

Types of controllers

The controllers that are available for urban irrigation can be grouped or categorised as follows:

1. **Stand alone controller** –Totally self contained unit that operates multiple stations, typically 6 to 50 valves. It may have provision for inputs from sensors (e.g. soil moisture or rainfall). Often one controller is dedicated to a single irrigated site.
2. **Master satellite controller** –The control functions are split between a central location (computer-based unit) and remote field locations. It has the capacity to control many stations (hundreds) and typically used on large sites such golf courses. Communication between the field and central control position is via land lines or wireless (e.g. radio or microwave).
3. **Central controller** – This has a large controlling capacity, which is centrally located in an area, district or large irrigated site. Distances between central and furthest control site may be 1020 km. The control unit at each site receives program instructions via telemetry. The system is capable of incorporating many input and output functions. It is well suited to local government, which has many remote sites (parks and ovals) to be controlled.

Controllers are discussed further in Chapter 10.

Monitoring and measurement

Incorporation of the measurement of the volume of water applied through the irrigation system is essential. Measuring instruments range from volumetric meters, which are manually read, through to electronic meters that continuously monitor the flow rate. Flow rate monitoring has the advantage of providing information about the operation of the system, as well as the distribution and application of water. Low flow rates may indicate blockages or equipment malfunctions and high flow rates may indicate pipe breakages.

Specifications and tendering

A well-prepared and thorough specifications document is an excellent aid in achieving a quality irrigation system. It also allows alternative system design to be compared and evaluated on an equal basis and minimum standards in equipment to be established. Independent irrigation design consultants generally prepare specifications and associated documentation as part of their professional services.

The following items should be included in the specifications document:

- scope of works
- warranties
- system water requirements
- system options
- water sources
- equipment quality – testing and standards
- inspection and permits
- marking out
- site plans
- maintenance manuals and training
- testing of system
- final inspection
- spare parts and tools
- pumping plant
- filtration equipment
- pipes and fittings
- valves – control, isolation and special
- sprinklers
- remote control valves
- control system
- installation
- site requirements – storage, clean up
- services – electrical, water
- materials list or bill of materials
- tendering requirements.

Specification detail

The documentation provided as part of the specifications will include plans as well as drawings of installation of all key components, including valve boxes (Figure 8.15)

Installation specification details, including drawings, are available from major irrigation company websites, including Hunter Industries, Rain Bird and Toro.

Installation and commissioning

The quality of equipment, the installation techniques and the competency of the irrigation contractor and expertise in the operation and servicing the system all need to be considered when selecting the irrigation company to supply and install the irrigation system.

All products, including pipes, fittings and valves, should comply with the relevant Australian standards. A list of Australian Standards relevant to irrigation is provided in Appendix 10. Further installation advice can be obtained from irrigation manufacturers, the *Urban Irrigation – Best Management Practices* (Irrigation Australia 2006) and also the *Irrigation Code of Practice and Irrigation Design Standards* (Irrigation New Zealand 2007).

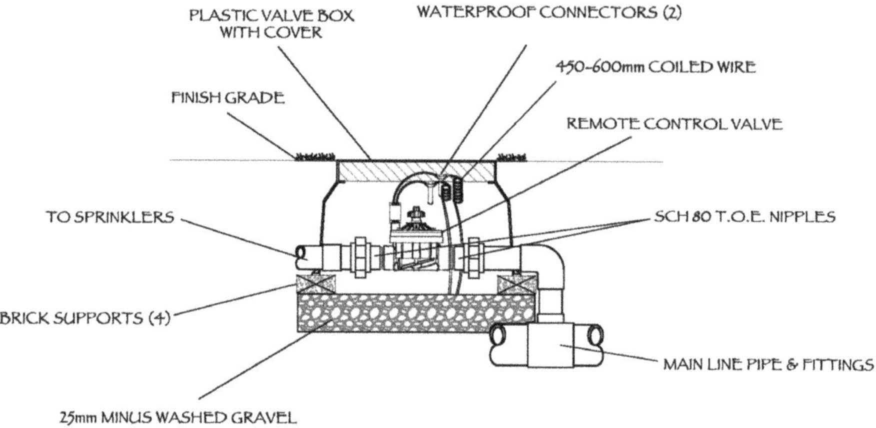

Figure 8.15. Specification details for installation of solenoid valve in valve box

A core requirement and duty of care in the installation of an irrigation system is the protection of underground utilities. A referral service, 'Dial Before You Dig' (www.1100.com.au), provides detailed information, including plans, on underground services in the proposed excavation area. Requirements in terms of clearances and digging techniques are usually provided by the asset owners.

An important stage in a new irrigation project is the formal acknowledgment that the equipment has been supplied and installed according to appropriate standards and is capable of performing to the required standards. In addition to visual inspections of equipment and installation techniques, specific tests should be carried out to test both the integrity of the complete system, the proper functioning of all equipment and the system application performance.

The following tests should be included:

1. Check that the pumping plant delivers the design duty (flow and pressure) at the designated efficiency.
2. Pressurise all mainlines and sub-mains to ensure that they are capable of holding the designated test pressure, under static conditions, for specified periods of time.
3. Check pressure variations along laterals, between laterals, along driplines and throughout the whole pipe network.
4. Check the flow variation between outlets (sprinklers, sprays and emitters) along laterals to ensure the system delivery rates are within acceptable limits. These are outlined in best practice guidelines in Chapter 7.
5. In the case of sprinkler systems, check the precipitation rate and the uniformity of application. Application uniformity coefficients, such as distribution uniformity (DU), should be determined. The measured field DU should be greater than 75%. In the case of drip systems, check the field emission uniformity, which ideally should be greater than 85%.
6. Check the integrity of the control and communication systems to ensure that all functions are operating correctly including all sensors (e.g. flow, rainfall, pressure and soil moisture) are operating according to design performance specifications.

Quotation variations for irrigation designs

On occasions, large variations in quotation values for the supply and installation of irrigation systems are reported. The reasons for large differences – greater than expected through commercial competitiveness – are usually due to differences in equipment and materials, specifications and quantities. These potentially have a significant influence on the overall performance of the system.

Examples include the following:

- Lower system capacity, in terms of flow rate. Smaller pumps, pipe sizes and valves are required.
- Wider spacing of sprinklers and sprays. Fewer heads are required, less pipework and fewer valves.
- Fewer control zones. Fewer valves, less piping and lower controller capacity.
- Lower quality products, including construction materials and performance specifications.
- Lower unit cost for key items such as sprinklers and valves.
- Lower standards of construction and installation. Excavation techniques and trench construction not sensitive to site requirements.

It is for these reasons that it is strongly recommended that the design specifications for irrigation system should include the performance to be achieved, as well as the quality and specifications of the system components.

Special irrigation design requirements

Designing for flexibility and efficiency

The requirements of an irrigation system are reasonably straightforward; delivery of a designated amount of water to the plant root zone, efficiently and at the right time.

The development of a design that meets these requirements for a specific site involves a process that optimises a number of design inputs that influence the final design outcome. These design inputs include water supply, labour, cost and agronomic considerations.

Although the need for uniformity of application and efficiency of delivery are now strongly promoted, there are also other factors influencing design. These mainly arise from a lack of security of water supply, water restrictions, lower water quality and a need for the irrigation system to be responsive to changing climate, water and regulatory environment. Irrigation designs are therefore required to incorporate flexibility in application and control.

To achieve the necessary degree of flexibility and adaptability, future irrigation system design will need to allow for the following:

- control zones that allow each vegetation area with different water needs – differing soil areas, differing surface conditions, differing microclimates, differing hydraulic conditions – to be separately watered
- precise hydraulic control and regulation so that all outlets on the system can operate at optimum hydraulic conditions
- outlets or applicators that allow uniform and effective delivery of water to the designated target areas
- supply capacity that allows peak demand to be satisfied (note that high flow rates allow greater flexibility in delivery, if time available for water access is reduced)
- water applied effectively, so that there is no risk of runoff (protection of the environment)
- monitoring, evaluation and reporting of water consumption
- monitoring of the operation of the system to alert of equipment failure or malfunction
- integration of sensors, rain and soil moisture
- use of weather data (daily ET) as an input in irrigation scheduling
- communication capability to allow remote two-way transmission of information and commands.

Key performance information required

The following is a list of the key information that needs to be readily available, or can be determined as part of the design process. This information provides the basis on which comparable designs can be prepared for a particular situation.

- peak system capacity (e.g. mm/week)
- site turf and landscape performance standards stated
- soil properties including infiltration rate (to avoid runoff)
- uniformity performance standards
- hydraulic performance, pressure variations allowances
- areas to be irrigated, including identification of 'no irrigation' areas
- identification of all hydrozones that require independent delivery and control
- installation requirements including depths, materials, construction requirements and re-establishment of ground and vegetation
- any site constraints such as pipeline locations, valve box locations that may have a significant impact on the cost.

Table 8.13 summarises the data required for design (input data) and the system performance data developed as part of the design solution (output data).

Table 8.13. Irrigation design data summary (example)

Design data input		**Irrigation system design performance data**	
Design input	**Value**	**Description**	**Value**
Site details			
Irrigated area	m^2	Annual application volume	kL, ML
Plant K_c or CF			
Peak irrigation requirement	mm/week	Design flow capacity	L/s
Soil type	Classification	System precipitation rate	mm/h
Soil infiltration rate – Initial	mm/h		
Soil infiltration rate – Steady	mm/h		
Root zone depth	mm		
Available water-holding capacity (AWHC)	mm/m	Irrigation depth	mm
Stored water	mm		
Management allowable depletion (MAD)	50%		
Irrigation depth (ID)	mm		
Performance standards		Uniformity of application – DU Design	>90%
		- Field DU	>75%
		Scheduling coefficient ($SC_{5\%}$)	<1.2
		Field Emission uniformity EU	> 85%
Hydraulics		Pipe network friction loss variation	e.g. 20%
		Manifold losses	e.g. 6 m (60 kPa)
		Lateral pipe friction loss	e.g. 10%
		Pipeline velocity	1.5 m/s max.

Irrigation design resources

Resources, in the form of design guides and manuals, are available from the major irrigation manufacturers including Hunter Industries, Netafim, Rain Bird and Toro.

The publication *Irrigation, Sixth Edition, 2011*, published by Irrigation Association (USA) (Irrigation Association 2011), is a very valuable resource and covers all aspects of design including equipment selection, hydraulics, control and installation.

Additional resources are available from Irrigation Association (USA) website:

www.irrigation. org/

Irrigation design resources, including publications and irrigation designer contact lists, are also available from Irrigation Australia Ltd (IAL) website: www.irrigation.org.au

Chapter 9

Achieving best practice – site studies

Main irrigation systems

Systems available

The main methods of irrigation that are available for urban irrigation include:

- sprinkler and spray irrigation systems
- surface microirrigation systems
- subsurface drip irrigation systems.

The selection and matching of a particular irrigation method to a horticultural situation is an important aspect of achieving high efficiency in water use. Each method of irrigation has performance characteristics and properties that may influence its suitability for use in particular situations. It is important to have knowledge of both the performance characteristics of the various methods and the nature and requirements of the site to be irrigated. The characteristics, benefits and limitations of the various irrigation methods are covered in Chapter 4.

In this chapter, each of the main irrigation methods is discussed with the aim of achieving high efficiency within a BMP framework.

Efficient sprinkler irrigation systems

Sprinkler irrigation systems have the major advantage of covering large areas. In many situations, sprinklers are the only real option in terms of applying water. Sprinkler technology has been developed so that good uniformity of application can be achieved. However, to ensure that high efficiency of application is achieved, there are numerous contributing factors that need to be satisfied. These include:

- correct sprinkler and nozzle combination
- correct layout
- correct hydraulic operating conditions
- maintenance of the system.

Sprinkler systems are very sensitive to variations in performance. A simple incorrect nozzle selection can result in marked deterioration in performance. The recommended strategy to maximise the chances of achieving high efficiency is to commence, with a sound design, prepared by a qualified designer.

The circular distribution of water from sprinklers and the distribution profile typical of sprinklers and sprays means that achieving high uniformity in square, rectangular and narrow areas is a challenge. The dependence on good weather conditions to achieve high efficiency of application is another limitation. If this method is chosen, weather conditions need to be monitored to achieve efficiency.

The maintenance of the system is critical. Sprinkler irrigation systems consist of numerous parts, some moving, which are at risk of malfunction due the harsh and demanding situations in which they operate. Also, like any piece of equipment, they are prone to wear.

Examples of sprinkler and spray systems that lead to inefficiencies include:

- excessive number of sprinkler heads on individual supply circuits
- excessive spacing of sprinklers
- weeping – lack of anti-drain valves in sprinkler heads, leaking wiper seals
- excessive pressure variation due to poor pipe sizing
- flooding due to excessive watering time programmed on controller
- uneven performance due to lack of hydraulic control
- malfunctioning of sprinkler heads due to lack of maintenance.

Efficient surface drip irrigation systems

Efficiency principles of drip irrigation

A key performance requirement of an irrigation system is that the water be delivered in a controlled way. Drip irrigation is based on delivering controlled amounts of water to the root zones of plants. The critical feature of a drip irrigation system is that the emitter is designed and operated to achieve reliable delivery of a precise, small amount of water, at each plant location, in a timely manner.

These systems have smaller diameter pipes and lower flow rates, than other pressurised irrigation systems. There are often hundreds, or even thousands, of components, which are mainly constructed from plastic materials that make up the system. Regular maintenance is critical.

Due to the sensitivity of drip systems to blockages, filtration equipment must be installed when low quality water (in terms of suspended solids) is being used. Provision should also be made for the flushing of drip laterals.

The efficiency of drip systems is founded on precision components, correct operating conditions and the maintenance of the system in good working order.

Point source and line source drip laterals

A characteristic of drip systems is that there is not a continuous distribution of water or wetting over the whole area. Water delivery is generally targeted to particular parts of the plant root system and soil volume. In the case of point source systems, the emitters are generally selected to water a proportion of the root system of individual plants. This may mean positioning from one to eight or more drip emitters around an individual plant. The number of drip emitters and dripper off-takes can vary greatly according to the planting layout.

Driplines in which the drip emitters are spaced at regular intervals are generally best suited to plantings, which have similar water demand and are at regular spacing. This arrangement produces a near continuous wetting pattern along the length of the dripline. Planting layouts need to be compatible with this soil wetting pattern. Driplines are used to provide total coverage watering (grid pattern) of landscape garden bed areas. In these cases, there may be a mixture of plants and a variety of plant size and densities. The approach is to water the complete area so that root development is encouraged over the whole target area. Caution is required in these situations, because there is the risk that individual small plants, with shallow root systems, that are not immediately adjacent to an emitter outlet will not be adequately watered and other areas where there are no active root systems may be over-watered. The use of point source drip systems would provide higher efficiency in these cases.

Efficient subsurface drip irrigation systems

The design and installation of subsurface drip systems is critical to their success. A thorough understanding of the wetting characteristics of the site soils, system hydraulics and product

Case study: subsurface drip – Tea Tree Gully, South Australia

Tea Tree Gully is located approximately 15 km to the east of Adelaide City GPO and has approximately 820 ha of public open space. The local Council has 430 irrigated parks and reserves and water consumption in the range of 700 to 900 ML.

The challenges facing the existing irrigation systems, include some irrigation designs operating below best practice, some sites with uniformity DU less than 50%, complex control systems and controllers located away from irrigation areas (Bradbrook 2005). The age of irrigation systems was also a challenge in achieving high efficiency.

Many maintenance problems have been experienced, including faulty wiring, unreliable solenoid valves and out-dated sprinklers. In addition to these maintenance issues, there has been a significant problem resulting from vandalism of equipment. It has been estimated that vandalism can represent up to 25% of the maintenance budget.

As a response to these challenges and the desire to achieve high efficiency in water use, the Council installed its first subsurface drip irrigation system in Tilley Reserve 1993. This was the first subsurface drip system installed on turf by local government in Australia. The field is a 1.0 ha soccer field (Figure 9.1).

Figure 9.1. SDI has been used on sports grounds by City of Tea Tree Gully, Adelaide, for more than 15 years

Tea Tree Gully has found that, although subsurface drip systems incurred a higher capital cost compared with pop-up sprinklers – typically in the range of 30% to 40% – this is outweighed by benefits. Reduced water consumption, maintenance and vandalism provide direct benefits to Council. The application rate was typically in the range of 4000 kL/ha (4 ML/ha).

The Tilley Reserve recorded a water consumption reduction of approximately 40%. Council believes that additional savings are possible through smarter scheduling, including use of ET-based control, soil moisture sensors and use of latest technology.

Subsurface drip systems require maintenance in terms of replacement of filter cartridges, incorporating herbicides to prevent root intrusion and regular flushing of the lines. Care needs to be taken not to perforate and damage the driplines when installing structures, such as soccer nets and shading frames, using anchoring pegs or carrying out ground renovation works on these playing fields.

Tea Tree Gully Council has highlighted that the design of a system must consider the soil properties and uniformity, turf maintenance and renovation requirements, root intrusion control, depth of drip pipe, and provision of fertigation points and turf management practices such as coring, verti-draining, aeration and soil compaction relief.

In the period 1993–2005, Tea Tree Gully Council installed in excess of 15 major subsurface drip systems, covering over 20 ha. These fields are used for a variety of purposes including soccer, school recreational activities, gridiron and community parks. In the majority of cases, significant water savings have been achieved and the maintenance requirements, particularly those associated with vandalism, have been significantly reduced.

performance is required to ensure that adequate and uniform wetting of the turf root zone is achieved. Subsurface drip has all the advantages of surface drip systems, including improved application efficiency, enhanced plant growth, improved use of fertilisers and chemicals, limited weed growth and reduced pumping energy requirements. In addition to these positive characteristics, subsurface drip systems have a dry or near dry soil surface (minimum soil evaporation), potentially no weeds, no exposed hoses or tubing that can be damaged, drier plant atmosphere (less disease because there is no surface water or wet soil) and an ability for water to be supplied continuously to the plant regardless of any field activities or use.

The delivery of irrigation water directly into the plant root zone in a controlled manner also reduces the risk of soil nutrients being leached to groundwater. This depends, of course, on the correct operation of the system.

The burying of the water distribution lines in the soil has some important implications. There is a much reduced margin for error in the design of the system, product quality and the management of this method requires higher skills than other forms of irrigation. Should a problem occur in any aspect or component of a subsurface drip system, then the turf or landscape planting is at risk.

Some examples of drip system efficiency risks

Although drip systems are potentially highly efficient, there are some circumstances which can adversely affect efficiency. These include:

- emitter blockage from low water quality (suspended matter), suck back of soil debris/particles and root intrusion
- poor pressure regulation – excess flow rates
- malfunctioning, due to mechanical damage – small plastic components are vulnerable; intentional (vandalism) or unintentional damage may occur
- poor water distribution, due to lack of knowledge of soil hydraulic properties or inappropriate soils (e.g. coarse fast-draining sands)
- using driplines (line source) when point source droppers are appropriate
- water draining below the plant root zone (Figure 9.2)
- over-watering, due to excess run times and poor timing of irrigation.

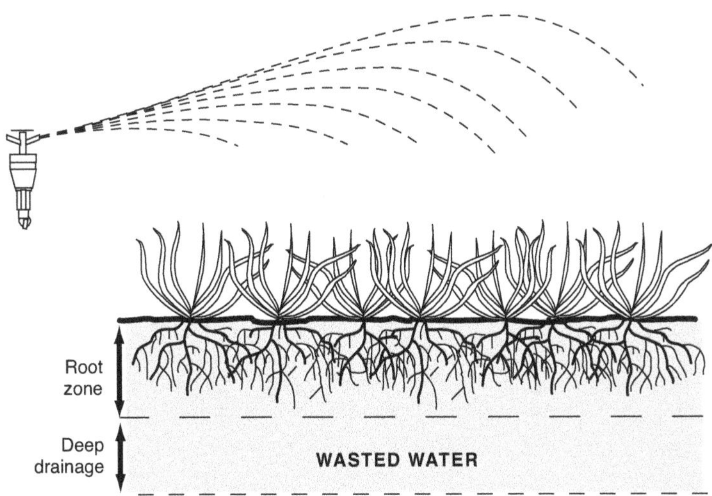

Figure 9.2. Over-watering of the root zone is a source of wastage and inefficiency for all types of irrigation systems

Irrigating turf

Achieving efficiency on turf

Turf surfaces need to be even and safe. The surface has very high demands in terms of use, wear and the need for rapid recovery (Appendix 6).

The control of the level of soil moisture, through irrigation, is a vital element in the overall management of quality turf areas. It is important to recognise that the effective irrigation of turf is a particularly demanding task, due to the special needs and characteristics of turf sites. These requirements have an impact both on the design and the management of the irrigation system.

Characteristics and requirements of turf surfaces

The special needs of turf are:

- **Uniformity** – Individual turf swards draw water from only very small lateral distances, the irrigation system must therefore apply water with a high degree of evenness or uniformity to ensure all plants are watered.
- **Limited root depth** – The active root systems of turf occupy only a comparatively shallow depth of soil (may be only 100 to 150 mm) and so, in order to avoid wastage through over-watering, the application depth must be precise. The limited root zone depth can impose significant constraints in the positioning of irrigation equipment so that it is not interfered with or damaged by soil cultivation (e.g. aeration) practices.
- **Responsiveness** – Turf responds very rapidly to changes in the soil moisture status and so, if inadequate levels of moisture are present in the root zone during periods of high water demand, then the turf will become rapidly stressed.

Turf irrigation systems must therefore be capable of applying pre-determined amounts of water to the turf root zone with a high degree of uniformity and with minimum wastage.

Irrigation challenges of turf

Turf has the following special irrigation needs:

- The water must be applied uniformly.
- There is often a shallow soil water reservoir, so application must be precise.
- Turf is sensitive to soil moisture status, so regular watering is usually required.

The turf/soil system

The role of water in maintaining a quality turf surface is critical. Many of the properties and processes involved in providing a surface of appropriate standard are directly influenced by soil moisture, water movement through the soil and water management.

The turf surface properties that are relevant in terms of irrigation are the: amount of thatch; amount of organic matter; organic content of the mat layer; surface hardness; compaction; infiltration rate and evenness or levelness of surface.

A quality turf surface requires a soil profile that will support healthy grass growth through the availability of nutrients, oxygen, soil water and a soil physical environment that will allow root development and extension. In terms of water-related properties, this means a soil profile that has adequate infiltration rate, good percolation rate, good water-holding capacity and good drainage characteristics.

The importance of adopting an integrated approach to water conservation has been highlighted by Carrow *et al.* (2004) who put forward a 10-point strategy for water conservation. A systems approach that values all aspects that impact on the site including grass species, management practices and education is required. These researchers pointed out the need to recognise that the turf/soil system is a dynamic one, changing in nature, both on a short-term and long-term basis.

Risks to quality of turf surfaces

Some of the water-related risks that can have a negative impact on the optimum conditions to provide a quality turf surface are:

- **Excess use and wear** – Causes soil compaction and reduced infiltration. There is a risk of waterlogging and irrigation inefficiency. Prolonged wetness leads to surface damage and also risk of disease.
- **Inadequate water applied** – Increased dryness results in increased hardness and risk of injury. These conditions are conducive to some weed species.
- **Excess water applied** – Causes soggy conditions and damages soil structure.
- **Excess thatch** – Reduced infiltration and ineffective irrigation.
- **Overdeveloped mat layer** – Excess organic matter in the top soil layer can lead to reduced infiltration and reduced percolation. Poor drainage is potentially a salinity risk if recycled water is being used and if the soil cannot be leached. The mat layer, which comprises decomposing roots and organic matter, plays an important part in the usability, wear tolerance and hardness of a turf surface. A balanced proportion of organic matter needs to be maintained.
- **Shallow, light irrigation applications** – These tend to cause shallow root systems, resulting in limited stored water and lack of drought resilience. Deep infrequent irrigations tend to encourage deep root systems, which provide increased reserves of water and nutrients.

All of these risk areas need to be managed to ensure that the required surface quality is achieved. Some of the strategies available to maintain a healthy turf/soil system include renovation practices, such as de-thatching, de-compaction, coring (hollow and solid), verti-draining and top dressing.

Turf irrigation methods

Virtually all types of sprinklers and sprays, including quick coupling valves (QCV), travelling sprinklers, manual systems and automatic pop-up systems, are used on turf. The characteristics of the various irrigation methods are outlined in Chapter 4. The main trends in turf irrigation are upgrading of sprinkler systems to achieve higher uniformities, improved control and the introduction of subsurface drip irrigation (SDI).

Irrigating sports grounds

Principles of irrigation efficiency on sports grounds

A sports ground is a relatively large area. Large volumes of water are required to maintain these facilities in a healthy condition, even those using water-efficient grasses.

The principles of high efficiency of correct depth of application, correct timing of irrigation and minimum losses and wastage that apply to turf generally are also directly relevant to sports grounds.

Because of the way in which these sites are used, the time available for irrigation is often limited. Also, there are usually restricted times in which supply is available: for example, 10.00 pm to 6:00 am, if water supply is a water agency main.

High irrigation efficiency is achieved through good uniformity of application and scheduling to match the irrigation to the climate, soil and use of the area. System maintenance is a key issue for efficiency and ground safety.

Characteristics of sports grounds

The relatively flat or even surface and regular shape of sports grounds differentiate these from many other urban horticultural sites. Examples of performance measures that may be used to assess a turf surface include:

- evenness – undulations, gradient
- trip hazards
- surface hardness
- traction
- compaction
- extent of grass coverage
- grass height
- soil moisture
- waterlogging.

Recommended performance standards, including methods of measurement, have been derived for community level sports fields in Australia (Holborn 2009). Performance limits for each component are included in this report.

Playing areas typically range in size from around 500 m^2 up to several hectares. The use includes training, organised sport and general recreation. The shapes of these grounds include rectangles, squares, ovals and circles, and combinations of these.

The surface layer, including the turf and the top 50 mm of soil, is very important in terms of surface performance. It is the contact layer that a user experiences. The nature of the soils used on sports grounds ranges from the native soils of the areas (clays, loams and sands) to soils, such as sandy loams, that are imported as part of ground construction and development and top-dressing purposes.

Irrigation challenges on sports grounds

The following features of sports grounds present challenges in terms of achieving high irrigation efficiency.

- The root zone depth of many sports grounds is often shallow: in the range 100–200 mm. Combining this with relatively low water-holding capacity of sandy soils, the soil water storage can also be low. This leads to the need for frequent shallow applications, which are not efficient (high proportion of application losses).

- Sports grounds are often subjected to concentrated use on parts of the ground, such training areas, including areas with lights, and goal squares. These areas require different irrigation design and management, to the other parts of the ground. The irrigation zoning should provide for these areas.
- The water supply to sports grounds is often from a mains potable supply. The flow rate and pressure available to be used on the ground are determined by the hydraulic conditions in the local main and the sizing of service off-takes, water meter and valves. In many cases, the flow rates do not allow appropriate irrigation depths to be applied, in the available time window.
- Irrigation equipment installed on sports grounds is exposed to harsh operating conditions. There is the wear and tear associated with contact (with sprinkler heads), the regular exposure to various types of ground maintenance machinery and a soil environment in which the sprinkler head operates can be severe (sand and grit can enter the turret, stem and seals).

Irrigation methods for sports grounds

The watering of sports grounds is typically achieved using pop-up sprinklers and, in some cases, mobile sprinklers. In recent times, there has been increased installation of subsurface drip systems. Portable sprinklers and various types of traveling sprinklers are also used. QCVs are used for the watering of selected parts of grounds, such as turf wickets, and as general water supply points.

Sports grounds sprinkler heads have been developed to incorporate properties and features that provide more robust and reliable operation and also achieve higher uniformity of application. Stronger materials, such as stainless steel, and new nozzle designs that achieve higher uniformity (DU) values are now used. Sprinklers should be mounted on articulated (swing joint) risers. The performance characteristics of the turf irrigation systems described in Chapter 4 on irrigation methods and Chapter 7 (BMPs) are relevant to sports grounds.

Strategies for high irrigation efficiency

The following strategies provide the basis for achieving high irrigation efficiency on sports grounds.

- **Auditing of irrigation systems** – Regular checking of the performance and determination of the uniformity of application is essential. Sprinkler uniformity is essential for high efficiency.
- **Deep root zone** – Encouraging deep root systems through both water management and soil management should be a priority. Deep, infrequent irrigation helps grass to establish roots deeper in the soil profile, rather than shallow root systems near the surface.
- **Maintenance** – Inspection of sports grounds will generally identify some examples of equipment not operating at optimum level. These include sprinklers set too low, titled sprinklers, nozzle assemblies stuck in the 'up' position and incorrect nozzles installed. Regular inspection of the functioning of equipment and rapid repair are the key to sound maintenance programs. It is very important to keep records of the inspection and maintenance and repair work carried out. In addition to assisting with the management of the systems, these records are required if a safety or injury issue arises from the use of the sports ground.
- **Pump and tank** – In situations where the mains water supply (pressure and flow) is inadequate to achieve an efficient irrigation, it is advisable to consider a pump and tank

system or booster pump (if allowed by the water authority). Water is delivered into the tank over extended periods and then pumped out to the irrigation system at optimum supply conditions. Often pump and tank systems operate at higher efficiency because of the higher pressure, constant pressure and ability to use higher flow rates. Variation in operating pressure is a consideration for all mains systems, even those that maintain adequate minimum supply pressures, because the pressure can vary considerably during the irrigation event.
- **Subsurface drip irrigation**– Consideration should be given to the installation of subsurface drip systems, which potentially have higher application efficiencies. Water losses due to evaporation, wind and surface runoff are eliminated or minimised. These systems need to be designed to match the hydraulic properties of the soil and the growth characteristics of the grass. They need to be operated to ensure that the water is delivered into the root zone and does not drain below this layer.

Safe turf surfaces and irrigation

The requirements for safe sports grounds and irrigation design guidelines are outlined in Chapter 8.

Golf irrigation

Principles of golf irrigation efficiency

Golf courses are recognised as large users of irrigation water. Both the irrigated area on a course and the numbers of courses contribute to the total water use. The need for efficiency is obvious, both in terms of the water used on the site and the water used overall by the golf industry. The total water demand of a golf course can be reduced or limited through irrigating only the areas that justify supplementary watering and by selecting plant species that provide the desired outcome or services, but use less water. High efficiency of water use is achieved through matching the species and management practices to the various outcomes required.

The conversion to warm season grasses in southern Australia has been a significant improvement in water use efficiency on golf courses. Plant species native or climate suited to the locality and soils should be selected wherever possible for non-play areas.

The preparation of golf greens is perhaps the most demanding in terms of quality, function and performance in environmental horticulture. Deficiencies are readily observed and shortcomings in performance quickly experienced by critical users. The demanding requirements mean that all inputs (fertiliser, water, mowing and aeration) need to be very carefully and astutely managed. This particularly applies to water.

Fairways are managed to reflect the different surface performance requirements in different parts of the fairway. Landing areas, approaches, surrounds and general fairway play areas can all be treated as having separate requirements.

There is the opportunity on golf courses to collect and reuse drainage water. Excess water draining through golf greens and bunkers, for example, can be collected and reused. These systems are very water and nutrient efficient.

Characteristics of golf courses

The scale of a golf course is an important feature in terms of water management. Typically a golf course covers an area of 20–30 ha. The irrigated area may be in the range of 30–50% of this total area. The diversity of vegetation, from trees through to intensely managed grasses used on greens, means that superintendents need to have a broad skill base to achieve the performance standards expected.

The range of microclimates around the course, from open exposed sites through to protected shaded areas, is another feature. In terms of soil types, the native soil is generally the dominate soil type to be managed, but areas such as greens – which have the potentially conflicting water property requirements of high infiltration rates and good water-holding capacity – require special management.

Golf course sites and surrounding urban areas are often used to harvest rainfall for irrigation and provide water for water features. Site water storage and drainage management are an integral part of the whole site water cycle management.

Irrigation challenges for golf courses

The main challenge in terms of achieving efficiency on golf courses is the scale of the facility. Large areas to be covered and limited time windows mean that delivery flow rates are high– often in the range of 30L/s to 80L/s.

The scale is also important in terms of maintaining optimum hydraulic conditions. Long distances and multiple delivery points require sound hydraulic design and the system to be operated so that each irrigated zone has a flow rate and pressure that is within the design performance specifications. Pipeline layout as a ring main is well suited to most golf course sites. This arrangement is hydraulically efficient and allows flexibility in delivery so that optimum operating conditions can be achieved in all areas of the course. Undulating terrain is another source of pressure variation within a golf course. Pressure regulation in the supply pipe network and valves is usually required to achieve acceptable operating conditions. The pumping equipment needs to be flexible in delivery so that variable amounts can be delivered at high efficiency. This often means multiple pumps, with variable speed drives, and control systems incorporating flow management programs.

Uniformity of application and precision in control is essential for good surface quality and high efficiency. Due to the shape and size of greens, achieving high uniformity (in excess of DU 75% in field tests) can be a challenge. The large width of fairways is demanding for pop-up sprinklers because achieving high uniformity over radial distances in excess of 25–30 m is a challenge. Wind exposure is another factor that can have an impact on the irrigation efficiency.

The range of microclimate conditions and varying turf surface performance and conditions requires a control system that can be programmed to meet the needs of individual zones and also has the capability of using daily ET data.

Irrigation methods

Golf courses have experienced a revolution, in terms of adoption of sprinkler irrigation technology, over the past three to four decades (Figure 9.3). Before the 1960s and 1970s, portable sprinklers and various forms of travelling sprinklers were common. There was the adoption of automatic irrigation systems with pop-up sprinklers arranged in groups or blocks, for example three to eight sprinklers supplied from the one remote control valve (solenoid or hydraulic). This was a major advance from manually intensive systems at that time.

The availability of valve-in-head sprinklers has been a significant advance for golf course irrigation. Control over the delivery of water from individual sprinkler heads provides flexibility that better matches the characteristics of each part of the course. The constraints of block design, where groups of sprinklers are operated together, are eliminated. The design of valve-in-head systems also means that optimum hydraulic conditions can be maintained throughout the course. This method of application has the potential for high efficiency when operated in conjunction with control systems that have the capability to operate individual heads according to individual programs.

The full range of irrigation methods are still used on golf courses. Not all clubs have the resources to be able to install the latest irrigation technology. The characteristics of the various

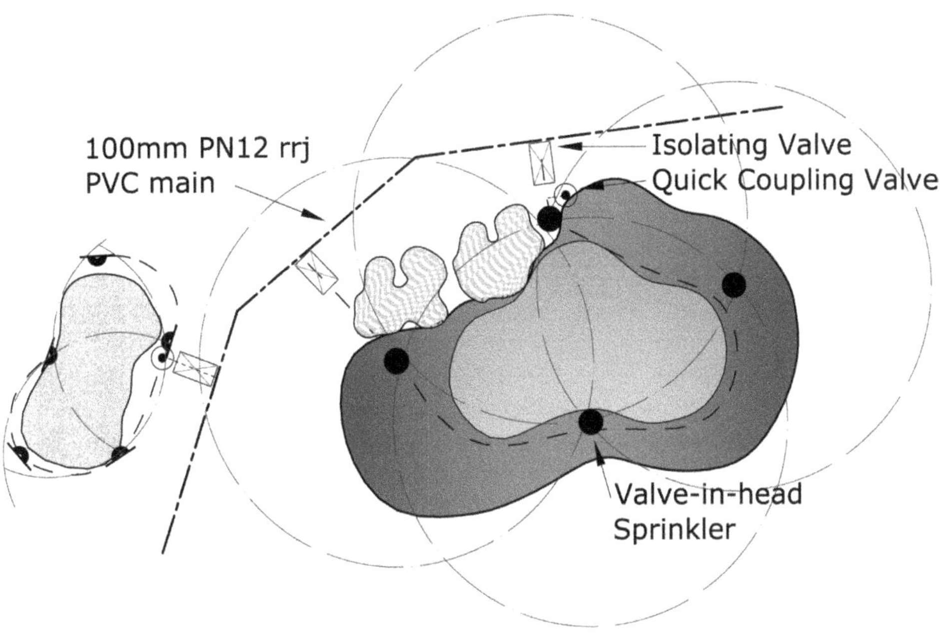

Figure 9.3. Sprinkler layout for effective coverage of golf green and tee (Source: Rainlink Australia Pty Ltd)

methods including travelling sprinklers, various types of manually operated sprinklers and automatic pop-up sprinklers are described in Chapter 4.

Strategies for high irrigation efficiency

Strategies for high efficiency include adoption of developments in application technology and improved scheduling.

- **System auditing** – Measurement of the application performance (DU) of the irrigation system and modification to achieve best practice (field DU_{LQ} >75%). Reference to sprinkler uniformity data, such as SpacePro (Center for Irrigation Technology, Fresno, CA), is recommended in making decisions about selection of sprinkler type, model and nozzle combination.
- **Zoning** – Zoning of sprinkler heads so that control of application to areas with unique water requirements is achieved. This may mean dual rows of sprinklers around greens to allow separate watering of surrounds and greens.
- **Nozzle replacement** – The wearing of metal nozzles can significantly affect the system performance. Long hours of use and water containing abrasive (sand) particles may contribute to wear and enlargement of nozzle orifices. The consequences include higher flows, lower uniformity, less than optimum operating pressure, excessive pipeline friction and increased pumping costs. In a study conducted by David Zoldoske of the Center for Irrigation Technology, Fresno, it was reported that golf courses replacing nozzles improved uniformity and reduced water consumption by an average of 6% (Zoldoske 2003).
- **ET control** – Irrigating according to plant water demand is a key part of irrigation efficiency. The irrigation decision making process should be based on the ET (evapotranspiration) rate of the golf course turf and landscape plantings.
- **Soil moisture sensors** – Soil moisture sensor technology has the potential to provide turf and landscape managers with up-to-date, accurate readings of the moisture level

within the soil profile. Intensively managed areas, such as golf greens, can use soil moisture sensors to provide feedback on rates of water depletion, soil drainage and irrigation effectiveness.

- **Monitoring, alarms and reporting** – The capacity to monitor and record the environmental conditions and the system operation is becoming increasingly important in golf course water management. Alarms that indicate malfunction assist with both equipment safety, prevention of waste and fault finding. Irrigation systems, which consist of many functioning components, require ongoing monitoring and regular maintenance.

The value of system and site monitoring is well illustrated at Royal Melbourne Golf Course, with its state of the art water source management and irrigation control. Multiple water sources are used, including stormwater, groundwater and potable water. All water sources are monitored, including groundwater levels, bore yields, storage levels and pumping rates. This is achieved through a supervisory control and data acquisition (SCADA)-based monitoring and control system. Multiple remote sensors provide feedback to the central control point so that the operation of the whole system can be optimised. This ensures that the irrigation can be carried out reliably and efficiently. Comprehensive reporting on the various water sources and system operation are key elements of this system.

The golf industry is well supported with resources to assist in the development of site-specific BMPs and water management plans to both achieve high efficiency and demonstrate environmental stewardship. One example is a template, prepared by the University of Georgia, which allows very comprehensive BMPs to be developed (Carrow *et al.* 2007). The Australian Golf Course Superintendents (AGCSA) (Website: www.agcsa.com.au) and the USA Golf Association have resources in water conservation available (Website: www.usga.org).

Irrigating racetracks

Principles of racecourse irrigation efficiency

Venues, such as racetracks, are highly visible and have to deliver specific performance standards. The properties of the turf surface directly influence the outcomes from the site (Figure 9.4). The sustainability of racetrack, in terms of water use, has become a major issue. The need for efficiency therefore becomes increasingly important. Achieving efficiency of application of

Figure 9.4. Turf surface needs to have full grass cover and evenness for racing

water requires attention to the scale and shape of the area. Watering needs to satisfy not only the turf evapotranspiration needs but also achieve the required soil conditions for rapid repair and surface performance.

Characteristics of racetracks

A racetrack surface is required to withstand heavy impact and intense use. In addition to the need for an even and uniform surface, there is the additional requirement that the surface readily repair from damage. Racetracks by their nature are long and also occupy reasonable large areas of turf, typically 3–6 ha.

In addition to the generic needs of turf, racetracks have additional requirements or constraints. These include the following.

- A firm consistent surface is required – over-watering cannot be tolerated (no waterlogging).
- An even surface is essential, which requires an even application of water to the soil root zone.
- The limited time window in which to apply the water often places significant constraint on the irrigation system, when issues such as training requirements, unfavourable weather conditions and staff availability are taken into account.
- They comprise areas and shapes, including curves and bends, that are difficult to effectively accommodate with the radial distribution pattern of sprinklers.
- The location and nature of the site and surrounds, particularly regarding the need for good visibility, mean that trees, which could offer some wind protection, are not grown. Wind is a significant factor in achieving effective application of water.
- Racecourses generate income from relatively few key event days. This places timeline pressure on track managers to prepare surfaces to the required standard and this may mean irrigating at times when atmospheric conditions are not ideal.

Irrigation challenges for racetracks

The challenges facing racetrack irrigation include the following (Figure 9.5).

- Sprinklers (usually half-circle) are required that can achieve a high degree of uniformity, when distributing water over long distances (e.g. 25–28 m).
- The curved track layout places constraints on sprinkler positioning for high uniformity.
- Sprinkler application rates need to be matched to soil infiltration properties.
- The system must be controlled and managed so that the correct depth is applied at the right time.
- Equipment must function effectively and operate reliably.

Irrigation methods for racetracks

The requirements of racetrack in-ground sprinklers are possibly the most demanding of all turf equipment. A consequence of the large distances to be covered is that the sprinkler heads will have high flow rates and consequently operate at high pressures. Ideally, they will be capable of performing effectively in moderate winds. Achieving a high degree of uniformity with single high-flow sprinkler heads is very difficult (Figure 9.5). The highest irrigation uniformity is generally achieved using low-flow nozzles mounted on a boom and closely spaced. Droplet size, droplet trajectory and closeness to the ground (of the nozzle all contribute to a more even application with booms. It is for this reason that mobile booms are sometimes used on racetracks.

Racetrack sprinklers must be extremely reliable. The maintenance of the irrigation system needs to ensure that the nozzles have a clear trajectory, the seals are effective, the retraction devices are working and the rotation drive mechanism are operating effectively.

Figure 9.5. To achieve uniform surface performance sometimes hand watering is required to supplement sprinkler system applications

Racetrack irrigation scheduling

Racetracks are often irrigated by applying water to a reasonably shallow depth (e.g. <5–7 mm). There are various reasons for this approach, including turf managers being cautious and not wanting to over-water, and limitations in equipment and systems that prevent deep applications. The travel speed of mobile irrigation equipment, flow rates, performance of pumping plants and pipework often limit the depth of water that can be applied.

Racetrack managers should be conscious of the consequences of shallow irrigation applications. They include:

- encouragement of shallow-rooted turf – drought resistance is limited and the surface has less ability to withstand stress
- water losses for each shallow irrigation application are higher
- limited opportunity to use nutrients in the shallow root zone.

Knowledge of the actual soil moisture level and changes under the varying influencing climatic and irrigation conditions is very useful to the irrigation manager. There is much to gain through the direct measurement of the soil moisture status of the racetrack.

Irrigating urban trees

Achieving water use efficiency for urban trees

Trees are critical elements in the urban landscape: they add value by providing function and amenity.

Large plants, such as trees, are high users of water. Trees need to take up large volumes of water to support their large mass and the extensive foliage area. Trees have the capacity to harvest a lot of water through extensive root systems (Plate 9.1).

Ideally, species are selected that are suited to the climate of the locality and provided with a growing environment that means supplementary water (additional to rainfall) is not required. However, most trees selected for urban situations require water for establishment. This may involve initial watering for several months or even for the first 1 or 2 years.

Figure 9.6. Mulching of trees provide benefits in terms of soil water retention and improved soil health

There are a range of reasons – including landscape design priorities, heritage, cultural, botanical and significance – that will require supplementary watering to ensure the tree maintains a healthy condition. Avoiding serious stress, as a result of low soil moisture, is a very high priority for authorities managing significant and high value trees in urban areas. Numerous preventative strategies have been implemented to keep trees in a healthy condition, including mulching (Figure 9.6). The amount of water required to prevent critical stress to a tree is small compared with the amount that a healthy tree, in good condition, might consume during hot summer periods when soil water is readily available.

Supplementary watering is required for many trees over the summer months, including, in some cases, natives and exotics, and the task is to deliver water efficiently in a sustainable way. Sites for urban trees range from open parkland sites where there is no restriction to root development and there may be limited competition from other vegetation, to street trees, where the available soil volume is very limited and the tree and soil is subjected to numerous pressures and stresses. These trees often need supplementary water to remain healthy (Figure 9.7). These trees can be subjected to a range of stresses including a polluted and potentially toxic atmosphere and, mechanical damage above and below the ground, as well as moisture stress. Moisture stress may arise through constrained root systems or lack of rainfall infiltration due to impermeable surrounds.

Trees in small containers obviously require irrigation (Figure 9.8). A calculation of days of storage is useful in assessing the water management of trees with confined root systems. This can be done using typical summer daily ET_c rates and the root zone soil volume, with allowance for water holding. Parkland trees may have 150 days of storage and a street tree may have only 10 days of storage.

City of Melbourne tree watering trial

Following periods of drought in the early to mid 2000s, the City of Melbourne investigated a number of watering options for mature trees. This evaluation is focused around various combinations of watering wells and watering trenches. The watering wells are commercially

Figure 9.7. Street trees are often growing in situations where tree root development is severely restricted and supplementary water is required

Figure 9.8. Trees growing in small containers require significant and frequent applications of irrigation water to avoid severe stress

available products from Rain Bird International and Hunter Industries. The watering trench system has been developed throughout the trial and is continuing to be refined.

The target of this trial is the delivery of reasonably high volumes of water (e.g. >100 L per watering point per day). Daily total delivery rates in the range 200–800 L are expected to be required at some trees locations.

Efficient water use requires that good use is made of rainfall and any stormwater that may be available to the tree site. Identifying the tree root system area (target area) that is to be watered is a very important part of the process. Also, recognition of the importance of the soil environment in determining the long-term survival and health of the tree is required. The irrigation water should be applied so that a significant portion (not necessarily the majority) of the root system is watered. This generally requires watering widely and deeply.

Characteristics of urban trees

Some of the characteristics that may be expected to be associated with urban trees include the following.

- The tree root distribution is highly variable, both laterally and vertically, in the soil profile.
- The physical, organic matter, water and nutrient properties of the soil varies.
- The tree roots are often in competition with other vegetation and in particular turf root systems.
- Compacted soils result in low infiltration rates – particularly on nature strips and traffic areas (pedestrians, machine and vehicles). Compacted soils restrict gaseous exchange and limit root development, and hence tree growth.
- Surface treatments (impervious paving) around street trees often limit access to root systems to source water.
- Some significant tree roots are located deep within the soil profile and effective water delivery is a challenge.
- Tree roots are sometimes constrained by street and roadway structures, which limits the water storage volume and catchment area.
- The tree response to applied water may take considerable time.
- Excavation works for construction and services often cause significant root damage to trees.
- The condition of a tree often reflects stresses or deficiencies experienced considerable time (sometimes months or years) previously.
- The amount of rainfall intercepted by the tree canopy can be significant.
- Trees need a spatially balanced root system required for stability under wind loads.
- Poor tree health may be due to stresses other than moisture, such as pests, disease, damage or environmental pressures or age.
- Trees can have high daily water demands that need to be satisfied.

All of these can have an impact on the water management of trees.

Irrigation challenges for urban trees

Urban trees present particular challenges in terms of irrigation. These include:

- The delivery of water to the root system requires a relatively large area to be covered.
- Compacted soils reduce the ability to deliver water effectively into the soil and deep within the soil profile.
- The risk of damage to irrigation equipment due to traffic (vehicle and human) and vandalism is high.

Figure 9.9. Installation of any buried equipment, such as watering wells, may result in significant root damage

- Soil conditions can be harsh in terms of risk of root intrusion and crushing of delivery pipes.
- Water penetration to deeper root layers generally requires slow application over long periods of time.
- There is a limited accessible area to apply water for some trees (e.g. street trees that have restricted ground spaces).
- The presence of roots within the installation area means there is a risk of root damage (Figure 9.9).
- Large volumes of water need to be applied.
- Non-homogeneous soils (e.g. rocks and solid debris) often result in non-uniform distribution of water.
- Competition from turf may limit the ability for effective tree watering.

Irrigation methods for urban trees

The choice of irrigation delivery method and the operation or control of the system is very important in the watering of trees. There are potentially numerous irrigation delivery options including:

- sprinklers
- sprays (short throw, high precipitation rate)
- microsprays
- surface drip (sub mulch)
- shallow drip (top 100 mm)
- subsurface drip (100–500 mm deep)
- water wells (e.g. 50 mm diameter × 300 mm deep) supplied by bubblers
- watering rings (perforated pipe and drainage material) supplied by bubblers or similar
- tree watering trench.

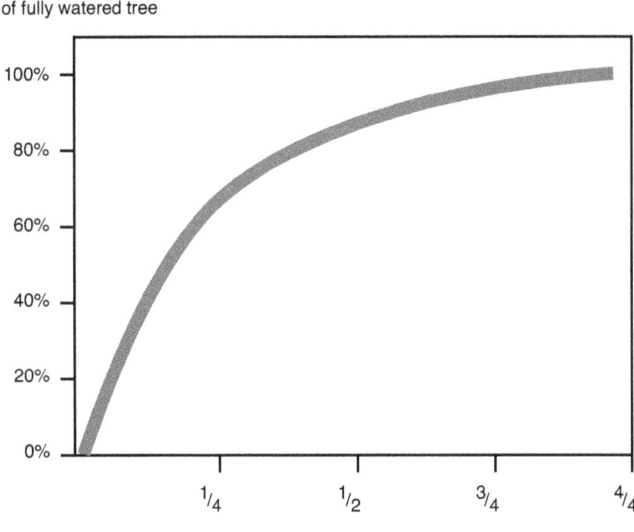

Figure 9.10. Tree root area watering (Source: adapted from Mitchell and Goodwin 1996)

Drip irrigation is well suited to the watering of mature trees. As a general rule, water should be applied slowly so that runoff does not occur, and deeply so that the soil volume, where the bulk of the roots are located, is watered, rather than just the top layer of soil. Delivery of irrigation water to the middle root zone, which may be at a depth of 300–400 mm, rather than just to the surface soil layers, is preferred. Shallow watering is commonly experienced with sprinkler and spray systems irrigating trees.

A subsurface drip system is one technique that allows direct application of water to the root zone. However, installing these systems alongside existing mature trees risks damage to tree root systems.

As discussed, the watering of large plants, such as trees, does not require the full root area to be watered. To maintain trees in a healthy condition, between 25% and 50% of the total ground root area can be watered to achieve good water uptake, around 80%, by the tree (Figure 9.10).

This graph is an adaptation of Figure 3.1 in Mitchell and Goodwin (1996), who reported on the relative water use of young apple and peach trees watered with drip irrigation. Trees respond to localised moist soil volumes in which nutrients and oxygen are available by producing increased numbers of roots in those areas (Figure 9.11).

Tree drip design and installation

The design of a drip watering system for a tree should take into account the root area to be watered and the total amount of water to be applied. These include the:

- type of drip emitter and drip supply line
- emitter discharge rate
- emitter interval along the drip supply line
- spacing between the drip emitter supply pipes
- installation requirements for the drip pipe
- operational requirements (e.g. run time and frequency).

The drip systems should ideally be designed to apply water from inside the tree dripline out to some distance beyond the dripline (Figure 9.12).

Figure 9.11. Concentration of roots of fruit (citrus) tree developed under localised water and nutrient application with drippers

Example – Drip watering of mature tree

Tree: Elm
Height: 9 m
Canopy width: 10 m
Drip layout design: three rings of dripline
Drip position: 3 m, 5 m and 7 m from trunk
Emitter interval: 400 mm
Emitter flow rate: 1.5 L/h
Total dripline length (three rings): 95 m (approx.)
Number of emitters: 230 (approx.)
Drip system delivery rate: 345 L/h
Water delivered in 2 hour run time: 690 L

Figure 9.12. Tree drip layout using three circles of dripline

Figure 9.13. Tree watering trench installation in roadway median (City of Melbourne)

Emitter requirements for subsurface drip watering of trees

The requirements of drip emitters to be used for the subsurface drip irrigation of trees include:

- operating very reliably and precisely in a mulch/soil environment
- being resistant to intrusion by roots and debris
- being resistant to ingress of debris through suck back into emitter outlet
- incorporating filtration to minimise or prevent blockage of emitters.

Watering wells can also be used to apply relatively large volumes of water into the soil volume around a tree. It is important to check the lateral distribution of water from the well position. A large tree may require six to eight wells.

The City of Melbourne developed the watering trench to allow delivery of relatively large volumes of water to the root zones of mature elm trees in confined root environments (Figure 9.13). The trench is approximately 1.2 m long, 400 mm wide and 300 mm deep. It is filled with small (5–7 mm) screenings (gravel) to provide storage and distribution. Water is supplied through a pressure compensating bubbler (1.9 L/min) and the system is designed to deliver in the range of 200 to 400 L per day for each trench.

Strategies for high irrigation efficiency for urban trees

1. Apply water so that the soil is moist to a good depth (e.g. >300 mm). This provides good soil water storage and provides reserves during dry conditions and builds resilience.
2. Sub-mulch drip systems are more efficient. Soil evaporation is minimised. The mulch generally enhances the tree soil environment as it breaks down.
3. Mulch around trees should be coarse mulch of mixed particulate size (with limited amount of fines) that allows rainfall to readily infiltrate. Mulch applications around trees should not be too deep (50–75 mm maximum).
4. Prepare the ground around the tree so that a natural water-holding basin is formed. If there is surface runoff in the vicinity, divert the water to the tree.
5. Ensure effective infiltration properties around the tree so that irrigation water does not run off.
6. Encourage infiltration (e.g. small trench filled with sand): this acts as an intercept for surface water and temporary storage.

7. Irrigation control and delivery to individual trees is required.

There should be a separate control zone covering the tree canopy and its immediate surrounds. There are many examples of lawns becoming waterlogged via over-watering, because the watering of a tree was not separated from the lawn irrigation system.

Irrigation scheduling for trees

In scheduling of irrigation for urban trees, the following need to be taken into account.

- Soil moisture levels are highly variable throughout the tree root system.
- The application may be to replenish the total soil water reservoir rather than to replace daily water needs.
- There is not one single overall refill point due the large variability in soil moisture in the root zone.
- Movement of water through the soil profile to replenish lower soil layers can take considerable time (e.g. days or even weeks).

When irrigating deeper-rooted plants, such as trees, the use of multiple sensors at varying depths is often recommended. Two sensors, one positioned at one-third of the root depth and one positioned at two-third depth, are suggested. The shallow sensor is used in the decision to commence irrigation and the deeper sensor is used to indicate that the irrigation water has reached the main water storage zone for the plant.

Watering trees during dry periods

Identification of stress in a tree is the first step in assisting trees in surviving extended dry periods, such as during a drought (Figure 9.14). Some of the indicators of stress are leaf curl, wilting/drooping, colour change (e.g. to yellow), leaf scorch (deciduous trees) and leaf drop. The key to the successful watering of trees is the application of sufficient volumes of water that wet the soil volume to a reasonable depth, (e.g. 300–400 mm). Emergency tree watering

Figure 9.14. A range of portable storage units has been employed to provide emergency watering of trees during drought

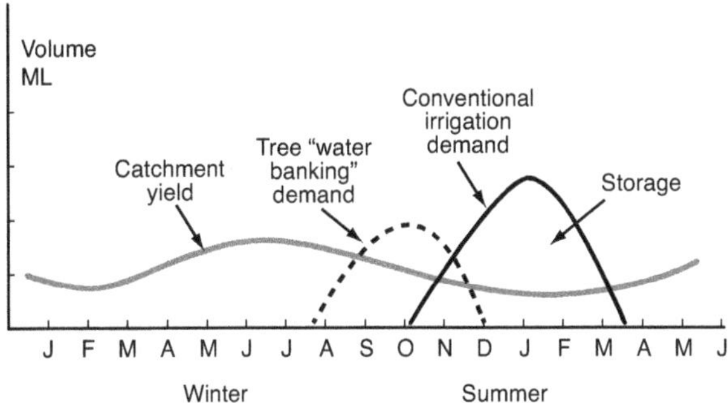

Figure 9.15. Stormwater yield can be used to replenish soil moisture at some sites as part of a water banking strategy

techniques should apply water in the same position at each application to encourage root growth in that soil volume.

Deep, infrequent watering is well suited to mature trees. In some circumstances, it is appropriate to wet up the soil volume before the period of high water demand: this approach is sometimes referred to as 'water banking' (Figure 9.15).

Irrigating garden beds

Achieving irrigation efficiency

The efficient irrigation of garden beds is a challenge both in terms of effectiveness of application and high efficiency of scheduling. Because of the diversity of plant size and shape and the density of plantings, effective water delivery can be difficult. The target is not the shrubs, not the mulch, not the leaves, but the root zone of the plants. Efficient irrigation is about delivering water into the root zone, where it can be beneficially used by the plants. The range in water demand within relatively short distances or areas makes precise scheduling a difficult in many garden beds. The scheduling of irrigation to cater for the needs of one or two individual plants should be reviewed. Options include a separate zone for watering or redesign of the planting.

Achieving high water use efficiency in a public garden involves adopting of a range of strategies. Experiences at the Royal Botanic Gardens Melbourne are described by Symes and Connellan (2004). Issues covered included the design of the irrigation system, effectiveness of water delivery (foliage penetration), water supply arrangements, maintenance of system, performance evaluation, technology options, soil management and staff skills.

Characteristics of garden beds

Garden beds are often used as landscape display areas. Maintaining plants in a healthy condition throughout the year so that the various colours, textures, forms are well presented is a priority. Garden beds are also used to provide protection and act as screens or barriers, both visual and physical. In these cases, the plants need to have the appropriate size and shape. Garden beds can be characterised by:

- irregular shapes and narrow planting areas
- mixtures of plants, in terms of species, size and shape
- multi-layered foliage
- dense foliage
- often large variation in water demand within one bed

Figure 9.16. Sprays need to be positioned and mounted, laterally and vertically, to avoid obstruction to distribution from foliage

- varying root zone depths and intermingled root zones
- irregular spacing of plants
- mulched surfaces
- tree foliage and large shrubs intercepting rainfall
- microclimates ranging from shaded and protected, to fully exposed (sun and wind)
- microclimates influenced by nearby structures, such as buildings, roads and paved areas
- non-homogeneous soils.

Irrigation challenges for garden beds

The issues that need to be considered in selecting appropriate irrigation applicators, designing irrigation systems and managing the irrigation of garden beds include:

- The foliage potentially interferes with the sprayed distribution of water (Figure 9.16).
- It is difficult to achieve uniform application with sprays in irregular shaped and narrow garden beds.
- Mulch absorbs rainfall and irrigation water.
- There is a risk of damage from maintenance activities and/or pedestrians.
- The plant response to soil moisture stress is variable, so the timing of irrigation is a challenge.
- Determining the appropriate irrigation application depth is a challenge, because of the variation in root zone depths and variation in species water requirements.
- There is a risk of vandalism to irrigation equipment and theft in public gardens.

Irrigation methods for garden beds

Sprays and sprinklers

Sprays, microirrigation equipment (microsprinklers, microsprays and drippers) systems and bubblers are used in garden beds. Sprays have the advantage of potentially wetting the whole area (foliage and soil) of the garden. Medium pressure sprays that have reasonable droplet size are recommended because small droplets are sensitive to wind drift and evaporation and do not have the energy to penetrate between foliage.

Precipitation rates are relatively high with sprays (fixed nozzles) – often higher than 20 mm/h – and so run times should reflect these high rates. If the soil infiltration rates are limiting (e.g. clay soils), the watering time window needs to accommodate cycling of the irrigation to allow infiltration.

Spray uniformity and efficiency

It is generally recommended that irrigation water should be applied uniformly. However, in the case of garden beds, effective application is more important. Using a spray with a high theoretical uniformity of application that does not penetrate between the foliage is not a sound practice. As a general rule, it has been found that stream- or jet-type devices are more effective than sprays in penetrating garden beds with dense foliage.

Microsprays

Microsprays have the advantage of providing complete coverage in relatively small areas of 1–2 m^2. They can be used to apply water under the canopy of tall plants. They are suited to soils that have reasonably high infiltration and percolation rates. However, droplet size is a significant factor in the performance of microsprays in garden beds. The small droplets are prone to drift and are also ready absorbed by finer organic mulches. Over pressurisation of microsprays appears to be a common problem in irrigating garden beds. The incorporation of pressure compensation (PC) in spraying type emitters is a significant advancement in terms of irrigation efficiency.

A study carried out on the irrigation of 50 homes in the Sydney region showed that garden watering can be inefficient (Maheshwari 2006). The average application rate was 8.7 kL/100m^2/month for the garden beds and 4.3 kL/100 m^2/month for the lawn areas. Microsprays, which tend to be used in garden beds, were found to apply the most water. The results indicated very significant over-watering of garden beds: approximately 200% of theoretical requirements. These results demonstrate the importance of human behaviour in irrigation efficiency. It was also found that manually operated systems applied more water than those operated with a controller. It should be noted that these results apply to relatively small irrigated areas (<50 m^2). Most experiences in urban irrigation indicate that residential properties with automatic irrigation systems used more water than those properties that are manually watered.

Drip irrigation

Drip systems are increasingly being used to water garden beds. The placing of driplines in garden beds under mulch is potentially a very effective water application technique. Achieving high efficiency requires attention to the selection of the emitter, the design of the system and the operation of the drip system. The risk of damage is real with drip systems. Anchoring of drip lines and tubing and having strategies that reduce risk of physical damage is essential. Efficiency is dependent on system reliability, and many garden and landscape drip systems are not maintained to provide effective application.

The total coverage of garden beds using multiple rows of driplines (referred to as a grid) is now relatively common (Figure 9.17). One characteristic of these systems is that areas that do not have active root systems are watered because the whole garden bed area is covered. This can lead to losses and inefficiencies.

If a significant proportion of the garden bed is not covered with vegetation – for example, if there is 40% or more open space – then consideration should be given to using point source emitters. The other option is to use dripline positioned around plants to avoid watering areas without plants.

Figure 9.17. Rows of dripline provide total area coverage of the garden bed

The discharge rate from dripline emitters is typically in the range of 1.0–3.0 L/h. Although this is a relatively low flow rate, the actual precipitation or application rate of dripline systems can be significant if short emitter intervals and close dripline spacings are used. A dripline system with emitters at 300 mm intervals, emitter flow rate of 1.5 L/h and spacing of 500 mm, has a mean application rate (MAR) of 10 mm/h. Operating a system, with this application rate, would apply 20 mm in a 2 hour run time. Because the emitter flow rate is low, these systems are often left on for several hours and so over-watering may occur.

The watering of trees and garden beds with drip irrigation is not as demanding as turf in terms of uniformity of application. There is not the underlying requirement to have even distribution of water. Non-uniformity can be tolerated, as long as the water is delivered effectively to the plant root zone and delivery is controlled. In the case of garden beds, both the spacing between emitters and the spacing between driplines is critical. Finer textured soils allow driplines to be spaced more widely apart. This has the advantage of the system requiring less pipe and fittings. With coarser soils, such as sands and sandy loams, closer spacing of emitters and driplines is required. This may require emitter intervals of 300 mm and pipe spacing of 400–500 mm. In a garden bed of 100 m length and 3 m wide (300 m^2), this represents a total dripline length of 500 m (0.5 km) or more.

The spacing of driplines in garden beds needs to be balanced between the high application rates, costs of closer spacing and the limited wetted area coverage of wider spacing. Although wide, (e.g. 1 m or greater) spacing may suit established shrubs, small plants, such as annuals, with shallow root systems require closer spacing. Also many garden soils are open, with high organic content. Good lateral distribution of water is often limited in these soils.

Table 9.1 provides a guide to the watering of landscape plantings using drip systems.

Strategies for high irrigation efficiency in garden beds

Scheduling

Garden beds should be irrigated using daily ET data. The timing of irrigation should be based on allowing the plants to be partially stressed and the soil reservoir depleted to a significant

Table 9.1. Guide to drip emitter numbers for nominated garden plant sizes

Plant canopy diameter (m)	Number of drip emitters (outlets) per plant
0.5	1
1.0	2
1.5	3
2.0	4
3.0	5
4.0	6–8

Notes:
Guide to dripper discharge rate: clay 1–2 L/h, loam 2–4 L/h and sand 4–6 L/h
The actual number of drippers required will depend on soil type and root area to be wet/covered.
These emitter numbers refer to point source (individual units) drip emitters and not dripline (in-line) products

level. Garden beds are generally characterised by plants with reasonably deep root systems and also good soil organic content. Watering should not be more frequent than once a week in many locations. If shallow-rooted display plants are part of the garden bed, then more frequent watering will be required in this area. These areas should be treated as a separate watering zone.

Garden beds often contain plants with strongly varying water demand, which can cause the over-watering of some species (Appendix 4). It is therefore preferable to group together (hydrozone) plants of similar water requirements.

Mulch

Mulch reduces water loss from the soil by providing a physical barrier to soil evaporation. If an organic mulch is used (such as fine bark, green waste or leaf litter), this layer can also absorb significant amounts of rainfall and irrigation. A coarse mulch, not deeper than 75 mm, that allows water (rainfall and irrigation) to infiltrate through the mulch is recommended. When used in conjunction with sprays and microsprays, organic mulches with high amount of fines result in considerable water losses due to absorption. Much of this water is lost due to evaporation into the atmosphere from the mulch.

Understanding the site water cycle – effective rainfall

There are numerous reasons why all rainfall does not reach the plant root zone. These include vegetation properties, ground condition and soil conditions. The proportion of rainfall that is delivered into the root zone, after all rainfall losses have been taken into account, is referred to as effective rainfall (P_{eff}). This is difficult to accurately determine without a full and detailed analysis of the site. It can, however, be estimated by taking into account some of the factors that will influence it. These include:

- rainfall above the amount that can be stored in the root zone will be wasted through deep drainage
- rainfall intensities greater than the soil infiltration rate will result in some runoff
- very light rainfall amounts may not result in a net addition of water to the soil root zone. It is likely to be lost by evaporation from vegetation and the soil surface. Rainfall less than 2 mm can be ignored because it is regarded as ineffective.

The estimation of P_{eff} should take into account the total amount of water that can actually be stored in the soil root zone, because rainfall in excess of this capacity will be wasted. Shallow-rooted turfgrasses growing in coarser soils will have a storage capacity typically in the range of 10–20 mm. Deeper rooting species may have a root zone storage capacity in the vicinity of 20–30 mm in coarse soils. An important characteristic of shallow-rooted turf is the limited ability to capture rainfall. Small root zone storages provide less opportunity to catch rainfall. Also, significant amounts of water can be stored on leaf surfaces and in thatch, which can contribute to the losses.

It is reasonable to assume P_{eff} to be 50% for many turf and garden situations, if a daily water balance analysis of rainfall and soil moisture is not possible. This assumption of 50% rainfall is reasonable for relatively shallow-rooted plants according to the Irrigation Association (2005). Analysis of rainfall entering the root zone of plants with a range of root zone depths and soil types indicated that the effective rainfall ranged from 44% to 68% for a range of soil types and soil depths. The average effective rainfall for sandy loam soils, with a root zone depth of 300 mm, was 50%. It was found that there was not a large variation for other soil types and depths.

Significant amounts of rain can be intercepted by overhanging foliage and is therefore not available to replenish soil moisture. The density of foliage, leaf form, leaf number, orientation and plant architecture all affect the amount of water trapped. The nature of the rainfall event (e.g. light drizzle compared with a heavy storm) result in varying proportions of rainfall interception. Some the intercepted rainfall may reach the ground by flowing down the trunk. Rainfall throughfall collection devices can be used to measure rainfall reaching the ground (Plate 9.2)

A study, which commenced in 2008 at the Royal Botanic Gardens Melbourne, in conjunction with Monash University, School of Geography and Environmental Science, is being undertaken to evaluate the amount of rainfall that passes through garden landscape foliage. Up to 80% of rainfall can be intercepted by foliage in a single rainfall event. Figure 9.18 shows that very significant amounts of rainfall are intercepted by foliage and will potentially not contribute to the soil moisture in the garden bed soils. There will be additional significant losses in mulch on the soil surface. Some of the intercepted rainfall will reach the ground via stem flow down the trunk.

Details on the rainfall throughfall project at Royal Botanic Gardens Melbourne are on the following website: www.rbg.vic.gov.au/horticulture

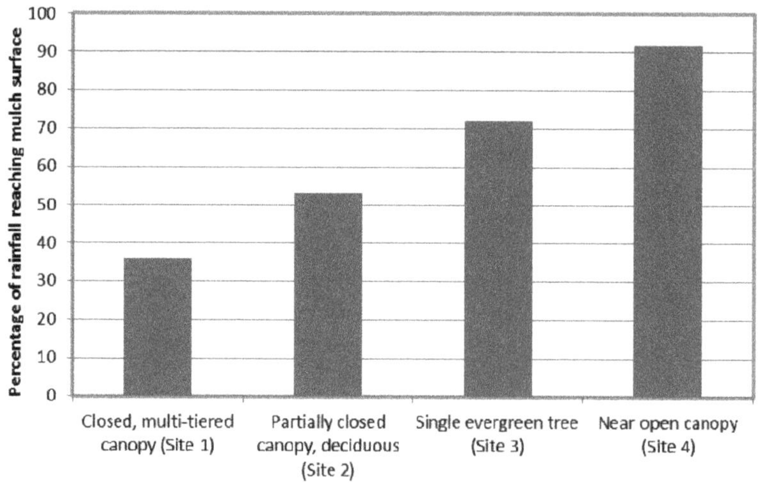

Figure 9.18. Proportion of rainfall reaching the ground at four sites in the Royal Botanic Gardens Melbourne

It is important to assess each site to determine the amount of rainfall that may be contributing to soil moisture. Soil moisture sensing technology is valuable for this purpose.

Demanding landscape sites for irrigation

Due to the nature and characteristics of some sites, particular attention to the delivery of water is needed to achieve high irrigation efficiency.

Narrow sites

A common problem is to water narrow garden beds and lawns so that no overspray occurs. The use of sprays on spaces that are less than 2–4 m wide is a challenge. These applicators tend to have small droplet sizes that are prone to drift and wind distortion, and the boundary of the wetted area is often poorly defined with sprays.

Watering narrow median strips and nature strips with sprays often results in water runoff and drift on to roads and paths. This is a highly visible waste of water. Although the risks can be minimised by selecting spray heads with lower pressures, wind and passing traffic can cause the spray to miss the target area. Drip systems, which readily accommodate narrow and irregular shaped areas, are well suited to these situations. The use of subsurface drip on narrow strips of lawn is an ideal solution, because it is efficient and removes the risk of water being deposited on non-target surfaces.

Irregular shaped areas

The effective coverage of irregular shapes with sprays generally results in the need for many applicators with short radial coverage to accommodate the changing shape of the garden bed (Plate 9.3). Overspray beyond the garden bed is commonly observed on irregular shaped garden and lawn areas. This is not consistent with good irrigation practice. The linear type water distribution achieved with driplines is well suited to garden beds where the width is narrow and shape irregular.

Sloping sites

The solution to the efficient watering of slopes is a combination of application technique and operation (run time and cycling of application).

Sprays on slopes

Sprinklers, sprays and drips can all be used on slopes with caution. The primary consideration with sprinkler and sprays is the reduced infiltration rate of the soil and the varied coverage. A guide to the maximum precipitation rates on slopes is presented in Table 9.2, which demonstrates the effects on increasing steepness of slopes in terms of reduced infiltration capacity.

Steep slopes readily cause runoff, so much lower precipitation rate systems should be used. Consideration should be given to the appropriateness of irrigating steep slopes with sprinklers. In many of these situations, using drip is more effective and efficient.

The spacing of laterals (assuming they run across the slope) needs to be adjusted for the reduced coverage up slope and increased coverage down slope. One piece of advice for sprays on slopes is to use shorter (e.g. 80% of recommended) spacing in the upper part of the slope, and wider (e.g. 120%) spacing, in the lower part of the slope. This requires assessment at the particular site.

Pressure regulation is a key consideration with systems installed on slopes. The availability of pressure regulation within the spray and sprinkler heads is a valuable feature for those to be used on sloping terrain. The availability of low precipitation rate, high uniformity, rotating

Table 9.2. Guide to maximum recommended precipitation rates for various slopes and soil types

Soil texture	Recommended maximum precipitation rate (PR) (mm/h)			
	0–5% slope Flat	5–8% slope Gentle slope	8–12% slope Medium slope	>12% slope Steep slope
Coarse sand	30	25	20	15
Light sands	25	20	12	10
Silt loams	15	10	6	5
Clay/clay loam	5	4	3	2

Notes:

1. This is a guide only.
2. Maximum recommended precipitation rates in mm per hour (mm/h) for sprinklers and sprays on ground surfaces with cover. If ground is bare, these values should be significantly reduced, particularly on sloping ground.

stream sprays provides an improved equipment option for many situations with inclined landscapes. The inclusion of anti-drain or check valves in sprinklers and sprays is also important in preventing drainage and seepage problems.

Ground slopes represent areas of varying water requirements and so zoning is very important. Control over sprinklers or sprays on each lateral installed on a contour (across the slope) is desirable. The monitoring of the water used on slopes is particularly important, because any pipe and equipment failures can create significant damage through washouts and erosion.

Irrigation control programs have to be able to operate multiple cycles so that water application in regulated and controlled amounts can be delivered. This reduces the risk of runoff.

Drip on slopes

Drip irrigation is well suited to slopes. The issues of ineffective infiltration, compared with spray systems, are minimised or eliminated and coverage is more readily controlled. However, the achievement of uniform coverage is still a challenge because the distribution of water down the slope is influenced more by gravitational forces than capillary forces. In other words, the water will drain down the slope to a greater extent than it will move up the slope. Understanding soil wetting behaviour is essential if a uniform application is required.

Pressure regulation is very important: in addition to the use of pressure-compensating emitters and sound hydraulic design, pressure regulation should be incorporated in control valves. As with sprays, wider spacings can be used down the slope on landscape plantings to take into account the drainage of water down the slope. It is also important that the zoning of drip systems be sufficient to allow at least the upper, middle and lower areas to be watered separately.

Irrigating coarse textured soils

At some locations, such as Perth, and in some situations, soils which have very high infiltration and percolation rates need to be irrigated. These soil properties have an impact on the method of irrigation and the management of the system. The main characteristics to consider are:

- the risk of soil solutions (leachate) draining below the root zone and contaminating groundwater
- the limited wetted lateral coverage with drip application
- the low AWHC of the soil, which limits the amount of stored water and results in the need for frequent irrigation.

Strategies that can be adopted to address these characteristics include:

- using high discharge rate drip emitters (e.g. 4 L/h or higher)
- monitoring of soil moisture in the lower part of the root zone

Figure 9.19. Drip irrigation of sandy soil, modified through the addition of soil fines, to improve hydraulic properties (Dubai)

- using fertilisers and plant health chemicals according to best practice. The risk of nitrogen and phosphorus contamination of groundwater and other water bodies should be assessed.
- giving consideration to improving the water properties of the soil (e.g. through adding organic material (Figure 9.19).

Windy sites

The first approach for windy sites should be to use, if possible, a technique that is not sensitive to wind, such as drip irrigation. If a spray type system is required, or already installed, avoid operating it in windy conditions.

Spray equipment selection for windy sites

Choose sprinklers and sprays that tend to produce larger droplets. This means using larger orifice sizes and avoiding microsprays. Sprays that have a fan type distribution are also prone to production of small droplets.

Typically a jet trajectory angle in the range of 20° to 30° is used because this provides the greatest distance on throw, under calm conditions. Select lower stream trajectory devices, such as in the range of 5° to 10° for windy sites. Note that these lower trajectory sprinklers and sprays do not achieve the distance of throw of the higher trajectory devices.

Sprinkler spacing and layout

It is recommended that the spacing of sprinklers be reduced to minimise the adverse effects of wind distortion. This may require spacing at less than 50% of wetted diameter (see Chapter 8). At sites, where winds are coming predominantly from a particular direction, the closer spaced sprinklers should be positioned, at right angles to the wind direction.

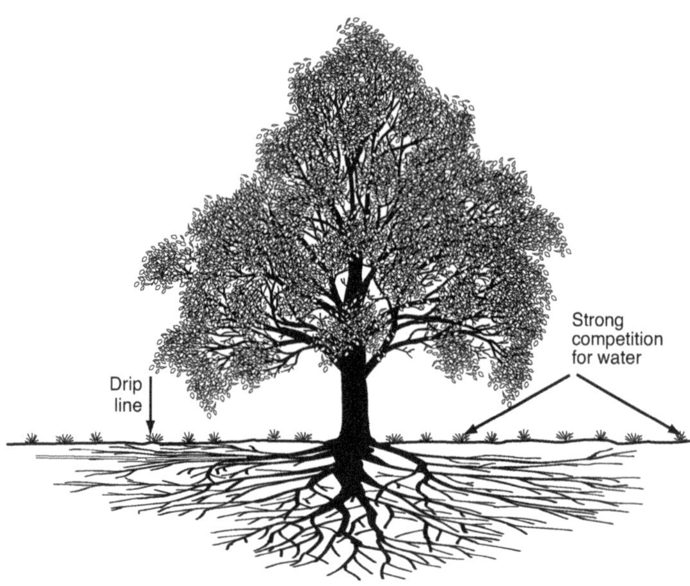

Figure 9.20. It is important to have separate control of irrigation to allow differential watering in areas of competition between trees and turf

Zoning and system control

The wind may have an impact on all of the system or only part of it. Each separate area that is either affected by the wind or can be used to respond to the wind conditions should be separately zoned. This provides flexibility in the operation of the system to allow each area to be controlled in response to the wind. Both wind speed and wind direction sensors can be used to reduce the risk of wind distortion and losses.

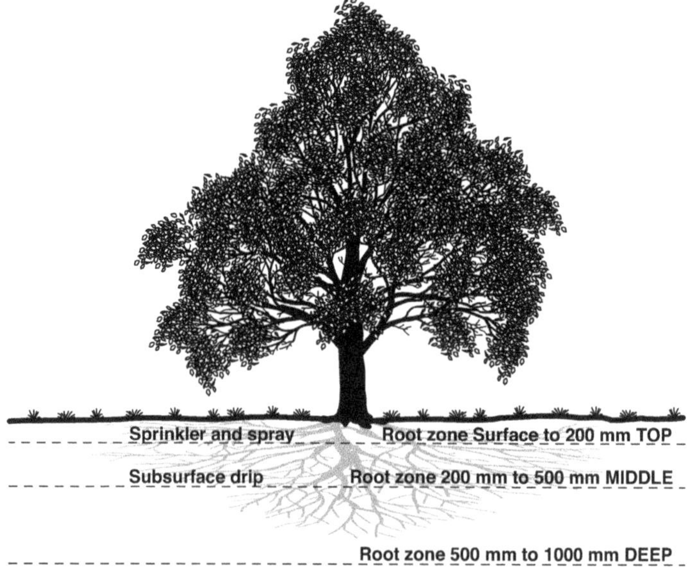

Figure 9.21. Subsurface drip irrigation (SDI) can be used to overcome limitations of sprays by only providing shallow watering for, say, the top 200 mm of soil

Competing landscape plantings

At times, there may be multiple and competing plant species extracting soil moisture from the same soil volume (Figure 9.20). There is the risk that one species will not receive adequate water while the other species may be over-watered. In terms of achieving efficiency, it is first necessary to identify these areas and then design the system so that the areas of competition are zoned separately.

A common site of conflict is in the treed areas that also include grass. In an attempt to keep water supplied to the tree and to allow the grass to grow, additional water is applied to the adjacent lawn area. This results in over-watering of the grass only areas. A novel solution applied to elm trees in Melbourne was the installation of two irrigation systems, one primarily designed to water the grass (sprays) and the other (SDI) to deliver water into the root zone of the trees (Figure 9.21).

Chapter 10

Strategies and technologies to achieve high efficiency

Water use efficiency and irrigation efficiency

Irrigation efficiency is a measure of the amount of water effectively delivered into the plant root zone, compared with the amount applied. It only measures the effectiveness of the irrigation system. Water savings achieved through reduced water demand by plants, are represented by the term water use efficiency. These terms and definitions are outlined in Chapter 1.

Irrigation efficiency can be broken down into two parts. These are the:

- effectiveness of application of water to the plant root zone, which is referred to as the application efficiency (E_a)
- timing of irrigation duration and event to meet plant needs, which is referred to as water management efficiency (E_{wm}).

The major factors that contribute to irrigation efficiency are outlined in best management practices section in Chapter 7. Underpinning the overall approach to achieving sustainability of irrigated turf and landscape sites, including turf, should be high water use efficiency and high irrigation efficiency (IE).

A target for IE should be in excess of 75%. This requires high application efficiency (E_a) in excess of 80% and water management efficiency (E_{wm}) in the range 90–100%.

Reduction of plant water demand

Plant selection

The first step in improving water use efficiency should be to reduce the actual plant water demand. This is achieved through informed plant selection, recognising that plants have different water use requirements. Species selection within the turfgrass family is an example of significant potential savings. For example, cool season grasses typically use 30% to 40% more water than warm season grasses.

The primary requirement is that the landscape provides the desired performance or outcomes and, if this can be achieved with species that have low water use rates, then this is a step towards sustainable water use for the site.

Selecting low water use and drought-tolerant plants

There are broad guidelines available for the selection of plants that have low water use rates and withstand some drought conditions. This advice includes selecting natives indigenous to the local area and plants that have inherent drought-tolerant properties and characteristics.

Plants that have adapted to survive in dry conditions and drought have adapted in a number of ways, including having:

- grey or silver foliage
- small narrow leaves
- fleshy water retaining succulent leaves and stems
- a hairy leaf surface
- summer dormancy
- stomatal closure.

Some of the adaptions that are characteristics of turfgrasses include rolled leaves, stomatal closure, thickened cuticles, and development of extensive root systems and rhizomes that can survive hot dry conditions. The various plant responses can be categorised as drought avoidance, drought tolerance and drought avoidance (Gibeault and Cockerham 1985).

There are plants lists available from botanic gardens, various societies and associations (e.g. Societies for Growing Australian Plants, Sustainable Gardening Australia and Greening Australia), nursery associations (e.g. Nursery and Garden Industries Australia) and water authorities in each state.

The publications *Water Use Classifications of Landscape Species (WUCOLS)* (Costello and Jones (2000) and *Landscape Plants for California Gardens* (Perry 1992) both provide valuable plant water use information for over 2000 ornamental plants. Both of these are based on Californian experiences and expertise. Plant lists are a starting point in the planning and development of a sustainable landscape. There are numerous factors to consider, that directly impact on the likely success or otherwise of the planting. These include the:

- climate of the area
- microclimate of the site
- site soils and geology
- topography – site aspect and drainage
- drainage patterns
- surface treatments – pavement, mulch
- site use – wear, compaction, risk of injury or damage.

The selection of a species to satisfy both the low water use criteria and specific site conditions and requirements may involve professional horticultural landscape designers or agronomy personnel, as well as water experts, to develop a solution that will be sustainable.

Appendix 4 provides a guide to plants with low water use rates (K_c 0.10 to 0.30) that will survive periods of low rainfall.

Establishment of low water use plants

Although some plants are categorised as low water use and drought tolerant, they will only perform in this way once they are successfully established. Supplementary water will be required for establishment.

The conversion from cool season grasses to warm season grasses will usually involve significant amounts of water over the establishment period of several months. The amount will depend on the particular approach, seeding, transplant or sod, the timing of the conversion and the weather conditions experienced at that time. Estimates of the amount of water required for sprigging establishment may be in the vicinity of 3–6 ML per ha over several months. Water requirement details for the establishment of warm season grasses are outlined in Chapter 5.

Management strategies to achieve best performance from warm season grass conversions are outlined in Holborn (2009). The impact of temperature on grass performance is evident. The warm season grasses are well suited to temperatures in the 25 to 35°C range, whereas cool season grasses are suited to 15 to 25°C. Cool winters result in the warm season grasses

Table 10.1. Water available for selected soil volumes

Soil/root system diameter (mm)	Soil depth (mm)	Water available (L)
200	200	1.2
300	300	4.2
500	300	11.8
1000	500	78
2000	500	314
3000	500	707

Note: Assume available water is 20% of soil volume

becoming dormant and losing colour and robustness. Over-sowing with cool season grasses is one approach to retain winter performance. This requires additional water to support the dual grass strategy.

Most urban trees, even low water use trees, require supplementary water over the first year or so after planting in order to establish themselves. Installation of temporary irrigation systems, such as drip irrigation, assists in achieving successful establishment.

Details relating to tree establishment, including volumes of water required, are included in articles by Dr P. May and G. Connellan on the Metropolitan Trees Pty Ltd website: http://www.metrotrees.com.au.

Access to adequate soil moisture is the single most important factor that will determine the successful establishment of many urban trees. The key strategy in tree establishment is to encourage the development of the root system to ensure maximum utilisation of the surrounding soil volume.

The amount of water that needs to be applied through irrigation depends on the size of the deficit between the water available in the soil through rainfall and the total water used by the tree. When a new tree is planted, the amount of water available from the soil/media is relatively small and hence frequent applications of water is required to ensure the root system, which may have been damaged, has access to water on a continual basis.

Table 10.1 shows that only very limited amounts of water are available from soil storage with newly planted trees (e.g. a soil root system 300 mm diameter, 300 mm deep, available water is 4.2 L).

It is therefore very important that water is supplied regularly to newly planted trees to ensure survival and maintain growth, and that the development and expansion of the root system is encouraged through providing a moist soil environment in which roots can explore.

Landscape design and irrigated area

Within many urban landscapes, there are opportunities to reduce the need for supplementary watering.

Large irrigated sites, such as parkland and golf courses, should consider the proportion of coverage of irrigated turf appropriate to the site, the environmental requirements and the available water resources. Limiting the total irrigated area to the key areas of play on a golf course is an important strategy to achieve high water use efficiency.

Plant management and maintenance

Mowing height

The manner in which plants are managed influences the water demand. Frequent, close mowing of grass results in a higher demand for water than higher, less frequent mowing. Lower mowing heights also tend to encourage shallower root systems that are less drought tolerant.

Figure 10.1. Perforation and penetration of turf surface aids in maintaining effective infiltration of rainfall and irrigation

Aeration and dethatching

Regular aeration and dethatching of grass is usually required for turf areas. Aeration assists with water penetration and dethatching minimises the amount of water absorbed by the thatch and subsequently lost through evaporation. Both of these assist in maintaining effective penetration and infiltration of water.

In a project, titled 'SurePlay', carried out by Queensland Department of Primary Industries and Fisheries Redlands Research Station and Queensland AFL, the benefits of using verti-draining to maintain soil infiltration rates at adequate levels during the summer months illustrated how capturing this rainfall could significantly reduce the need for potable water (Figure 10.1) (see website: www2.dpi.qld.gov.au/horticulture/16834.html) (Bransgrove and Henderson 2005).

Root zone depth

Maximising the potential water storage in the soil should be a key water management strategy. This is particularly important for shallow-rooted plants, such as many cool season grasses. Encouraging deeper root systems is strongly recommended. Deep infrequent irrigations are advised, rather than shallow frequent applications (e.g. an application of 20 mm, rather than 5 mm). The deep watering ensures water reaches the lower parts of the root system and also provides the increased opportunity to capture rainfall. Allowing the root zone to dry out encourages root development in the lower part of the root system and the resilience of the plant is enhanced.

The principles of soil and irrigation management to achieve deeper root systems have been promoted for some time. McIntyre and Jakobsen (1992) detail irrigation management strategies to encourage deep rooting in turf.

Reducing supplementary water through rainfall optimisation

The utilisation of rainfall to minimise the need for supplementary watering should be considered early in a site development planning process. The principles of water sensitive urban design (WSUD) can be used to develop the site so that surface runoff water can be retained on the site and landscapes designed so that plant material is compatible with the natural soil moisture regimes. Stormwater runoff from adjacent surfaces can be diverted to the areas requiring supplementary water. This will reduce the need for supplementary water for the site. The diversion of roadway stormwater to rain garden tree pits is an example of this approach.

Mulch

The application of mulch is a very effective water conservation strategy. Water loss from the soil by evaporation is minimised or eliminated, and the germination and growth of weeds, which also waste water, can be greatly restricted. The term 'mulch' can include many different materials, such as plastics, paper, wool, stone, rocks, crushed rock, screenings and wood chips. The main types of mulch in use in public open space areas are organic mulches, both fine and coarse.

The potential benefits of mulch are:

- water saving through reduced surface water evaporation
- improved soil condition (break down of organic mulches)
- preventing compaction around key plantings (e.g. trees)
- modifies soil temperature
- enhanced aesthetic qualities.

Mulches have a good reputation as a water-saving technique and rightly so. However, if used in conjunction with inappropriate irrigation systems, mulch will not provide the savings expected. Some organic mulches (fine particle mulches) have the capacity to intercept and hold considerable amounts of water and do not readily allow water to pass through. They act as a barrier. Both light rainfall and irrigation water applied as small droplets from sprays and microsprays can be prevented from entering the underlying soil by a layer of fine mulch. The behaviour and properties of organic mulches, with regard to water penetration and water retention, are strongly influenced by particle size and mulch depth (Handreck 2008).

Recommendations and observations from this work include:

- Mulches should have more than 80% of their particles larger than 5 mm.
- A 50 mm thick layer of mulch is usually adequate, rather than the general recommendation of 70 mm to 100 mm.
- Fine organic mulch, if used, should be applied at shallow depth, say, 50 mm.
- Mulches allow maximum transmission of water if they are water repellent.
- Water savings of mulches are in the range of 10% to 20%, and not 70%, as often quoted.

Although some irrigation application equipment can be placed either above the mulch or below it, installing below is more efficient, if it is practicable. Drip irrigation systems can be covered by the mulch and work very efficiently. The irrigation system is then out of the way and the mulch maintains moist soil conditions, without water lost by evaporation from the soil surface.

If irrigation equipment is used above the mulch, then spray outlets with very high precipitation rates should be selected. High precipitation rate sprays (greater than 20 mm depth per hour) or bubblers, which produce a localised wetting by using high flow rates, are recommended. Drippers can be used above mulch layers, but it is important to select drippers with relatively high flow rates, (e.g. 4 or 8 L/h) so that there is enough flow to encourage water to drain through the mulch.

Organic mulch presents a uniform and aesthetically attractive surface. Being an organic material, it will break down in time and be incorporated into the upper layer of soil. Organic mulches generally improve soil properties, such as water-holding capacity, and encourage microbiological and worm activity.

Improving soil water properties

Water-holding capacity

It is advantageous to be able to maximise water storage in the soil profile. If the soil has only small amount of soil water available, in relation to the amount regularly extracted, then frequent replenishment (irrigation) is required. The water-holding properties of soils, such as sands, that have low water-holding capacity can be improved with the addition of organic matter. In the case of sandy loam soils, the addition of organic matter can increase the AWHC from around 150 mm/m to 200 mm/m. These improvements are, however, very site specific.

There are a number of synthetic materials, such as water crystals (e.g. cross-linked acrylamide) and gels, which are marketed as improving soil water-holding capacity. They are generally not used on large-scale irrigated areas. Many of these are chemically active or reactive in the soil environment, and so have a limited life. Soil temperature and soil salinity both affect the water availability properties of these substances. Repeated cycles of wetting and drying reduce the amount of water held by the crystals. According to Handreck (2008), they are not an effective technique in modifying soil to reduce plant water demand in the long term.

Infiltration rate – cultivation practices

A low soil infiltration rate, (e.g. <6 to 8 mm/h) can present significant restrictions on the design of sprinkler and spray irrigation systems. The main losses associated with low infiltration rate soils are runoff losses due to precipitation rates being higher than soil infiltration rates. The precipitation may be rainfall or irrigation. In addition to water lost from the intended targeted area, the ponding of water on the surface can cause problems. There may be losses due to evaporation of water from the wet and saturated ground. Also, if the wet conditions are prolonged, there is the risk of disease and damage to the soil, through the saturated conditions.

The production of thatch within turf is another significant contributor to inefficiencies. The thatch can prevent water entering the root zone. In addition to being a physical barrier to movement, the thatch can act as a temporary storage for the water, which is then lost through evaporation. Several techniques are employed to maintain soil infiltration of turf areas at acceptable levels. These include de-thatching, coring using hollow and solid tines, slitting, de-compaction and verti-draining.

Wetting agents and hydrophobic soils

Some soils exhibit water repellency (hydrophobic) properties (Figure 10.2). Various chemical treatments are available to improve wettability and infiltration properties of these soils.

The reason for the water repellency is not fully understood, but it is understood that a complex organic substance, which behaves similar to wax surfaces, forms on the surface of soil particles. The occurrence of hydrophobic conditions tends to be more common in sandy soils and also during extended dry soil periods when there is a high organic content. Sandy soils have large particles and provide large surfaces. These soils naturally have low water-holding capacity and maintaining infiltration and good water-holding capacity is important.

One popular treatment of hydrophobic soils is through application of wetting agents, also called surfactants. The surface of hydrophobic soils exhibit very low adhesion forces. This means that the cohesion forces of the water are greater and keep the water in a droplet form

Figure 10.2. Water repellent soils (hydrophobic) reduce efficiency of water use (Burnley Campus, University of Melbourne)

rather than spreading out and infiltrating into the soil micropores. Surfactants reduce the cohesion forces of water. Soil wetting agents (SWAs) are available in both solution and granular forms. SWAs can be applied through the irrigation system or applied separately. SWAs are a temporary treatment for a soil condition that is likely to continue to occur. The chemical needs to be continually applied at regular intervals. SWAs do not change the structure of the soil. SWAs are an effective means of improving infiltration and hence contributing to increased application efficiency of many large turf areas and garden beds.

Improved irrigation application efficiency

Nozzle replacement

The wearing of nozzles can significantly affect the system performance. Long hours of use and water containing abrasive (sand) particles may contribute to wear and enlargement of nozzle orifices. The consequences include higher flows, lower uniformity, less than optimum operating pressure, excessive pipeline friction and increased pumping costs. In a study conducted by David Zoldoske of the Center for Irrigation Technology, Fresno, it was reported that golf courses replacing worn nozzles improved uniformity and reduced water consumption by an average of 6% (Zoldoske 2003).

Replacement of damaged or missing nozzles, in the field as part of regular maintenance programs, with incorrect nozzle sizes leads to reduced uniformity. The landscape irrigation maintenance program should include a policy that replacement nozzles must be matched to the original design sizes, unless otherwise approved by the irrigation system designer.

Conversion kits are available for some older sprinklers that allow a new sprinkler assembly to be fitted to the original sprinkler body. This allows updated sprinkler nozzle technology to be employed in older irrigation pipe circuits.

Optimum operating conditions (hydraulic and atmospheric)

Optimise hydraulic operating conditions for outlets

Maintaining optimum hydraulic conditions in each part of the irrigation network can be challenging. Significant pressure and flow variations can occur as a result of design, equipment and maintenance issues. Undulating terrain will also affect pressure distribution. The incorporation of pressure reduction devices in sprays and the use of pressure regulation valves contribute to efficiency by ensuring that the design duty is achieved in all parts of the system.

Sprinklers that incorporate automatic shut down if nozzles are damaged, interfered with or removed are now available.

Regular testing of the system hydraulics, flow and pressure should be part of irrigation maintenance. The use of a pressure gauge to check an irrigation system is a very valuable tool. Control systems that include system monitoring, such as flow and pressure, are a valuable aid to site water management.

Valve-in-head sprinklers

Control over the operation of each individual sprinkler allows each part of a large landscape area to be separately watered. It provides a high degree of zoning of water application. This feature is incorporated into reasonably large sprinklers, typically those with flow rates in excess of 60 L/min (Figure 10.3).

Although these systems are more expensive than the traditional block design, where groups of sprinklers are operated simultaneously, they do provide greater opportunity for both efficiency and achievement of landscape performance appropriate to each area. They are commonly used on premium sports facilities and golf courses. Valve-in-head sprinklers have several key benefits. These include:

1. The application can be matched to the specific needs of each part of the turf area or landscape.

Figure 10.3. Incorporation of independently controlled valve in sprinkler provides greater precision in application and flexibility (Source: Rain Bird Australia Pty Ltd)

2. Virtually any combination of heads can be operated to optimise the application under varying atmospheric (e.g. wind) conditions.
3. The operation of individual heads allows hydraulic management (e.g. flow) to be optimised.

A typical application of a valve-in-head sprinkler is shown in Figure 10.3.

The adoption of irrigation designs using part-circle sprinklers that prevent overspray into rough areas of golf courses is another example of advances in water use efficiency. This approach provides controlled watering of the turf and eliminates supplementary watering of the species (e.g. native grasses) in the adjoining non-turf rough areas. An example of this approach is at the Settlers Run Golf Course, Victoria, where reported savings are in the order of 100 000 to 200 000 L per irrigation cycle are estimated through the prevention of overspray from fairways. The elimination of supplementary watering also minimises the ingress of fairway creeping grasses into the roughs and native grass areas.

Zone control – dual sprinkler design

Another example is the use of dual sets of sprinklers around greens of golf courses. One group waters the green and the other group waters the surrounds, which can have very different terrain and microclimate conditions.

Maintenance and minimisation of waste

Maintenance programs

Irrigation systems require regular maintenance. These systems are made up of many wearing and vulnerable parts. Pop-up irrigation systems are a particular issue. Sprinkler heads may not lift to the required operating position or they may become stuck in the high position, and be subsequently damaged by mowers and machinery. The correct functioning of valves also needs to be constantly monitored to ensure that flow and pressure is correct.

Flow and pressure sensors, strategically positioned throughout the system, allow the functioning of the system to be monitored (e.g. using data loggers) and action taken to alleviate problems. Monitoring of the system also provides valuable information (e.g. water volumes) that can be used to evaluate the performance of the irrigation.

Each organisation should have a formal maintenance program in place that prompts scheduled maintenance activities and records replacement items so that the overall performance of the system can be monitored and reviewed. The frequency of checks should be determined in consultation with equipment suppliers and industry best practice. A guide to frequency of maintenance tasks is presented in Table 10.2.

A valuable aid in conducting maintenance of irrigation systems is to have a checklist of the items or aspects of the site that need to be observed. It is important that a record of all works is completed and continuously updated.

Wastage – low head drainage

Water can be wasted following shut down of a sprinkler or spray line because water will drain to the lowest part of the pipe system. Incorporation of low head shut down valves in sprinkler and spray heads eliminates this source of wastage. Prevention of waterlogged areas around sprinkler heads protects the quality of the turf surface.

Irrigation applicator/outlet matched to the site

High uniformity, low precipitation rate rotating stream sprays

A significant development in irrigation technology has been the availability of a spray that incorporates a moving multi-stream as the method for the distribution of water. This

Table 10.2. Irrigation system maintenance task check list

Nature of task	Frequency of task
Controller settings	Weekly
Pump	Monthly
Water treatment	Weekly
Filter	Weekly
Fertigation	Weekly
Master valves	Annually
Remote control valves	Monthly
Sprinklers	Biweekly
Emitters/drippers	Weekly
Pressure regulation devices	Annually
Flushing valves and laterals	Monthly
Weather station and sensors	Bimonthly
Records of checks and repairs	Recorded in daily diary. Monthly report prepared

technique of delivering water is more effective than fixed nozzle sprays. These rotating stream sprays are suited to small areas (Plate 10.1).

The distribution of water in a series of moving streams is fundamentally different to water application with fixed sprays. Water applied in a moving stream results in less water applied and a more stable radial distribution of water. Also, the delivery of water in a stream is able to achieve greater distance of coverage for the same flow rate.

The performance required for small grass areas is a sprinkler or spray that will:

- effectively cover short distances
- apply the water uniformly
- apply the water at a rate that is less than the infiltration rate of the soil.

An important system design consequence arising from the use of rotating stream type sprays is that the required flow rate is lower than fixed spray type sprays. There is a lower supply flow rate required to service an area. Typically the flow rate is 50 % lower than for fixed spray systems. This means either more area (more spray heads) can be simultaneously watered or multiple areas or zones can be watered from the same supply at the same time. This provides greater flexibility in the layout and operation of the irrigation system. There is, however, a longer time required to apply a given irrigation depth.

Fixed sprays are high application rate devices. The PR rate is often in the range of 30 mm/h to 50 mm/h. For many soils this results in runoff, unless multiple cycling of irrigation programs is adopted.

The rotating stream type spray is more complex device and considerably more expensive. It does, however, have significant performance (DU and PR) advantages and provides the opportunity for high efficiency, in watering of small and challenging areas.

The radial distribution profile for the fixed nozzle spray and rotating stream spray is presented in Figure 10.4. The sharp peak of high application of the fixed nozzle spray means that is difficult to achieve an overall even application. It also means that in that part of the coverage, the actual precipitation rate is very high.

Table 10.3. Comparison in performance of rotating stream and fixed nozzle (fan type) spray

	Fixed fan spray type		Rotating stream type spray	
Test conditions	Precipitation rate (mm/h)	Uniformity (DU %)	Precipitation rate (mm/h)	Uniformity (DU %)
200 kPa Calm, <5 km/h wind	20.8	49.0	12.0	70.3

Source: Connellan and Denman (2009)

The differences in performance, DU 70.3% compared with DU 49.0%, of rotating sprays compared with fixed sprays, both in terms of precipitation rate and uniformity of application, are presented in Table 10.3.

In a trial conducted in California, the benefits of changing from fixed spray nozzle to multi-stream, multi-trajectory rotating sprinklers were demonstrated (Solomon *et al.* 2007). The average uniformity for the fixed sprays was DU 44% prior to the conversion and increased to DU 70% after the conversion. This represents an improvement from a system rated as poor to a system rated as very good. Due to the lower precipitation rates, it was necessary to operate the sprinklers 1.7 to 2.3 times longer to deliver the same depth of application.

Adjustable trajectory nozzles

The ability to adjust the vertical alignment (e.g. from 7° to 30°) of the sprinkler nozzle has application in situations where trees and vegetation obstruct the sprinkler stream. Windy conditions and sloping ground are also situations that can potentially be accommodated through on-site adjustment of the nozzle.

Sprinkler head design for high efficiency

The following is a list of the features that may be incorporated into sprinkler heads (rotors) and sprays to assist in achieving precise and effective application:

- stream break up with minimum fine drops that evaporate or are wind sensitive
- pressure compensation for constant flow at varying pressure

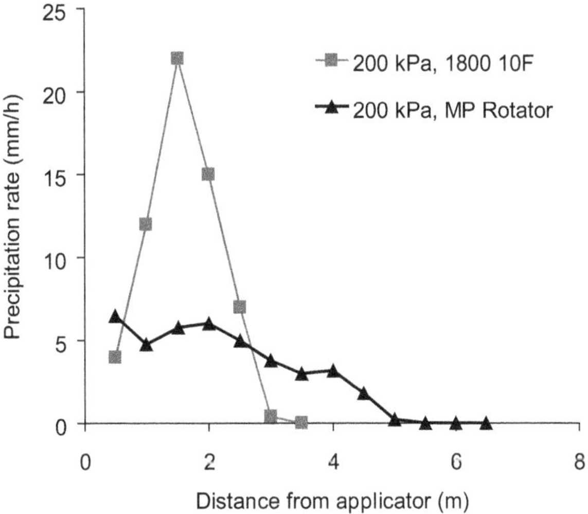

Figure 10.4. Radial distribution of fixed nozzle (■) and rotating stream sprays (▲) (200 kPa)

- pressure regulation
- anti-drain/check valve
- flushing of supply lines and laterals
- effective and reliable seals
- strong and reliable retraction
- readily serviced and adjusted from the top.

Improved irrigation water management efficiency

Evapotranspiration (ET) scheduling

The rates at which plants use water (the evapotranspiration rate) vary according to the weather conditions. It is therefore important to time irrigation to match the actual water demand. The ET value can be obtained from several sources including an on-site weather station and Bureau of Meteorology data.

In recent years, controllers have been developed to incorporate ET data as an input reading. The readings are then processed within the controller and used to make scheduling decisions. These controllers are referred to as ET Controllers. This technology is considered to be a major potential contributor to improved irrigation efficiency in the future.

A range of irrigation technologies have been identified in the USA as having the potential to aid in water conservation and increased efficiency. A program, referred to as SWAT (Smart Water Application Technology) has been initiated, by the Irrigation Association, USA, and is now being actively promoted (see the website: www.irrigation.com).

Weather station

Irrigating according to plant water demand is a key part of irrigation efficiency. The weather station is a tool that can provide a breadth of information relevant to the water management of the whole site. The collection of current weather information by a weather station and the storage of this data for later use have many applications in the management of an irrigated landscape. These include:

- aiding irrigation scheduling through ET data
- providing rainfall records – historical data and site water management including water harvesting
- identifying plant health risk conditions
- providing fire risk information
- monitoring of weather conditions (e.g. wind) that may reduce the effectiveness of application of water (irrigation) and chemicals (spraying)
- monitoring of the soil environment – temperature and relative humidity
- continuous monitoring and recording of local climate conditions to identify trends and changes that may impact on the landscape planning and future management of the site.

In terms of fire risk, the monitoring of some key weather parameters, such as air temperature, wind speed and wind direction, provides valuable information regarding atmospheric conditions and risks associated with fire around the property.

Daily ET control is virtually real-time control. It uses the latest up-to-date weather information. The next step is to base current decisions on future weather. For example, if it is expected to rain in the next couple of days, then delay irrigation. If it is going to be unusually hot and dry, then additional watering may be required.

The Bureau of Meteorology is developing a service, known as ACCESS, which will provide predictions on daily evaporation. An irrigator will receive a 7-day forecast, presented graphically, of daily rainfall, temperatures and evapotranspiration.

Soil moisture sensors

Value of SMS

Soil moisture sensor (SMS) technology has the potential to provide turf and landscape managers with up-to-date accurate readings of the actual moisture level within the soil profile. This knowledge is extremely valuable when making decisions about water management in urban areas and can achieve very significant increases in water use efficiency.

Urban horticulture is lagging behind production horticulture and viticulture in the use of this technology, where moisture sensors are an integral part of the irrigation management. The overall adoption rate of SMS in the urban sector has not been high. This technology, however, has the potential to provide high-quality information necessary for good water management. There are numerous soil moisture sensor products available.

Soil moisture sensing techniques

There are broadly two types of soil moisture sensors on the market: sensors that measure water content and sensors that measure water tension. The water content of the soil can be measured using the soil density, electrical and thermal properties of the soil.

The types of *water content* sensors that are potentially suited to urban landscapes include:

- capacitance
- electrical resistance
- dielectric constant
- thermal dissipation
- neutron particle scattering and absorption.

The types of *water tension* sensors that are potentially suited to urban applications include:

- tensiometers
- gypsum blocks
- granular matrix (gypsum)
- ceramic block.

As a group, water tension sensors are sometimes referred to as porous type sensors. Tension type sensors are well suited to urban turf and landscape situations because the reading is directly related to the amount of effort that the plant is exerting on the water in the soil (Figure 10.5). A limitation of water content type sensors is that they require calibration for each particular soil type (Figure 10.6). Tension type devices do not require calibration, but knowledge of the likely response of the plant to the tension values is required.

The movement of water through the soil can be monitored using a mechanically operated device called FullStop (Stirzaker and Hutchinson 2005). The detector, referred to as a wetting front detector, is positioned at the bottom of the root zone and visually shows when water reaches this point with a pop-up indicator. This system is very simple in operation and also allows samples to be taken of the water reaching the detection point for analysis and monitoring.

A comprehensive description of soil moisture sensors, including the methodology used in measurement, calibration, data handling, maintenance, potential limitations and positive points, is presented in the publication *Soil Water Monitoring* (Charlesworth 2000).

Requirements for urban soil moisture sensors

Soil moisture sensors to be used in urban areas need to meet a range of specific requirements (Figure 10.7). These include:

- provide accurate and rapid measurement of soil moisture in a wide range of soil types (e.g. sands, loams and clays)
- work effectively in relatively shallow soil depths (e.g. 50–200 mm)

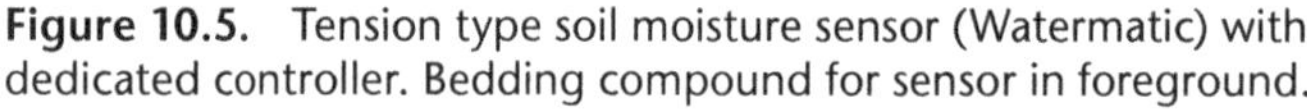

Figure 10.5. Tension type soil moisture sensor (Watermatic) with dedicated controller. Bedding compound for sensor in foreground.

Figure 10.6. Portable dielectric probe type sensor (Theta). Readings are recorded on a hand-held data storage device.

- have calibration that is not adversely affected by soil nutrient levels, soil salinity and temperature
- have an output signal capable of effectively controlling irrigation
- have electrical stability, so that lightning and power surges do not cause failure
- be robust, reliable and have a reasonable life (e.g. >5 years).

For open space managers, particularly of turf, the sensor needs to provide representative and accurate readings of the soil moisture level (tension or water volume), be reliable and be suited to turf management practices at the site. These requirements are quite demanding because the irrigation and soil conditions can vary considerably and the root zone is relatively

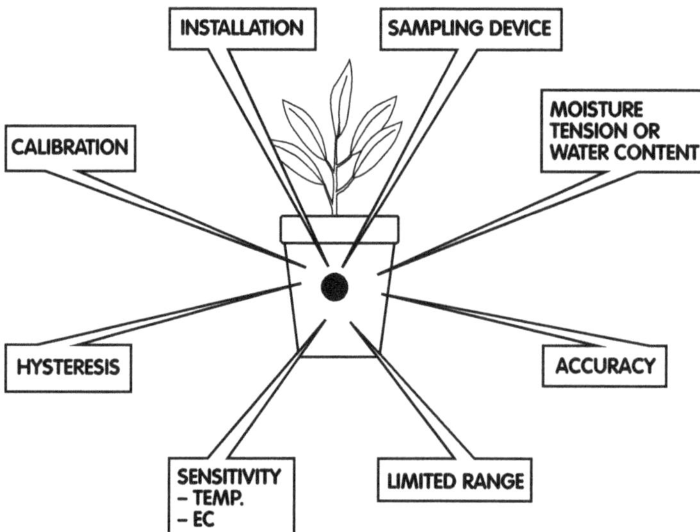

Figure 10.7. Generic characteristics of soil moisture sensors

shallow, so the sensing device needs to be in close proximity to the surface (which results in a risk of damage from ground machinery/equipment).

Water saving potential of soil moisture sensors

Soil moisture sensors can save significant amounts of water. In a turf field trial in Melbourne, conducted on sports ovals in the Camberwell area a tension type sensor (Watermatic) was found to provide water savings in the range of 40–50% (Connellan *et al.* 1992).

In a trial conducted by the University of Western Australia (Pathan *et al.* 2003), soil moisture sensors were found to save 25 % in the irrigation of turf plots. These savings were achieved when compared with the water applied to turf irrigated according to industry recommendations. The soil moisture sensor represented savings of 100 L/m^2 for the summer of 2002–03.

Soil moisture sensors provide greater savings than rain switches. A soil moisture sensor, if working correctly, takes into account the net contribution of any rainfall or irrigation in adding moisture to the soil and also monitors the effect of the plants extracting moisture from the soil. The periods of cancelled irrigations will be longer because the soil moisture sensor responds to both the climate conditions and water use characteristics of the plants.

Information available from SMS

The actual data that are available from SMS depends on the type of sensor, location and number of sensors installed, but may include:

- the soil moisture level at specific depths and locations
- identifying active root zones – through comparison of water extraction rates, as reflected by a decrease in soil moisture in different layers in the root system, it is possible to gain an appreciation of the absolute rate of extraction for a layer and to identify which parts of the root system are active relative to other layers
- the soil water reservoir for the whole root zone – the extent of water distribution and total amount of water available in the root zone
- the effectiveness of irrigation – monitoring of soil moisture, particularly in the surface soil layers, readily shows the entry of water into the soil and down the soil profile. Through measurement of the volume of water delivered to the site a water balance can be carried out to evaluate the efficiency of application. In some cases, hydrophobic soil conditions may be identified.
- the soil conditions including EC and temperature.

Effectiveness of rainfall

Monitoring surface soil layers shows how much rainfall is beneficial. The contribution or otherwise of small rainfall events (e.g. 2–4 mm) can be evaluated and also the effect of high-intensity events, where significant losses may occur.

Drainage characteristics of the soil

Understanding the movement of water down the soil profile, by monitoring changes in soil moisture levels in particular soil layers over time, is valuable in appreciating the water behaviour of the soil. Low drainage rate soils and potential drainage losses below the plant root zone can be detected. The recommended position of soil moisture sensors in the soil is shown in Figure 6.13.

Royal Botanic Gardens (RBG) Melbourne complex landscape SMS trial

The availability of up-to-date, detailed knowledge of the extraction of soil moisture from the different soil layers in garden beds is extremely valuable in the water management of complex landscapes. In a collaborative project between the Royal Botanic Gardens Melbourne, Sentek

Figure 10.8. A capacitance type soil moisture sensor probe transmitting readings in real time to host website (Royal Botanic Gardens Melbourne)

Pty Ltd and the University of Melbourne, multiple soil moisture installations have been used to monitor the soil moisture of selected garden beds. This information is continuously relayed at frequent intervals to a host website from where it can be viewed and analysed by the project partners.

The project, entitled '*Soil Moisture and Irrigation Methods in a Complex Landscape*' began in May 2007 with the installation of multi sensor (five sensing depths per probe) capacitance type probes (EnviroSCAN), supplied by Sentek Pty Ltd. Four garden beds are being monitored (see Royal Botanic Gardens Melbourne website: www.rbg.vic.gov.au/horticulture/environmental-management/water-conservation) (Figure 10.8).

The trial has produced some key outcomes, including gaining an understanding of the moisture extraction pattern of deep-rooted plants, such as trees, and the use of this in the scheduling of the irrigation so that water application is matched to the needs of the plants. The availability of soil moisture data through the internet allows the actual soil moisture conditions to be monitored in real time and direct reference made to the condition of the plants. In periods of high temperatures and high evaporative demand, this information allows informed water management decisions to be made. The soil moisture data generated allows key indicators to be used to aid in the water management of the landscape. The evapotranspiration stress index (ETSI) has been developed as part of the project and is continuing to be refined. The ETSI is a measure of the effort being exerted by the plant, driven by evaporative demand, to extract soil water relative to the water available in the soil. Several other significant benefits

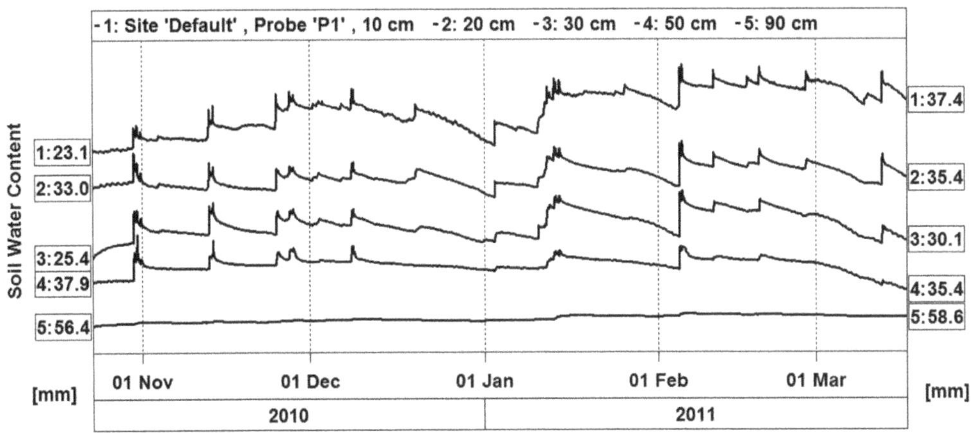

Figure 10.9. Multi sensor probe graphical display of readings in garden bed, Royal Botanical Gardens Melbourne (Sentek Pty Ltd)

have been achieved through the project including the tools to assess the effectiveness of irrigation and rainfall. Also, in some situations, the soil moisture data can be used to indicate hydrophobic conditions.

The printout of soil moisture readings recorded continuously at five depths provides a comprehensive insight into water behaviour within the soil (Figure 10.9). The daily extraction at the various depths by vegetation can be observed. In addition, it also shows how the water from irrigation and rainfall is used. The progressive extraction from deeper root zones is also evident.

In terms of making scheduling decisions, the balance of water remaining in the soil volume can be monitored and irrigation initiated to avoid significant plant stress (Figure 10.10).

Figure 10.10. Soil moisture sensor (Enviroscan) transmitting readings in real time to host website (Royal Botanic Gardens Melbourne)

Environmental sensors

Rain shut off devices

A common source of wastage in irrigated urban areas is the operation of sprinklers and sprays, following or during rainfall. Technology, such as rain shutoff devices, is a simple and effective way of reducing this potential wastage.

The water-saving potential of rain sensors is influenced by the total rainfall, the pattern of rainfall, the scheduling program of the irrigation controller and the operating characteristics of the particular sensor in use. In relatively dry periods, the savings achieved with rain sensors will be small.

In a Smart Water Fund project carried out at the Burnley Campus, University of Melbourne, rain sensors were evaluated (Figure 10.11). Based on analysis of 15 years of rainfall records for Melbourne, rain sensors were estimated to provide savings in the range of 8–10 %. In a wet year, it would be expected to be higher (e.g. ~12 %) and in a dry year lower (~ 4 %). In this trial, significant variations in the switching on and off of the sensor were recorded. Both accuracy and reliability of sensors are critical to achieving efficiency of water use.

Salinity, pH and temperature sensors for water and soil

The monitoring of both the water source and the soil environment is part of best practice water management. The monitoring of the water quality is essential when alternative water sources are used. This information is required both for the management of the water storage and the potential impact of the irrigation water on the site. The measurement of salinity, pH, temperature and nutrient levels of water is available. Potential risks can be identified, including the likelihood of algae occurring.

The monitoring of soil conditions is part of the ongoing management of an irrigated site to ensure that the soil is healthy and the environment is protected. Soil salinity (EC measurement) is a risk in terms of soil health and plant toxicity.

Figure 10.11. Selection of rain sensors used in trial at Burnley Campus, University of Melbourne, to assess switching performance during rainfall events

Soil temperature is valuable in monitoring a range of plant responses, including germination and the threshold growth and dormancy temperatures for warm season and cool season grasses. The timing of fertiliser applications can also be assisted with soil temperature data.

Wind sensors

Wind speed needs to be monitored to determine, if conditions are acceptable, for the spray application of water. When recycled water is being applied with sprinklers, there is usually a requirement that the system be shut down when a threshold wind speed is recorded, such as 10 kilometres per hour (kph).

Irrigation controllers

Controllers – from basic to 'smart'

The automatic operation of an irrigation system to pre-set times and programs is a major advance on any manually operated controlled system. Reliability and precision are two of the key advantages. The development of the 'smart' controller is now extending the control capability to incorporate all relevant information in the irrigation decision-making process. This allows high efficiency to be achieved.

The term 'smart' controller applies to controllers that incorporate weather-related information or soil moisture sensor measurements. This type of controller ranges from a controller with provision for a single sensor input through to a controller that accesses evaporation data in real time and allows multiple sensors as inputs into the irrigation decision-making process.

Controller developments

The modern irrigation controller allows the integration of all of the factors and systems that may have an impact on irrigation decision making. Features offered by manufacturers include:

- ET calculation
- rainfall measurement (totals and instantaneous rates)
- pumping system optimisation
- hydraulic management including flow control optimisation
- computer screen with interactive map graphics incorporating GPS-based mapping data
- on screen system representation of equipment operation and interactivity function
- individual valve and head status report
- two wire control of multiple (potentially hundreds of) solenoids using decoders
- compatibility with latest software and operating platform.

An irrigation controller is a data management system. It has the capacity to acquire or collect data; it can display data, process data and carry out supervisory control functions to meet the requirements of the program. The capacity of the controller to manage information is becoming increasingly important, not only from the aspect of improving the quality of the decision making but also in terms of reporting on the use of water.

Precise and reliable application of water

To achieve high efficiency, each irrigation zone needs to be watered according to the water requirements of the vegetation and the site at that time. The actual irrigation station run time, the number of activations per complete irrigation cycle and the frequency of watering all need to be customised for each hydrozone. The desired run time needs to be precisely programmed, from 1 minute to hours, to accommodate both sprinkler and microirrigation systems. Global adjustments of programs are required to allow proportional reduction in run times during periods of reduced water demand.

The activation of each station needs to allow heavy soils and sloping ground, which have low infiltration rates, to be watered through repetition or cycling. The controller also needs provision for input from a range of environmental sensors so that water is not wasted following rainfall.

A controller needs to be able to withstand electrical supply irregularities and be able to function effectively following power failure through the incorporation of a non-volatile memory to save the operating program. Many irrigation controllers on the market today have all of these water efficiency features and additional capabilities including flow management and wireless communication.

Central control

Features

A single control system that operates multiple sites from a single location has advantages not only in terms of labour saving but also in the sophistication that can be applied to the decision-making process. The modern irrigation controller is required to not only operate pumps and valves, but also it is an information management system. The control system needs to be able to integrate information from a range of sources and execute multiple programs to achieve high efficiency of water use.

The acceptance of central control by organisations that manage multiple irrigation systems is evidence of irrigation technology delivering improvements in resource use efficiency and irrigation efficiency.

Some of the key functional developments in recent years include:

- increased flexibility in programming of individual stations
- cycle and soak to enable the plant the best opportunity to take up the water
- pump station monitoring and the ability to react to pump or station failure
- switching programs for multiple water supplies, where, for example, slightly more saline water, or less desirable water, can be used to irrigate fairways and the better quality reserved for the greens
- chemical dosing, such as applying fertilisers or wetting agents and for balancing pH levels
- for multiple site systems, global programming for operations such as quick shut down that saves not only water resources but also energy and manpower resources.

Water savings in the range of 20–40% have been reported following implementation of central control. The ability to command many remote sites and execute individual programs without the need for personal attendance is a major advance.

Controller and sensor communications

Irrigation control involves the transfer of large amounts of data using multiple pathways, both in receival and transmission modes. Ensuring that the output readings of sensors and weather stations are compatible with the irrigation controller is essential. The formatting of the readings so that they are readily understood by the user is also important. The selection of the most appropriate communication system needs to consider the following aspects of data transfer and the specific requirements of the site. These include:

- data format
- frequency and volume of data transmission
- power availability at weather station site (mains or solar)
- robustness of communication system (e.g. lightning risks)
- capital cost of system
- operating cost (significant with some telephone systems)
- security of data transmission.

The transfer of data to the host processing unit (central computer or controller) can be achieved in a number of ways, including via:

- a hard wired connection e.g. cable, fibre optic
- a telephone landline with modem access
- the internet via mobile phone networks GPRS e.g. 3G, 4G
- satellite
- radio transmission e.g. UHF.

Wireless technology

The expansion of wireless communications is providing opportunities for much-expanded monitoring and control of irrigation and water management systems. Environmental sensors, including soil moisture, temperature and salinity (EC), can now be positioned anywhere on a site and the data transferred back to a central receiving point in real time. The flexibility provided by this technology means that not only can more remote locations and conditions be measured but also there is much greater flexibility in the sensing of the system operation and site conditions.

Hand-held and mobile remote control devices

The operation of irrigation controller functions using a hand-held remote has numerous advantages for the irrigation manager. A primary benefit is the time saved in travelling to and from the controller position. The ability to turn on and off sprinklers, for brief periods of time, allows heads to be located and operation checked for correct functioning. They are a valuable aid in the conduction of irrigation audits. The programming and communication capability of hand-held devices, such as iPhones, iPads and androids, and irrigation remotes, is greatly facilitating remote access and management of irrigation systems.

Monitoring, alarms and reporting

The capacity to monitor and record the system operation and environmental conditions is becoming increasingly important. The capability of comprehensive monitoring system flow rates so that pipeline breaks, missing sprinklers and faulty valves can be detected and shut down minimises water wastage. The automatic monitoring and checking of electrical systems assists in fault identification and reliable operation of the irrigation system.

Alarms that indicate malfunction assist with both equipment safety, prevention of waste of water and fault finding. The correct functioning of sprinklers, sprays and valves also needs to be constantly monitored to ensure that flow and pressure is correct. Flow and pressure sensors, strategically positioned throughout the system, allow the irrigation to be monitored and action taken to alleviate problems.

Monitoring of the irrigation system hydraulics also provides valuable information (e.g. water volumes) that can be used to evaluate the performance of the irrigation. Monitoring and measurement of water use is a key part of all good irrigation management.

Monitoring of the irrigation system components, such as solenoid valve current draw, can provide valuable information to assist in planned maintenance programs.

Smart metering

Monitoring the flow through an irrigation system in real time provides the operator and manager with data that greatly increases the capacity to achieve high efficiency. This is one feature of a smart meter (Figure 10.12). Ready access to an internet site that hosts data from a smart meter provides additional advantages in terms of water use information management. Analysis and reporting of water use including peak flows, average flows, time of event and

Figure 10.12. A smart meter provides real-time continuous monitoring of system flow rates and volume totals

accumulated volumes, can be readily carried out, with the sharing and remote access of the information adding to the overall quality of the management process.

The selection of the appropriate meter type and size is critical to achieving sound water management. The range of flows over which a meter is accurate is limited. An accuracy of ± 2% should be achieved. In situations where a range of flows is expected careful attention needs to be paid to the performance of the meter under consideration.

Pump stations

The provision of an optimum water supply from a pumping plant involves consideration of not only hydraulic efficiencies but also energy efficiencies. Due to the long hours of operation and the total amount of energy used, any inefficiency in the system has a major impact.

The availability of pump sets, which allow high hydraulic efficiency to be achieved for a range of irrigation duties, is important. The flow and pressure requirements of irrigation systems are becoming more complex, because of the increased range of watering regimes. The availability of variable frequency drives (VFD) and the use of multiple pump units means that pump performance can be more closely matched to demand and high energy efficiencies achieved.

Pump station output can be accurately matched to the hydraulic demand (often, highly variable) in the field, thereby saving energy and wear and tear on pumps. Most significantly, the impact on irrigation systems when starting and stopping can be softened, because of the ramp up and ramp down capability.

In selecting the right pumping unit for an application, it is important that the efficiency of the electrical motor and also any associated control equipment is taken into account in determining the overall energy efficiency. Pump controls can now have advanced graphical interfaces for better understanding of the operation of the pump station executing functions such as remote monitoring, reporting and fault alarms.

Pump station communication and the internet

The remote access to pumping plants provides managers, operators and service personnel with the tools to assess the functioning of the equipment and the ability to make programming changes. Although this facility has been available through telecommunication modems, the linking of the pump unit control to the internet greatly increases accessibility, processing and interpretation of the information. The pump operation can be viewed and analysed from anywhere and the management of the information performed at a much higher level.

Water storage management

Reducing water storage losses

The amount of water that can be lost from water storage through both seepage and evaporation is potentially significant. Seepage losses can be eliminated through the use of impermeable liners (Figure 10.13). This is relatively expensive, but, because the value of water increases and the stored water becomes more important to the sustainability of the site, this option is now receiving increased attention.

Several techniques are available to reduce evaporation. These include suspended covers or netting, such as shade cloth, floating covers and also chemical barriers. The losses from evaporation can be in excess of 30% in many cases. An evaporation rate of 600 mm over the irrigation storage period represents a loss of 600 mm in depth from the storage. In a 10 ML storage with a surface area of 2500 m^2 and a depth of 4 m, this represents 1.5 ML of evaporation loss.

There have been various methods investigated to reduce evaporation losses. These include shading the storage and using various floating barriers including plastic covers and polymer coatings.

The National Centre for Agricultural Engineering (NCEA) and the CRC Irrigation Futures have investigated liquid monolayers (polymer) and other evaporation reduction techniques. (Craig *et al.* 2005). The monolayers are sprayed or pumped on to the storage and are readily dispersed over it. It was found that they reduce evaporation in the range 0–40%, compared with shade structures, which can reduce losses by 70–90%. Although monolayers are simple to apply, they may only last a few days in some cases (Craig *et al.* 2005). The stability and effectiveness of these products are continually being improved.

Water quality is also important in terms of efficiency. The presence of algae in water storages presents a number of problems not only for the storage but also for the systems in which the water is used.

The following are some treatment options to reduce risk of algae:

- Minimise vegetation, suspended solids, nitrates and phosphates – maintain the catchment in a clean condition free of weeds and plant seeds, with a stable bank (not readily erodable) and avoid contamination (e.g. fecal matter) from animals.
- Reduce phosphorus – exclude sediments that release phosphorus, especially under low oxygen conditions and select and use low phosphorus fertilisers.
- Use aerators – these have several potential benefits in terms of improving the quality of the storage: destratifying the storage and remove bottom sediments, increasing dissolved oxygen levels, enabling controlled removal of organic sediment.

Capacity to achieve high efficiency

The competent management and maintenance of an irrigation system requires a high level of skill and expertise. Organisations should continually pursue opportunities to advance the skills of staff involved in irrigation. In addition to educational institutions providing training, many irrigation companies provide a good range of training opportunities for staff.

Figure 10.13. Irrigation buffer storage with liner to eliminate ground seepage and netting to protect from bird contamination

An organisation's capacity to achieve high-quality water management depends upon:

- putting policies and objectives in place to support the efficient use of water
- obtaining a comprehensive, up-to-date and accurate database of information on all water-related issues/aspects including water consumption, site details, usage type and patterns, climate and water infrastructure
- developing systems and procedures to support implementation of water management strategies
- employing and training appropriately skilled personnel to deliver water efficiency outcomes. This includes system management and system maintenance.

The capacity of the organisation should be objectively assessed and steps taken to overcome any gaps between the performance required and the current capacity level (gap analysis). The skills required in order to be a competent irrigation manager include:

- a knowledge of plant soil relationships
- understanding the concept of irrigation scheduling
- an ability to select appropriate irrigation schedule for the site
- understanding system performance in terms of flows, pressure and precipitation rate
- an ability to evaluate the performance of the system
- knowledge of system maintenance and repairs
- an ability to read and action plans, specifications and manuals
- understanding safe operating procedures and OHS regulations
- an ability to record water and energy use and operational details.

Investment in water management training is an ongoing requirement, because, in addition to the need for core skills, the technology is continually changing and generally becoming more complex. A simple cost–benefit analysis can be carried out to evaluate the training through comparing the cost of the training and the reduction in water use or increase in efficiency.

According to Peter Symes, Curator, Horticulture, the development of staff expertise in water management has been fundamental to the achievement of improved water use efficiency at the Royal Botanic Gardens Melbourne (Symes 2000). In 1995, the Royal Botanic Gardens Melbourne arranged for Burnley Campus, University of Melbourne to facilitate water management training. After implementation of the training, there was a dramatic increase in efficiency as evidenced by a reduction in the irrigation index from around 3.5 to 1.2. This was achieved without any noticeable reduction in landscape quality.

Training in irrigation and water management is available from Irrigation Australia Ltd Website: www.irrigation.org.au. Certificate courses in urban horticulture, landscaping and irrigation are available from TAFE colleges.

All major irrigation manufacturers provide training in their specific product areas and some provide more general training in irrigation maintenance and water management.

Chapter 11

Evaluating and benchmarking irrigation system performance

Performance evaluation is essential

Evaluation requirements

The community has high expectations that water used in urban open space areas is used efficiently and in an environmentally responsible way. The performance of turf and landscape irrigation systems is generally not well understood or appreciated. It is important for the irrigation industry and the community to have performance standards that accurately describe current irrigation practices and provide a basis for encouraging the pursuit of high efficiency. If an area is to be irrigated, then it should be done efficiently.

Basically, the evaluation of the performance of irrigation systems is an essential component of good water management of urban areas. So, how do you assess the quality of the system and the quality of management of the irrigation? The four requirements for an effective and efficient irrigation system are that water should be applied:

- uniformly
- with minimum losses due to evaporation, runoff and deep drainage
- to the correct depth to meet the site vegetation needs (Figure 11.1)
- at the right time (taking into account rainfall and climatic conditions).

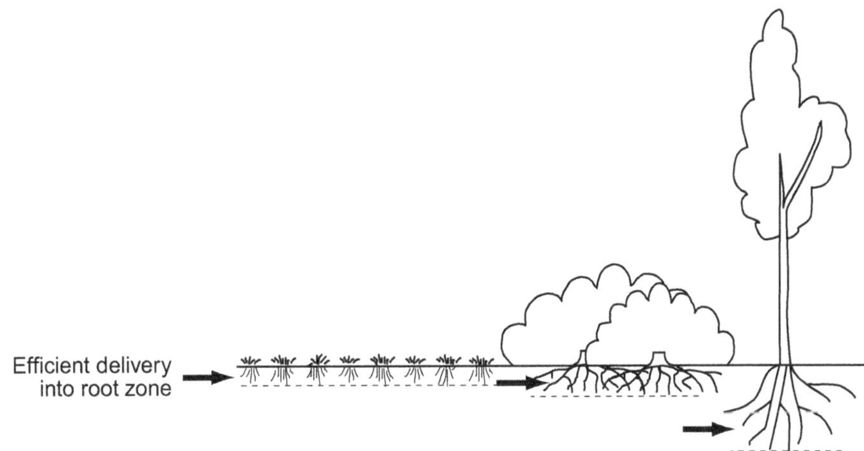

Figure 11.1. Effective delivery into target root zones is a core requirement of efficiency

The minimum information requirements in the assessment of the performance of an irrigated site are:

- the water consumption used over irrigation season
- the water budget for the site
- the operating effectiveness of irrigation components (e.g. valves, sprinklers and controller)
- the hydraulic operating conditions – pressure and flow
- the uniformity of application or uniformity of delivery
- the precipitation rate of system (PR)
- a site-specific irrigation schedule
- achievement of turf and landscape performance quality targets
- a maintenance program (documented and recorded).

Benchmarking of urban irrigation

The term 'benchmarking' is used in a variety of ways within the irrigation industry. The context here will be for the determination of quantitative measurements that can be used to achieve improved performance.

The key elements of performance benchmarking are:

- measuring performance quantitatively
- comparing with best practice in industry and identifying gaps in performance
- identifying strategies to improve performance
- fostering a culture of ongoing improvement
- demonstrating and reporting water use performance.

Performance benchmarking has been demonstrated to be a powerful tool in agriculture, production horticulture and viticulture (Skewes and Meissner 1997). Benchmarking can be used as an aid in reaching high standards of urban water management. To implement effective performance benchmarking for turf and landscape, it is essential to have performance indicators that are appropriate, clearly described and precisely defined.

Role of irrigation system evaluation

Outcomes and benefits

The evaluation process should be carried out within the framework of a water management plan (see Chapter 12). The irrigation audit will:

- show you the current operating efficiency and uniformity or evenness of the system
- identify any weaknesses in the system
- establish key system performance parameters, such as pressure, flow rate, precipitation rate and uniformity
- provide basis for developing an irrigation schedule for the site – how much water to apply and when to apply it.

The following are the potential benefits of an audit:

- potential water savings and associated cost savings
- higher quality plants – turf and landscape
- savings in time and labour
- improved scheduling based on field measurements
- better utilisation of fertilisers – as a result of reduced water losses
- improved management of a valuable resource – water.

An audit of an irrigation system is a very powerful tool in achieving high standards of irrigation efficiency. The process involves initially carrying out a visual inspection of the system and then conducting a series of tests.

Irrigation performance indicators

The assessment of the performance of an irrigation system should provide the manager with a numerical or quantitative measure that will allow the effectiveness and efficiency of irrigation to be evaluated and referenced to an industry standard or benchmark. In the case of sprinkler irrigation systems, it is important to consider both the operating effectiveness of the system at any point in time and the management of the irrigation over a longer period, such as the irrigation season. There are therefore broadly two types of performance indicators: one type evaluates the effectiveness of application of the system and the other evaluates how well the system was managed, over a designated period (month, year or irrigation season) in terms the amount of water applied compared with the amount that should have been used.

The various aspects of performance can be grouped according to those that reflect the effectiveness of the irrigation system and those that reflect the water management of the site.

Irrigation system operating performance

These indicators include the:

- uniformity of application for sprinkler and spray systems
- evenness of delivery from microirrigation emitters
- precipitation rate of sprinkler and spray systems
- programmed or scheduled irrigation depth
- flow rate and flow variation of sprinklers, sprays and emitters
- pressure (design operating value) and pressure variation
- application effectiveness
- capacity of the system to meet peak water demand.

Total water consumption performance

These indicators include the:

- water consumption efficiency (irrigation index)
- total irrigation depth (expressed as mm)
- total water consumption volume (ML, kL)
- irrigation management efficiency
- application rate (ML/ha, kL/1000 m^2)
- overall irrigation efficiency (IE)
- water savings compared with a reference period (e.g. month, past year or designated year).

Uniformity of application indicators

Coefficients of uniformity

Individual sprinklers and sprays do not apply water uniformly along the wetted radius. Overlap of sprinkler distributions is therefore required to achieve even application. Tests that provide a measure of the variation in application depths of planned and existing irrigation systems are therefore essential in assessing performance. There are numerous indicators that have been developed to measure the evenness of application of water to a surface, such as the ground.

The initial analysis and description of the performance of sprinkler system was carried out by JE Christiansen in the 1940s and described in the publication *Irrigation by Sprinkling* (Christiansen 1942). He proposed the measure of uniformity, which became known as Christiansen uniformity (CU). The indicators that are in common use with sprinkler irrigation systems are the distribution uniformity coefficient (DU), Christiansen coefficient of uniformity (CU) and the scheduling coefficient (SC). The determination of the various indicators is outlined in the following sections.

Distribution uniformity (DU)

The uniformity measure most commonly used in evaluation of turf and landscape systems is the DU, which compares the average of the lowest 25% of test can readings with the average of all readings. It is referred to as 'field DU' as it is based on measurements taken in the field. The term 'design DU' is used to describe target DU values set as part of the design process and is based on desktop sourced data. A DU of 1.0 or 100% would indicate that the application was perfectly even. In practice, this does not happen.

As an example, if the average reading of the five lowest-reading cans, out of a total of 20 cans, is 15 mm, and the average of all readings is 20 mm, then the field DU is 0.75.

The DU value can be calculated in a number of ways. The number of lowest readings used in the calculations can vary. A $DU_{25\%}$, also referred to as the DU_{LQ} , where LQ refers to the lowest quarter (25%), is the most commonly used DU term. This is the term mainly referred to in this publication. Other DUs, such as $DU_{10\%}$ and $DU_{5\%}$ are used in uniformity analyses and $DU_{50\%}$ is used in run time determinations.

The value of DU is presented both as a percentage and a decimal. Strictly speaking it should be a decimal because it is not a true efficiency term. However, due to the widespread use as a percentage, both terms will be used.

The value of DU is calculated using the following expression:

$$DU_{LQ} = \frac{V_{c25}}{V_{cav}}$$

where:

V_{cav} – average value of all catch can readings.

V_{c25} – average of lowest 25% of readings

The industry standard for DU is a minimum of 75% or 0.75.

The following is an example of the calculation of DU_{LQ}. The readings have been obtained from a test where 20 catch cans were positioned in a grid pattern between four sprinklers. The catch cans, which are available from IAL, for auditing can be read as either volume (mL) or depth (mm).

To determine the uniformity of application, it does not matter whether the reading is a volume or a depth. The uniformity value calculated will be the same. It is, however, convenient to measure in a depth, because this allows the direct conversion to net precipitation rate (PR_N).

Example DU calculation

The test can readings for DU calculation are given in Table 11.1.

Average of all readings: (920/20) = 46.0 mL

Average of lowest 25%: (190/5) = 38.0 mL

$$DU_{LQ} = \frac{38.0}{46.0}$$

$$DU_{LQ} = 0.826$$

Christiansen coefficient of uniformity (CU)

The CU is widely used in the broader irrigation industry and in the nursery industry. The CU value is calculated by determining how much variation there is for each can reading from the average reading. The total variation from the average is calculated by adding up any difference, whether it is above or below the average value. This total variation is used to determine the overall non-uniformity of the application. It takes into account both over-watering and under-watering.

Table 11.1. Test can readings for DU calculation

Can position	Volume (mL)	Lowest 25%
A1	45	
A2	53	
A3	47	
A4	38*	38
A5	41	
B1	48	
B2	58	
B3	52	
B4	40	
B5	49	
C1	40*	40
C2	55	
C3	50	
C4	38*	38
C5	40	
D1	57	
D2	47	
D3	35*	35
D4	48	
D5	39*	39
TOTAL	920	TOTAL: 190

Note: * Indicates one of the 25% lowest readings. The number of readings is 20, so number of 25% lowest readings is five.

$$Cu = \left(1.0 - \frac{\sum V_d}{V_{cav} \times n}\right) \times 100\%$$

V_{cav} – average value of all can readings
$\sum V_d$ – total of variation of each reading from the average
n – number of can readings

A CU value of 100% would represent a perfectly uniform or even application of water. Due to the mathematical method of calculation of CU, it will always have a higher value of uniformity than DU for the same set of readings. To avoid confusion, it is therefore important to specify which particular uniformity coefficient is being used. The industry standard for CU is a minimum of 84%.

The turf and landscape industry tend to use DU_{LQ} as the uniformity measure for the field evaluation of a system. In situations where both over-watering and under-watering measures are valuable, then CU is a valuable tool. CU is often used by irrigation designers as part of the process of achieving an optimum design solution.

The following example uses the same readings as in the previous DU_{LQ} calculation.

Example – Calculation of CU from can readings

The test can readings for CU calculation are given in Table 11.2.

Table 11.2. Test can readings for CU calculation

Can position	Can reading (mL)	Difference: can reading to average of 46 mm
A1	45	1
A2	53	7
A3	47	1
A4	38	8
A5	41	5
B1	48	2
B2	58	12
B3	52	6
B4	40	6
B5	49	3
C1	40	6
C2	55	9
C3	50	4
C4	38	8
C5	40	6
D1	57	11
D2	47	1
D3	35	11
D4	48	2
D5	39	7
Total	920	**Total difference: 116**

Number of readings: 20
Average of all readings: (920/20) = 46 mL
Total differences between average reading and actual reading: 116

$$\text{CU} = (1.0 - 116/920) \times 100$$

$$= (1.0 - 0.126) \times 100$$

$$= 0.874 \times 100$$

$$\text{CU} = 87.4\%$$

Note that CU value is higher than DU, even though the same numbers (readings) are being used. The differences between the two measures are generally greater with poorer or lower uniformity values.

Scheduling coefficient (SC)

Although DU and CU provide an overall measure of uniformity of application, the SC term is used to identify and quantify specific areas within the pattern where the application is low (or high) or in other words dry (or wet). It is mainly used in the design of irrigation systems to ensure that the absolute variations in water depth are not excessive. The SC is both a uniformity coefficient and can be used to provide a time adjustment factor to ensure that the dry or under-watered areas receive an adequate depth of application.

$$SC = \frac{\text{Average of all can readings}}{\text{Selected low readings (dry area)}}$$

There are several versions of SC. The Center of Irrigation, Fresno (Zoldoske *et al.* 1994) states that the SC is calculated using lowest readings for a contiguous area (1%, 2% or 5%). The SC is valuable in assessing uniformity performance, as part of irrigation design, by indicating the degree of dryness of the driest part of the application.

Analysis of sprinkler applications, based on laboratory testing of sprinkler nozzles, allows dry areas to be identified. Potential weaknesses in layouts can be avoided by selecting sprinklers, nozzles and layouts that meet SC performance standards. The design of an irrigation system that meets both DU and SC performance criteria is recommended. The particular SC used is $SC_{5\%}$ and the maximum recommended $SC_{5\%}$ is 1.2. In some situations, this value may be lower.

Access to computer analysis during the design stage allows dry areas to be identified and determinations made. Identifying 'dry areas' in the field is more difficult. However, noting the highest and lowest readings measured in the field provides a measure of overall variation. The calculation of the ratio of highest to lowest provides a visible measure of the degree of variation that exists. It is not uncommon for the ratio to be in excess of 2:1. In the example above, the ratio is 1.66 (58/35) in an application with a DU_{LQ} of 0.83. The corresponding field $SC_{5\%}$ calculated from these can readings is 1.31 (46/35).

Run time multiplier

Adjustment of run time

When water is applied with sprinklers and sprays, there is some variation in depth of water applied across the wetted area. To ensure that the low or 'dry' areas receive an adequate amount of water, the run time of the system can be increased. This also means that wetter areas receive more water. This time adjustment is described using various terms including run time multiplier (RTM), scheduling coefficient (SC) and scheduling multiplier (SM). The discussion here will cover RTM and SM because these are the terms mainly used for the scheduling of irrigation based on audit results.

A general expression to determine the value of run time adjustment is:

$$\text{Run time adjustment factor} = \frac{1}{\text{Uniformity measure}}$$

The lower the uniformity value, the higher the required run time adjustment. There are a number of ways of calculating run time adjustment values. Historically the most common term has been the following:

$$RTM\,(DLU_{LQ}) = \frac{1}{DU_{LQ}}$$

where:

$RTM\,(DU_{LQ})$ – run time multiplier using DU_{LQ}

When water is applied to the ground with sprinklers, there is some redistribution of water in the soil. The degree of non-uniformity on the surface is not as great within the soil profile. DU_{LQ} therefore tends to over-estimate the actual soil moisture non-uniformity. Mecham (2001) proposed that DU_{LH}, calculated using the lowest 50% (half) be used rather than DU_{LQ}.

The results in Table 11.3 show that the redistribution of soil moisture within the soil profile was not as variable as the application of the water to the ground by the sprinklers.

The results show that the distribution uniformity of the water in the soil is approximately 20% higher (sandy loam 75%, compared with 57%), than the measured application distribution.

Table 11.3. Comparing soil moisture in similar sprinkler zones of different soil types

Soil type	Sprinkler type	Catch can DU_{LQ}	Soil moisture at 12 cm DU_{LQ}	Soil moisture at 20 cm DU_{LQ}
Sandy loam	Rotor	57%	75%	77%
Silty clay loam	Rotor	68%	86%	87%

Source: Mecham 2001

Because it is the distribution of soil moisture that is important for turf and landscape water management, the use of an adjustment factor that more accurately reflects the soil moisture distribution in the soil profile is logical.

$$RTM_{LH} = \frac{1}{DU_{LH}}$$

RTM_{LH} : Distribution uniformity (DU) using lowest 50% of readings

Statistically the value of DU_{LH} is higher than DU_{LQ} and so RTM_{LH} is lower than RTM_{LQ}. The recommended run time adjustment time is therefore less.

More recently, another term, scheduling multiplier (SM), has been described in Irrigation Association (USA) Landscape Irrigation Auditor (Irrigation Association 2011). The following expression is used to calculate SM:

$$SM = \frac{1}{0.4 + (0.6 \times DU_{LQ})}$$

where:

SM – scheduling multiplier

DU_{LQ} – distribution uniformity using lower 25%

The SM term has the advantage of reflecting, to a degree, the redistribution of water in the soil and also limiting the increased run time required as the uniformity of the system decreases.

Comparing RTM (DU_{LQ}) and SM:

If DU_{LQ} is 0.70

$$RTM_{LQ} = \frac{1}{0.70} = 1.43$$

$$SM = \frac{1}{0.4 + (0.6 \times 0.70)} = \frac{1}{0.82} = 1.22$$

A value of RTM_{LQ} of 1.43 recommends an increased run time of 43% and for SM 22%. The amount of recommended reduced run time is very significant.

The use of DU_{LH} in the calculation of RTM has also been studied by Dukes *et al.* (2006). They found that DU_{LH} more closely represented the actual soil moisture distribution in the soil. They also pointed out that soil moisture tends to be redistributed more at increased depths. Shallow root plants, for example in the top 100 mm, are more prone to non-uniformity than deeper rooted plants.

It is important to recognise that every step should be taken to improve the performance of the system rather than adjust the run time. Also, it can become impractical to apply high run time values because they may not fit within the time window for irrigation.

Although SC and DU are both uniformity coefficients, the fundamental difference between the two is that DU uses the lowest readings as a group of numbers, whereas SC uses a contiguous area of low readings or 'dry spot' as the basis of the calculation. DU provides a more general measure of non-uniformity whereas the SC focus is on extreme values (usually low or dry) for identified sections or parts of the application test area.

The adjustment of the run time for turf is warranted, so that dry areas are brought up to an acceptable standard (Plate 11.1). In the case of garden beds, there will not be the same direct consequences if some areas receive less water than others. In many cases, the areas over-watered will encourage root development in that part of the soil volume, and the water will be used by the plant. Woody plants, for example, have a good capacity for exploiting soil water within the vicinity of the plant. The adjustment of the run time for these situations should only be made if there are significant consequences arising from the unevenness of the application. In reality, it would be recommended that the irrigation system be corrected, rather than to increase the run time to try and overcome the problem.

Key uniformity performance indicators and best practice

The following section describes the key irrigation performance indicators for uniformity and best management practice recommendations currently in use.

The industry standards for uniformity under operating conditions are: field DU 75% and CU 84%. It is recommended that SC 5% should not exceed 1.2 for turf and landscape situations. Both DU and SC need to be defined in terms of data sets there are to be used in calculation of the respective values.

The above uniformity standards refer to sprinklers and sprays operating in the field. Other uniformity standards are applied for design. Designs are based on theoretical uniformity values based on test data collected inside a test facility, where there is no wind. Field uniformities will be lower because of wind, variation in sprinkler performance and possible hydraulic variability.

Design DUs are sometimes specified as being minimum DU 85% or DU 90%. This ensures that the system has been designed with the potential for high uniformity of application.

The type of equipment also influences the uniformity that can be expected in the field. As a general rule, spray heads used to water small areas and in many cases irregular shapes do not achieve the higher level of uniformity of sprinklers covering a larger area. The distribution characteristics of sprays with fixed nozzles limit the uniformity compared with the rotating steam distribution of sprinklers. Guidelines for benchmarking uniformity values should recognise these differences.

Table 11.4 illustrates the difference in equipment. It also indicates the achievable and uniformity target ranges for sprinklers and sprays.

These values illustrate the relative low uniformity that is achieved with fixed nozzle type sprays.

Some organisations have adopted minimum DU standards for both existing and new systems. The City of Salisbury, South Australia, (Charlton 2003) includes a policy requirement that systems should be redesigned, upgraded or replaced if field DU is less than 0.75 (75%), and that new systems have a design DU greater than 0.85 (85%).

It is important for the determination of the status of system performance and communication with stakeholders that the system be able to be rated in terms of quality. Table 11.5 provides

Table 11.4. Target uniformity (DU) ranges for sprinklers (rotors) and sprays

Sprinkler/spray type	Achievable DU	Target DU
Sprinklers (Rotors)	0.75–0.85	0.65–0.75
Sprays (fixed nozzle)	0.65–0.75	0.55–0.65

Source: adapted from Irrigation Association (2011)

Note: Although the target value should be DU 0.75 for sprinklers and DU 0.65 for sprays, higher values are considered achievable.

Table 11.5. Performance rating categories for sprinklers and sprays

	Field distribution uniformity (DU_{LQ}) performance rating categories for sprinkler and spray irrigation equipment				
	Excellent	**Very good**	**Good**	**Fair**	**Poor**
Fixed spray	0.75	0.65	0.55	0.50	0.40
Sprinklers (gear and impact drive)	0.80	0.70	0.65	0.60	0.50

Source: GreenCO and Wright Engineers (2008)

a guide to the field DU values that allow a performance rating to be applied to sprinkler and spray systems. These values indicate the lower uniformity values that are achieved with fixed nozzle type sprays.

A similar quality rating system can be applied to microirrigation systems. Table 11.6 provides a guide to the field EU values that allow a performance rating to be applied to microirrigation systems.

Uniformity and efficiency

The achievement of high efficiency with sprinkler irrigation systems depends on a high degree of uniformity of application. It is not possible to achieve high efficiency with sprinklers that have poor uniformity. If a system with poor uniformity is operated for a set period of time, some areas will be under-watered. This results in low efficiency because some parts of the soil root volume do not receive the programmed amount of water. If the irrigation system is operated so that the under watered areas receive adequate water, then other areas will be over-watered and water is wasted, and again low efficiency occurs.

In the case of microirrigation systems, high efficiency depends on each plant receiving a designated amount of water. If there is a large variation in discharge in emitters, then water is wasted because some plants are under-watered and some are over-watered. Drip systems require a high degree of evenness in the amount of water delivered by each emitter in order to achieve high efficiency. Drip uniformity of delivery is measured using the term EU, which is discussed later in this chapter.

There are some urban landscape situations that are particularly challenging in terms of high efficiency. Sprays are often used in garden bed situations, where foliage is usually of varying size, height and density, and the plant architecture is highly variable. In this case, the uniformity of application (overhead) is not a core requirement for high efficiency. The ability of the spray to penetrate the foliage and achieve effective coverage throughout the garden bed will be the key factor in achieving high efficiency. Also, garden bed plants usually have the capacity to exploit uneven distribution of water in the soil.

The DU term, although not strictly an efficiency measure, is a valuable education message in improving irrigation system performance.

Table 11.6. Performance rating for microirrigation systems

	Emission uniformity (EU) performance rating categories for microirrigation systems		
Type of zone	**Excellent**	**Very good**	**Good**
Microspray	0.80	0.70	0.60
Drip – standard	0.80	0.70	0.65
Drip – pressure-compensating	0.95	0.90	0.85

Source: GreenCO and Wright Engineers (2008)

Auditing of irrigation systems

What is an irrigation audit?

Conducting a field test to evaluate the current performance of an irrigation system is the basis of an irrigation audit.

In the case of sprinklers and sprays, a number of catch cans are placed throughout the system area to record the depth and distribution of water (Plate 11.2).

There will be some initial costs for both the audit and the potential repair or upgrade of the irrigation system.

Conducting an irrigation system audit

The steps in the process can be summarised as follows:

1. Select a suitable site (e.g. an area with potential for improvement).
2. Obtain information on the site, vegetation, soil and climate.
3. Obtain information on the water supply and irrigation system (plans and documentation).
4. Test a representative part of the system.

Some common irrigation problems

Experience has shown that irrigation systems may fail to meet the requirements of good irrigation practice for numerous reasons, but the most common tend to be:

- damaged or malfunctioning sprinkler and spray heads
- sprinkler distribution ineffective (heads too low or stream obstructed) (Figure 11.2)
- mixed heads and incorrect nozzles on same pipe circuit
- excessive sprinkler head spacing (poor design)
- excessive irrigation run times (too long)
- irrigation system operated too frequently.

Figure 11.2. Catch cans being used to measure uniformity and effectiveness of garden bed sprays

The need to improve the performance of sports grounds was highlighted by Bransgrove and Henderson (2006) in a study of 11 fields in Queensland that were mainly used for Australian Rules Football. The DU_{LQ} ranged from 0.51 to 0.72, with the majority of the systems in the range of 0.55 to 0.65. On seven of the nine fields tested, 33% or more of the sprinklers were not operating optimally. This study highlighted the importance of maintaining individual sprinklers heads and correct operating pressure.

Preparing for an audit

Background information

Conducting an audit of an irrigation system requires establishing an accurate record of the system, the site and the vegetation. The foundation to building a quality irrigation management program is a detailed and accurate plan, which not only includes records of locations of important features and water infrastructure, but also reference to accurate location and description of installed equipment. The make, model, size of components (sprinklers, valves, etc.) should be recorded. If a plan is not available, then the preparation of a detailed sketch is required or arrangements made to prepare an accurate plan of the site.

Details of the water supply and control equipment are particularly important – pump or water meter, controller, master valves, and so on. It is also critical that details of control programs for each control station are noted so that recommendations can be made on the appropriate run times of the system, as tested, to meet the needs of the vegetation at the particular site.

The soil water properties for each irrigated zone need to be determined (Figure 11.3). The key properties are infiltration rate (IR) and available water-holding capacity (AWHC). Identifying the soil type allows these properties to be obtained from a reference table (see Chapter 6). There are also techniques available, using instruments, to measure infiltration rate in the field.

Figure 11.3. Soil samples provide valuable information on root depth, soil type and soil moisture conditions. Picture shows darker (organic) topsoil overlying finer clay.

Site test conditions – weather and hydraulics

An audit should be carried out under conditions, which provide a fair representation of the normal performance of the system. The climate conditions, in particular wind, should be within acceptable limits during the test. A maximum wind speed of 10 km/h can be used as a guide. The wind conditions during the test should be recorded and if they are considered to have an impact on the test results, this should be noted.

Preliminary visual check of system or 'walk through'

Stage 1. Overall effectiveness of system

The first step in the evaluation process involves assessing the overall effectiveness of the system.

The following aspects should be checked:

- the appropriateness of the irrigation method for the particular situation
- the design of the system, in terms of layout and spacing (e.g. sprinkler spacings, drip emitter intervals, dripline spacing)
- the zoning of the system to meet separate microclimate, plant water, topography and soil characteristic requirements
- evidence of over-watered areas (e.g. damp areas, ponding or dead plants)
- evidence of erosion and runoff
- evidence of under-watered areas (e.g. stressed or dead plants)
- potential soil issues that may impact on effectiveness of application (e.g. compaction
- presence of thatch, which may impact on effectiveness of infiltration
- mulch depth and condition (e.g. if there is no mulch, assess and recommend application, if appropriate).

Stage 2. Checking the condition and operation of the system outlets and key components

The next stage involves checking of individual outlets to assess the ability of the system to deliver water effectively. This evaluation can be conducted by observing the operation of the outlet (sprinkler, spray or drip emitter) and checking the effectiveness of the delivery.

Functioning of sprinkler and spray systems

- Operate the system to observe the water flow and trajectory from the head.
- Check that pop-up units function in terms of lift, retraction and sealing.
- Check the vertical alignment and depth of sprinkler heads (Figure 11.4).
- Check that the sprinkler rotation is uniform and duration is within manufacturers' specifications.

A comprehensive and detailed approach to checking sprinkler irrigation systems is presented in Barrett *et al.* (2003).

Functioning of drip systems

- Operate the system to observe the delivery from the emitters.
- Check the functioning of flushing valves.
- Check for damage or kinking of poly pipes.

Stage 3. Application performance evaluation

There are two types of evaluation of the application performance:

- checking the uniformity of application for sprinkler and spray systems
- checking the evenness of delivery of drip emitters and microirrigation equipment.

Figure 11.4. Tilted heads are a common problem with pop-up sprinkler systems

Some of the aspects and items that might be identified during the 'walk through' include design and hydraulic aspects and maintenance.

Design and hydraulic aspects

These include checking the:

- amount of overlap with sprinklers and spays
- regular spacing and pattern of sprinklers
- stream break up – misting or dry areas.

Maintenance

This includes checking the:

- correct nozzles are fitted to each sprinkler or spray head
- direction of spray – overspray on to non-target areas
- vertical alignment of the head
- water stream clears surrounding grass
- rotation speed is even
- seals are effective – no leaking
- effective reseating of sprinkler head following operation (Figure 11.5).

Any problems observed should be identified, according to position on the ground, and referenced to the controller station operating that particular sprinkler or outlet. This information should be recorded and noted on the plan as part of a maintenance record of the irrigation system.

Not all problems will be able to be fixed before the test. The audit may indicate system deficiencies (problems), such as incorrect sprinkler spacing or low operating pressures, that may involve major works or system design changes.

Figure 11.5. Sprinkler heads stuck up after operation are a hazard

Carrying out an audit of a sprinkler system

Audit test equipment

The following equipment is required to carry out an irrigation audit:

- catch cans – 20 to 40 catch cans. IAL catch cans (Figure 11.6) or cylindrical cans or containers are suitable (diameter: 70–150 mm, height: 50–100 mm, straight sides)
- bucket and tubing – to measure sprinkler flow rate
- pressure gauge and pitot tube
- stopwatch
- measuring flask
- measuring tape
- recording sheets for can readings and test details (customised worksheets are available as part of the IAL Irrigation Efficiency Course (IEC) training program).

Full details on conducting audits are presented in the *Irrigation Efficiency Course Resource Manual* (Irrigation Australia 2010). This replaces the Certified Landscape Irrigation Auditor (CLIA) training program manuals. For further details, see the IAL website: www.irrigation.org.au.

Setting out the cans

To carry out a test, it is advisable to use at least 20 cans to obtain a fair representation of the system performance. Greater number of cans will allow larger areas to be simultaneously tested (Plate 11.3). Also, for any selected test area, the higher the number of cans, the greater the accuracy of the measurement.

It is helpful to allocate each can a position number, so that the location of each particular reading can be readily identified. Irrigation Australia Ltd catch can markings include both volume (mL) and depth (mm) of water. It is most convenient to read in millimetres (depth).

Audit Test Procedure Summary for Sprinkler and Spray Systems

Step 1. Record details of sprinkler make, model and nozzle size.
Record sprinkler design operating pressure.
Measure spacing of sprinklers.
Step 2. Set cans out in a regular pattern.
Check that cans are vertical and the tops level.
Step 3. Commence test, noting pressure and weather conditions.
Note start time.
Record water meter reading.
Step 4. During test continue to monitor pressure and weather conditions.
Wind speed and wind direction are important.
Step 5. Finish test, noting time accurately.
Record water depth/volume in each can.
Record water meter reading.
Collect readings in a methodical manner.
Step 6. Calculate DU
Calculate Precipitation Rate (PR)
Step 7. Prepare base schedule
Step 8. Prepare report based on test results.

Can layout

The cans should be spaced so that they are in a grid within the area bordered by the sprinklers. The aim is to test a representative part of the system. In setting out the cans, the options are to test a large part of the system or concentrate on testing a smaller part. If a large area of the

Figure 11.6. Audit catch can allows direct reading of depth (mm) and volume (mL)

Figure 11.7. Containers of various sizes can be used to measure uniformity of application and the system application rate

system is tested, the distance between cans is generally too large. As a guide, can spacing should not be greater than 2 to 3 m, because significant variations in application can occur over short distances. If spacings are too wide, these variations will not be recorded.

Cans should not be positioned too close to individual sprinkler and spray heads (a minimum distance of around 1.5 m is suggested). The trajectory of the spray may mean that some water is not collected by the can if they are too close.

The can layout shown in Figure 11.8 shows the test area surrounded by four sprinklers. A larger area (e.g. six sprinklers) could be tested using more cans. The test would need the irrigation controller at two stations, numbers 6 and 7, to be operated to achieve a sample of the system application. If four sprinkler heads are operated from one valve, then positioning the cans between the four heads provides a good representation of the system and the test, in this case, can be carried out with only one station operated. In other cases, it may be necessary to operate one sprinkler lateral and then operate the adjacent lateral to obtain readings of the full system coverage of that area.

In most cases, setting out the cans between a group of four sprinklers is the most convenient. Cans can be spread out over more sprinklers, but the increased spacing may not pick up significant variations in application.

Cans should be spaced in a regular pattern of rows and columns, which should be clearly identified, (e.g. rows A to E, columns 1 to 5). Record the rows and columns in relation to sprinkler positions and station layout by preparing a sketch of the area to be tested and the layout of the test cans (Table 11.7).

Recommendations regarding the number of cans to be used and the spacing of cans are presented in the publication *Irrigation Evaluation Code of Practice,* Irrigation New Zealand (2006).

Test conditions

Note the date, time, air temperature, cloud/rain conditions and wind (both speed and direction) during the test. Record the time the pressures were checked.

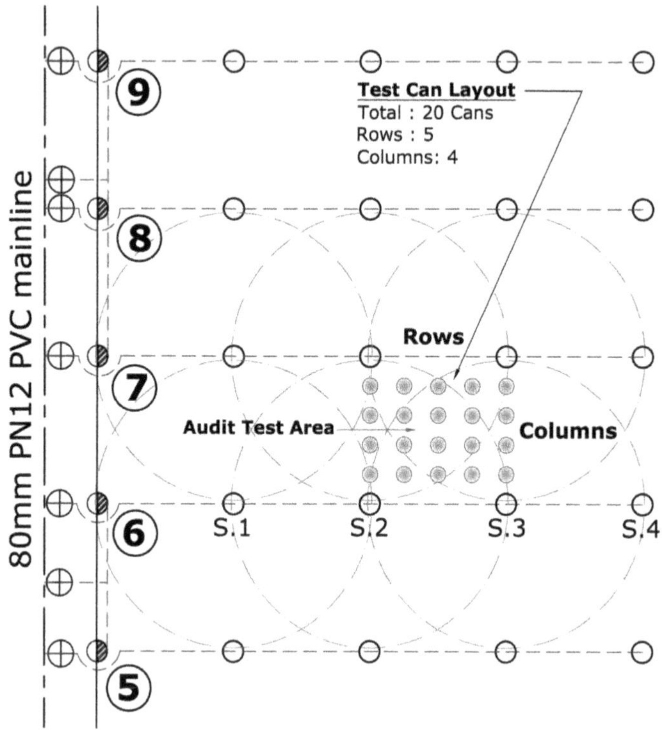

Figure 11.8. Test can layout between four sprinklers operating on two different laterals on a sports ground (Source: Rainlink Australia Pty Ltd)

Test duration

The test should be conducted for enough time to obtain accurate readings. Generally this will be 20 minutes or so for sprinklers. Shorter times may be warranted for sprays that have higher precipitation rates.

Sprinkler full rotations

The rotation speed (time for one complete rotation) of the sprinklers should be noted during the test. There should be at least six rotations of the sprinkler during the test.

Table 11.7. Sample catch can layout and recording table

	Auditor: .... Audit date: .... Test conditions: Wind speed .... Wind direction .... Temperature .... Time of test				
	Site name: .... Test area designation: .... Controller station number:				
Rows	**1**	**2**	**3**	**4**	**5**
A					
B					
C					
D					
E					

Table 11.8. Summary table of audit test data

Category	Description
Test site	Site name and location
	Contact person, contact details
Site details	Irrigated area
	Vegetation type, areas, condition
	Zone microclimates, plant densities
	Soil type, properties, root zone depth
Test conditions	Weather: wind speed, direction, temperature
	Date of test, time of test start, duration of test
Water source	Type: meter or other source
	Product identification, size
	Hydraulic supply conditions: flow and pressure
	Meter reading; start, finish
Controller	Manufacturer, model, number of stations
	Program; days, start times, run times
	Sensors: rain, weather station
	Features; water budgeting
System review – sprinkler	Manufacturer, model, nozzle/s sizes
	Design operating conditions; pressure and flow
	Layout and spacing
	Design precipitation rate
	Sprinkler rotation time; number of rotations
System review – drip	Type, manufacturer, discharge rate
	Emitter interval spacing, lateral spacing
	Pipe and tubing sizes and material
	Valves: flushing, air release, pressure regulation, filtration

Table 11.8 provides a list of the type of information that needs to be collected as part of the audit process.

Evaluating uniformity of delivery of microirrigation systems

The evaluation of the uniformity of delivery from outlets, such as drip emitters and microsprays, requires checking of selected outlets, within the system. The measure of uniformity of delivery is the EU and it is calculated in a similar (mathematically) manner to DU. In this case, a volume of water is measured from a number of outlets and the variation in volumes delivered used to gauge the degree of non-uniformity of the system (Figure 11.9).

Calculating EU

$$\text{EU} = \frac{\text{Average of lowest 25\% of readings} \times 100}{\text{Average all readings}} \quad \%$$

As a general guide, laterals close to the system supply point, in the centre and at the furthest point from supply can be used. Within any lateral, as a minimum, emitters at the start, midpoint and end should be used. Table 11.9 shows the results of a test carried out on a drip system.

Figure 11.9. Collection container, measuring cylinder and stopwatch to check drip emitter flow rate

Analysis of results

Total number of readings: 20
Number of readings for 25% lowest: 5
Total of all readings: 604 mL
Average of all readings: 30.2 mL
Lowest 25% readings: 25, 26, 27, 29, 29 Total: 136
Average of lowest 25%: 27.2 mL
EU = 0.90 or 90.0 %
Emitter flow rate: 30.2 mL per 1.0 minute equals 1.81 L/h

Table 11.9. Results of evaluation of a drip system

Test details
Site: garden bed
System: 6 dripline laterals
Emitter specified delivery rate: 2.0 L/h
Recommended operating pressure: 150 kPa
Emitter interval (S_{ei}) 400 mm. Lateral spacing (S_{ls}) 600 mm
Measurement: Volume (mL) delivered in 1.0 minute

	Start of lateral (mL)	Quarter (mL)	Halfway (mL)	Three quarters (mL)	End of lateral (mL)
Lateral 1	35	33	32	30	30
Lateral 2	32	32	29	30	29
Lateral 3	33	32	30	31	30
Lateral 4	29	27	29	25	26
Totals	129	124	120	116	115

Comment on results

An EU of 0.90 or 90% represents good uniformity of delivery in the field. Operational EU values are lower than the design values. Some recommendations suggest minimum operational EU of 0.75, but, ideally, a drip system should be able to perform at a level of EU 0.85 to 0.90.

The average flow rate from the emitter is 1.8 L/h. This is within 10% of the specified delivery rate (2.0 L/h). This slightly lower flow rate may be due to low pressure. This can be checked as part of the overall assessment of the system.

The evaluation of SDI systems is difficult in terms of checking for uniformity of delivery from emitters and distribution of water in the soil. Flow meters that allow individual laterals to be monitored and pressure gauges can be used to assess the hydraulic performance. Comparing actual flows with manufacturer performance specifications provides a reasonable measure. Portable soil moisture sensors can be used to check surface soil layer moisture while installed sensors can provide only a representative sampling of the performance of the system.

Pressure testing

An accurate pressure gauge is an extremely valuable tool for the evaluation and monitoring of irrigation systems. Pressure is the heart rate of the irrigation system performance. Part of the audit test will involve checking the actual sprinkler operating pressure and pressure variation throughout the system (Figure 11.10).

Some of the key information that can be provided through pressure measurements includes:

- checking that sprinklers, sprays and emitters are operating at correct (optimum) pressure
- determining the pressure variation along the lateral – is it acceptable?
- the pressure variation between stations and outlets in different parts of the system
- the amount of pressure loss due to friction in the system
- the pressure loss across valves, filters and special fittings.

A pitot tube gauge (a small diameter tube inserted into the water stream) can be used to check nozzle pressures (Figure 11.11). The positioning of the pitot tube inlet is critical to obtaining a reliable reading. The pitot tube inlet should be positioned slightly in front of the sprinkler nozzle. If the pitot tube is inserted into the nozzle and restricts flow, it will produce an elevated pressure reading.

When using this method, it is important to note that the nozzle pressure will be lower than the inlet (base) pressure to the sprinkler head. Irrigation systems are most commonly designed on sprinkler inlet pressure and so this difference needs to be taken into account when analysing a system. The pressure loss or difference is typically in the range of 40 to 60 kPa.

The system pressure can also be checked at strategic points, such as at the downstream side of solenoid valves. This can be done using rapid connect Schrader valves (car tyre valves) and tubing. A data logger is valuable in providing continuous monitoring of system pressure.

Key scheduling information to be collected

The following is the key information to be obtained during an audit, in order to analyse the performance of the system and prepare the base schedule.

- plant water requirement
  - plant material type (e.g. grass – warm season)
  - plant species coefficient (K_c or K_s).
- irrigation requirement
  - precipitation rate (PR_N)
  - distribution uniformity (DU_{LQ}).
- scheduling requirements

Figure 11.10. Solenoid valve pressure being measured using quick connect hose and gauge

  - root zone soil type
  - active root zone depth (RZD)
  - management allowable depletion (MAD)
  - run time per cycle (time until ponding or runoff) (TRO).
- flow rate
  - flow rate per head (Q) (L/min, L/s)
  - number of heads per station
  - flow rate (L/min) for station (QS).

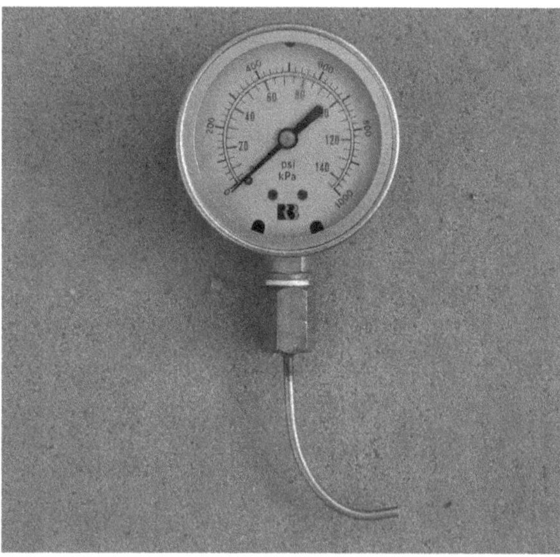

Figure 11.11. Sprinkler nozzle operating pressure can be measured using a pitot tube

Table 11.10. Summary of possible causes of low DU

Possible cause of low DU	Solution: services/technology
Excessive spacing of heads	New system design and replacement
Poor precipitation profiles	New sprinkler heads
Incorrect pressure	Design, pressure regulation
Blocked nozzles	Maintenance
Misaligned head assembly	Maintenance
Wind	Install shut off wind speed sensor

Analysing audit test results

Uniformity readings

The two key performance readings that an audit will provide are the evenness or uniformity of the application (DU) and the average precipitation rate (PR). The first step is to compare the measured uniformity values with the industry standards – DU_{LQ} greater than 0.75 or 75% and CU greater than 84%. If the values are significantly lower, then possible causes should be investigated (Table 11.10). Low uniformity values may be due to either a poorly designed system or the system not functioning correctly. A check of the sprinkler or spray performance and measurement of sprinkler and spray spacings will provide details needed to ascertain if there are flaws in the basic design. The wetted diameter of sprinklers should be checked to see that they meet the required spacing guidelines for high uniformity.

If the system is shown to meet basic design guidelines, and it has low uniformity value, then other aspects should be assessed. Each sprinkler and spray should be checked for pressure, flow rate and distribution. Individual readings within the test area should be reviewed. A pattern of low or high readings may indicate specific problems within the system. For example, one sprinkler may have a damaged nozzle, which distorts that part of the pattern.

If the DU is low, there could be a range of reasons, including those associated with carrying out the test, as distinct from those that indicate a problem with the system. Check the following:

- Was the test a true reflection of the performance of the system?
- Was the system tested at a time that allowed optimum hydraulic conditions (e.g. testing during the day may result in pressures lower than those experienced at night, when the system would normally be operating)?
- Were the atmospheric conditions (wind) suitable for testing? Wind speeds in excess of 10 km/h (2.8 m/s) are considered unsuitable.
- Did the layout and positioning of catch cans, including those close to sprinkler heads, provide an accurate measure of the performance of the irrigation system?
- Were adjacent overlapping sprinkler heads operated to provide an accurate measure of the system?

In reporting the value of DU obtained in the audit, it is valuable to comment on the results in the following contexts:

- compared with industry best practice
- compared with the industry generally
- compared with the organisation's performance standards – do they have any (this information should be in the water management plan)?

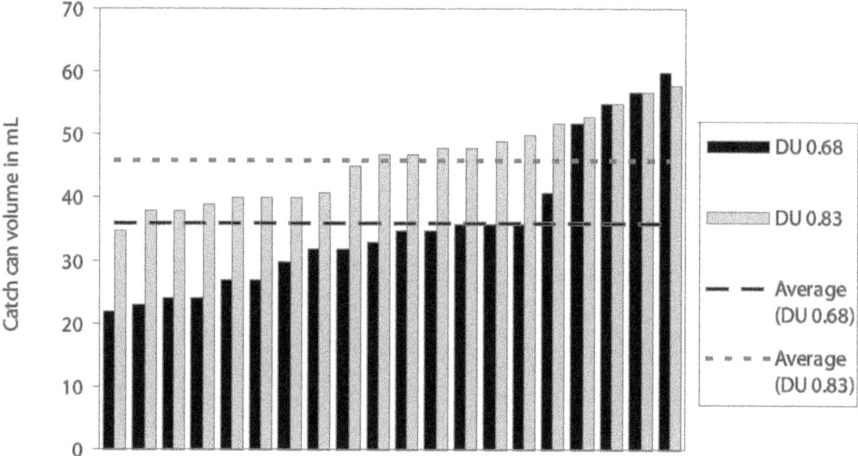

Figure 11.12. Presentation of catch cans results; two uniformity results

Recommendations for tackling low uniformity (DU_{LQ})

The following are some suggested actions that should be taken following recording of low DU:

- DU 0.65–0.75 (65–75%): repair or adjust to achieve acceptable performance
- DU 0.50–0.75 (50–64%): upgrade system components (sprinklers, new nozzles, etc.)
- DU <0.50 (50%): replace system.

Illustrating non-uniformity

The presentation of catch can results graphically provides a powerful message to stakeholders regarding the range of readings that occur. Figure 11.12 shows the results from two tests: one meeting industry uniformity standards and the other not.

Precipitation rate

The PR is used to check that the design of the system and current head setup are correct. The actual PR should be less than the soil infiltration rate. If the PR is too high, it will generally be apparent through runoff or ponding during the test. There are not a lot of options available to change an existing irrigation system design. It may be possible to change nozzle combinations, or even the sprinkler or spray heads, but this would need to be checked with the product manufacturers or system designer.

An alternative approach is to apply the required depth of water, in several small amounts, rather than one continuous application. The controller is required to initiate multiple starts for each irrigation station. For example, 15 mm may be applied in three lots of 5 mm, with each irrigation event separated by a period of 1 hour. This is referred to as cycling.

Calculating precipitation rate

The theoretical precipitation rate of the system can be calculated from the flow rate of the sprinklers or sprays and the spacing in the layout. This rate is referred to as the gross precipitation rate (PR_G). It assumes all of the discharge water is deposited on the surface. The calculation of the PR_G for a sprinkler irrigation system is outlined in Chapter 8. This value provides a check for the rate measured during the tests. The PR is expressed in mm/h.

The field precipitation rate can be readily obtained from the audit test readings. In this case, it is a measure of the actual water incident on the surface. It is referred to as the net precipitation rate (PR_N). If the can readings are in mm, then the average of all readings provides a value of precipitation for the period of the test.

Example

Number of readings: 20
Total of all readings: 160 mm
Catch can readings: average: 8 mm
Duration of test: 20 mins (divided by 60 to convert to hours)

$$\text{Precipitation rate: } (PR_N) = \frac{8 \times 60}{20}$$
$$= 24 \text{ mm/h}$$

Calculating PR_{NET} from can volumes

If a round can is used (not calibrated in mm), rather than the funnel type that gives a direct reading, then the depth of water can be determined by dividing the volume of water by the area of the top of the can. To calculate the average PR_N, the average depth of water of all the cans is calculated, taking into account the duration of the test.

The following expression allows the net PR_N to be directly calculated from the can readings, where the volume is measured.

$$PR_N = \frac{V_{cavg} \times 60\,000}{T_t \times A_c} \quad \text{mm/h}$$

where:
V_{cavg} – Average volume in test cans (mL)
T_t – Test run time in minutes
A_c – Area of can in mm^2
Note: 60 000 converts mins to hours and mm^2 to m^2

Example

Catch can diameter: 120 mm
Average volume in cans: 42 mL
Test run time: 20 minutes

$$\text{Area of top of can } (A_c)\text{: } \frac{\pi \times d^2}{4} = \frac{22 \times 120 \times 120}{7 \times 4} = 11314 \text{ mm}^2$$
$$PR_N = \frac{42 \times 60\,000}{20 \times 11314} = 11.1 \text{ mm/h}$$

Pressure check of outlets

A pressure check is needed that not only the minimum operating pressure is available, but also that the pressure variation is within acceptable limits (e.g. ± 5%). Table 11.11 presents results from pressure checks on two sprinkler laterals.

Processing of results – Lateral No. 2

The degree of variation is determined using the highest and lowest readings. The difference between the average of the highest and lowest, called the mid-point pressure, is used as the reference value.

Table 11.11. Pressure reading check of sprinkler laterals

Site details System: Pop-up sprinklers Number of laterals: 12 Design pressure at sprinkler: 320 kPa		
Sprinkler position number	**Lateral no. 2**	**Lateral no. 4**
1	350 kPa (highest)	340 kPa (highest)
2	345 kPa	340 kPa
3	340 kPa	335 kPa
4	335 kPa	330 kPa
5	330 kPa	330 kPa
6	325 kPa	320 kPa
7	320 kPa (lowest)	315 kPa (lowest)

Example

Highest: 350 kPa
Lowest: 320 kPa
Average or mid-point: 335 kPa
Difference: mid-point to Lowest: 15 kPa
Percentage variation: (15/335) × 100 = 4.5%
The pressure variation along the lateral is ± 4.5%, which is within acceptable limits (e.g. ±5%).

System water flow balance

It is important to check that the field performance of the system meets the system design capacity. The availability of water meters or flow meters allow this check to be readily made.

Example

Drip system: six zones each of 80 m of dripline
Emitter interval: 300 mm
Emitter delivery: 1.5 L/h pressure compensated (PC)
Station run time: 60 minutes

Design flow rate

Number of emitters: 80 m/0.30 m = 266 emitters
Zone flow rate: 266 × 1.5 L/h = 399 L/h or 6.65 L/min

Design volume

Zone volume: 399 L/h × 1 h = 399 L

System check

Initial meter reading: 09875.150 kL (m^3)
Final meter reading (after 60 mins): 09875.600 kL (m^3)
Volume delivered in 60 mins: 450 L

Comment on results

The measured volume is 51 L higher than the design volume. Consider checking for leaks, that the installed product matches the specifications, the length of dripline and possible pressure variations (although emitters are PC), and the discharge from selected emitters.

Preparation of base schedule

The preparation of a base schedule should be included as part of an audit of an irrigation system (the procedure is outlined in Chapter 6).

Water consumption analysis

Comparing water consumption totals

Water consumption can be reported and evaluated in a number of different ways. A simple technique is to report the total water consumption in megalitres (ML) or kilolitres (kL). The presentation of historical water consumption totals is valuable to demonstrate trends from year to year. If a reference year (e.g. 2007–08) is adopted, then 'savings' can be highlighted relative to the reference year. Although this allows annual trends to be reviewed, it does not give an indication of efficiency or allow comparison with other enterprises.

Reference can be made to the climate data for the site. In the Bureau of Meteorology 'Climate Averages' rainfall reports for a locality, deciles 1, 5 and 9, are presented. These indicate the likelihood of the rainfall amounts occurring. The decile 9 (90%) rainfall indicates that the rainfall in 90% of years is expected to be less than this amount. Table 11.12 provides an example from Brisbane Regional Office.

For this location, for 90% (decile 9) of the years, it could be expected that rainfall will not be greater than 1602.1 mm. In a dry year (which may be defined as a 1 in 10, decile 1, or 10%), it would not be expected that the annual rainfall would be less than 719.9 mm.

The reporting of water consumption as a water application rate (WAR), expressed, for example in ML/ha, is a more informative than a total volume. This allows comparisons within the organisation and between businesses. There are now several units used for WAR in

Table 11.12. Rainfall statistics including deciles for Brisbane

Site: Brisbane Regional Office BOM Station number: 040214 Period of records: 1840 to 1994		
Rainfall statistics	**Annual rainfall (mm)**	**Year of event**
Mean	1149.1	–
Highest	2242.4	1893
Lowest	411.5	1902
Decile 1	719.9	
Decile 5	1106.7	–
Decile 9	1602.5	–
Highest daily rain	465.1 mm	21 Jan 1887

Source: Bureau of Meteorology website (www.bom.gov.au)

environmental horticulture including kL/1000 m^2, L/m^2 and ML/ha. The calculation of the WAR is straightforward.

$$\text{WAR} = \frac{\text{Total consumption (e.g. annual)}}{\text{Irrigated area}}$$

Example

Site: sports ground.
Total water consumption: 6.6 ML/year.
Irrigated area: 1.5 ha.

$$\text{WAR} = \frac{6.5}{1.5}$$
$$= 4.33 \text{ ML/ha/year}$$

Water consumption should preferably be expressed in terms of efficiency of water use, rather than gross values. Water use efficiency takes into account the actual water needs of the site and the weather conditions experienced by the site during the irrigation season (variation from long-term data, such as a dry or wet year). Explaining variations in consumption and providing reasons for excess applications is a key part of reporting. Water consumption reporting should be seen as an essential part of sound water management planning.

Efficiency measures and terms

There are a number of terms that can be used to determine the efficiency of irrigation. The irrigation index (Ii), irrigation efficiency (IE), application efficiency (E_a) and water management efficiency (E_{wm}) have been discussed. A calculated crop coefficient (K_c) value can also be used to provide a measure of efficiency. Details on efficiency terms and calculations are outlined below.

Irrigation index

The irrigation index (Ii) is a very useful measure to evaluate water consumption performance and also to gauge progress in terms of improvement in overall efficiency. The Ii compares the depth of water actually applied with the estimated depth of water required over the complete irrigation season (Kah and Willig 1992) (Figure 11.13). This relatively simple measure provides the irrigation manager with a visible, readily understood measure of how well, or how efficiently, the system is performing and how the performance compares with other sites. An irrigated area that is being well managed would have an Ii value of 1.0 or less. If the Ii value is

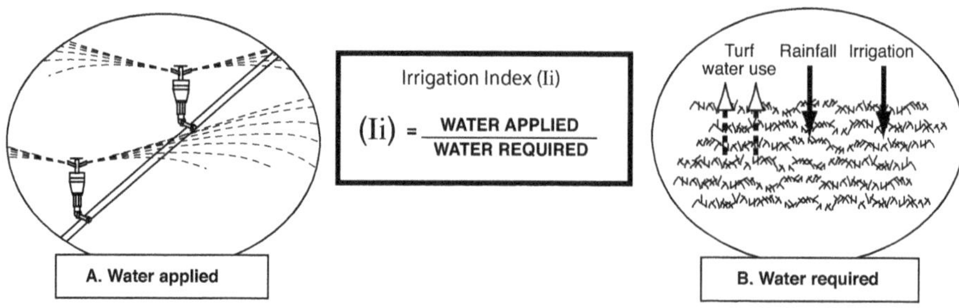

Figure 11.13. Elements of the irrigation index

greater than 1.0, it would suggest that there is some wastage of water. The Ii can be defined in the following way:

$$\text{Ii} = \frac{V_A}{AIV_R}$$

where:
V_A– Water applied to site
AIV_R – estimated annual irrigation water required

Determining water applied (V_A)

The volume of water applied can be determined from water meter readings or from irrigation controller reports.

In situations where the total depth of irrigation is known, the volume can be determined using the following expression:

$$\text{Volume} = \text{Area (m}^2) \times \text{Applied depth (mm)} \quad \text{L}$$

Note: 1 L is equal to 1 mm spread over an area of 1 m^2.

It is important to keep records of meter readings not only at the start and end of the irrigation season, but also on a regular basis throughout the season. This assists with the monitoring of the site, the equipment and irrigation scheduling. As a minimum, monthly readings are required.

Net water requirement (NWR)

The amount of water that needs to be deposited, by the irrigation system, in the root zone, to satisfy plant growth, is the net difference between the plant water use (ET_c) and the amount supplied through rainfall (P_{eff}).

$$\text{NWR} = (ET_c - P_{eff})\ \text{(mm)}$$

The accuracy of this technique depends upon the frequency of calculating the net water requirement. This can be done on a daily, weekly or monthly basis. Daily analysis provides the most accurate measure, but monthly analysis provides a good overall guide to efficiency.

Effective rainfall

The proportion of rainfall that is actually used by plants, after all rainfall losses have been taken into account, is referred to as effective rainfall (P_{eff}). As a guide, it is assumed to be 50% for turf areas, but for garden landscape sites it may be lower (e.g. foliage interception). Each site needs to be individually assessed. Details of the effectiveness of rainfall are outlined in Chapter 9.

Annual irrigation volume requirement

Due to inefficiencies, the irrigation system needs to apply more water than the estimated NWR. Some water may be lost due to wind and evaporation, some may drain below the root zone and there is always some unevenness in the application. The irrigation system application efficiency (E_a), which takes into account these losses, can range from very low values up to around 90%. Achievable or minimum acceptable system efficiency (e.g. 75%), can be selected to provide a reference performance standard for sprinkler systems.

$$AIV_R = \frac{\text{NWR (mm)}}{E_a} \times \text{Area (m}^2) \quad \text{L}$$

The determination of the Ii can be carried out using a relatively limited amount of data and the application of simple analysis techniques. In a study of sporting ovals in Melbourne (Keig 1994), the value of Ii was determined to be in excess of 2.0 for two ovals. This represents a total water use rate double that should have been used on those sites. There were therefore plenty of opportunities for water conservation at these sites.

Irrigation efficiency (IE)

The following is the definition of IE.

$$IE = \frac{\text{Water beneficially used by plants (output)}}{\text{Water applied to area (input)}} \times 100$$

It is very difficult to measure water actually used by the plants. It is for this reason that the IE is commonly based on the water delivered to stores in the root zone. It is assumed that it is all used beneficially by the plants.

An alternative approach is to use the estimated water required.

$$IE = \frac{\text{Estimated water required over season (output)}}{\text{Water applied to area } (V_A) \text{ (input)}} \times 100$$

In this case, the estimated water required does not include an assumed efficiency value. It considers the plant water demand and effective rainfall only.

Estimated net water required (NWR) is determined using the following expression:

$$NWR = (ET_c - P_{eff}) \text{ mm}$$

NWR is calculated for each month and summed to obtain an annual value.

Efficiency can be determined using depths (mm) or volumes (KL or ML).

Using the data presented above, the IE can be determined.

$$IE = \frac{\text{NWR summed for each month (mm)} \times \text{Irrigated area } (m^2)}{\text{Total water applied } (V_A)} \times 100\%$$

The efficiency can be calculated using either the volume of water or the depth of water (mm). The depth of water applied is calculated directly from the volume (L) and the area (m^2).

Crop coefficient (K_c) for site

Efficiency can also be expressed and described in terms of a crop coefficient (K_c) value for the site. K_c is defined in the following way:

$$K_c = \frac{\text{Depth of water applied}}{ET_o}$$

In the expression presented here there is no allowance for rainfall. Both the total rainfall and effective rainfall are sometimes used in this calculation. Effective rainfall will be used in the following expression:

$$K_c = \frac{\text{Depth of water applied}}{(ET_o - P_{eff})}$$

Calculation of efficiency terms

The following is an outline, using a sports ground, for the determination of the Ii, IE and crop coefficient (K_c).

Table 11.13. Climate data used in efficiency determinations

Month	Historical monthly ET_o (mm)	Actual monthly ET_o (mm)	Historical monthly rainfall (P) (mm)	Actual monthly rainfall (P) (mm)	Actual effective rainfall (P_{eff}) (mm)	Net water requirement (NWR) (mm)	Irrigation requirement (IR) (mm)
October	166	170	38	26	13	55.0	73.3
November	202	196	26	58	29	49.4	65.9
December	222	220	24	32	16	72.0	96
January	253	262	17	18	9	95.8	127.7
February	217	214	18	22	22	63.6	84.8
March	186	172	20	46	32	36.8	49.1
Total	1246	1234	143	202	101	372.6	496.8

Table 11.14. Site water consumption

Month	Calculated monthly water required (AIV_R) (kL)	Actual monthly water consumption (kL)	Difference in water consumption (kL)
October	1099.5	1270	+ 170.5
November	988.5	1224	+ 235.5
December	1440.0	1415	–25
January	1915	2310	+ 394.5
February	1272	1240	–32
March	736	990	253.5
Totals	7452 kL	8449 kL	+ 997

Example

Site: sports ground – cricket
Area: 1.5 ha (15 000 m^2)
Ground rating: local sports, community ground
Grass: couch
Crop coefficient (K_c): 0.40
Soil: sandy loam
Irrigation application efficiency (E_a) (Assumed): 75%
Climate and water consumption data are presented in Tables 11.13 and 11.14.

Calculation of efficiency values

The following calculations show how each of the efficiency measures are determined.

A. Irrigation index (Ii)

$$Ii = \frac{V_A}{AIV_R} = \frac{8449 \text{ kL}}{7452 \text{ kL}} = 1.134$$

B. Irrigation efficiency (IE)

$$IE = \frac{\text{NWR (mm)}}{\text{Depth of water applied } (ID_A)}$$

$$\text{Depth of water applied } (ID_A) = \frac{8\,449\,000 \text{ (L)}}{15\,000 \text{ (m}^2\text{)}} = 563 \text{ mm}$$

$$IE = \frac{372.6}{563} \times 100 = 66.2\%$$

C. Crop coefficient (K_c)

$$K_C = \frac{ID_A}{(ET_o - P_{eff})} = \frac{563}{(1234 - 101)} = 0.5$$

Guide to interpreting irrigation index (Ii) values

The results presented for Ii indicate that the site was over-watered by 13% (Ii = 1.13). The reasons for this over-watering may be due to excessive run times, excessive frequency or

Table 11.15. Guide to irrigation index values

Irrigation index	Performance rating	What does it mean?
>1.3	Poor to very poor	Very wasteful irrigation, much overwatering
1.1-1.3	Medium	Significant over-watering
0.9–1.1	Good to very good	Watering close to optimum
0.7–0.9	Medium	Under-watering, landscape quality may be affected
<0.7	Poor to very poor	Serious under-watering, landscape at risk, sports surface may be unsafe (too hard)

deficiency in the application performance of the system. The calculation of the monthly water required uses both actual evaporation and actual rainfall data. An irrigation index of 1.0 is the best practice target value. Table 11.15 is a guide to the interpretation of Ii values.

RBG Melbourne experiences in use of irrigation index

The improvement in water management is clearly demonstrated by the experiences of the Royal Botanic Gardens Melbourne, where the Ii values were reduced from 3.5 in 1995–96 to around 1.0 in 1999–00 (Symes 2000) and has been maintained around this level (Figure 11.14). For further details, see the irrigation management plan, Royal Botanic Gardens Melbourne website (www.rbg.vic.gov.au/about/water).

Discussion on efficiency terms

Each efficiency term has its own characteristics and limitations.

The Ii term highlights the efficiency of management of the irrigation system (E_{wm}). It makes assumptions about the application efficiency (E_a). For example an E_a of 75% is often assumed. This is not the actual measured value and the field value may be lower or higher. This value is aligned to an industry best practice value. It is a sound basis on which to determine a

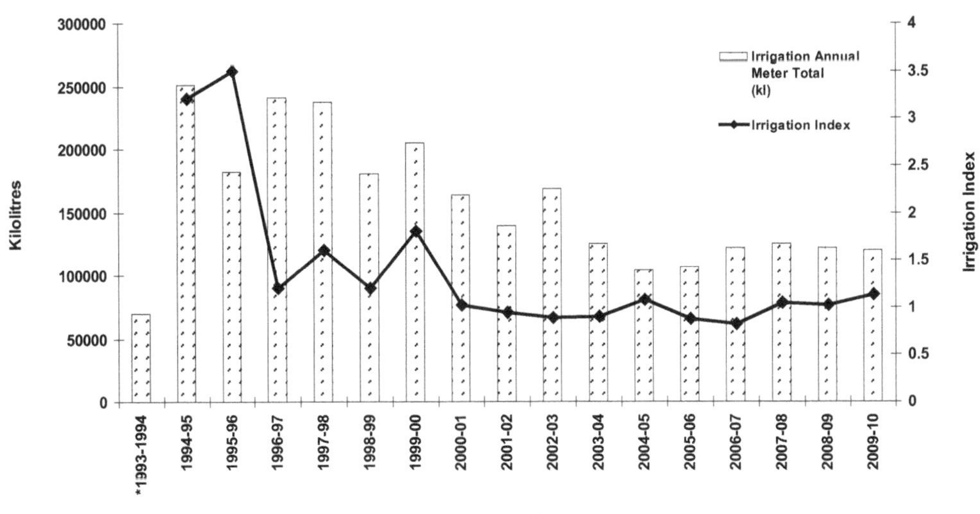

Figure 11.14. Trends in water consumption and irrigation index (efficiency) at Royal Botanic Gardens Melbourne

value for Ii. If Ii is, for example 1.15, then it is indicating that the management of the system has resulted in over-watering by 15%.

Due to the way IE is calculated, it includes both the application and water management efficiency in the calculation. The overall low value of 66% determined in above example takes into account both the application efficiency and the water management efficiency. To establish the reasons for the low efficiency, an audit will provide an insight into potential contributing factors. It does highlight the need to have high levels for both E_a and E_{wm}.

K_c is an indirect measure of efficiency because it needs to be referenced to another K_c value to gain an appreciation of how well the area was watered. The calculated K_c value (0.5), based on water used, is higher than the assumed value (K_c 0.4). This value also indicates over-watering.

Previously it has been stated that DU is not strictly an efficiency measure. It does, however, provide a field measure of the capacity of the system to apply water uniformly.

The surest way to achieve high efficiency in the field is to ensure that the system has the capacity to apply water uniformly (high DU and EU) and to manage the system according to current soil moisture levels. This may be done through ET_c or soil moisture measurement, or both.

Monitoring the performance of the vegetation and site conditions, the irrigation can be adjusted in a controlled way so that the amount of water is close to what the plants need. This is referred to as an adaptive management approach.

Irrigation efficiency – Irrigated Public Open Space (IPOS)

The *Code of Practice – Irrigated Public Open Space* (SA Water 2008) outlines the procedures for the continuous monitoring and adjustment of irrigation applications. This approach should be used to achieve high efficiency in systems where the irrigation schedule is based on long-term historical climate data.

A water budget (base irrigation requirement, BIR) is determined for a selected period (usually a month). This value is based on the long-term values of evaporation and rainfall and takes into account the crop coefficient for the plant and performance of the irrigation system. This is the amount to be applied by the irrigation system and is programmed into the controller.

During the irrigation season, the requirements of the plants will depend upon the weather conditions experienced. They may be more or less demanding that the long-term average amounts. The actual amount that should have been applied is determined using the daily weather data recorded during the selected period. This amount is the irrigation requirement (Ir). The difference or variation between BIR and Ir indicates how much over- or under-watering there has been, compared with the predetermined water budget. Some difference is expected, because the weather conditions experienced in a particular year will be different to the long term average data on which the BIR is based.

The BIR provides a reference value, which is typical of that particular month, and is a sound strategy to adopt in planning a water budget or schedule for the coming irrigation period. The daily adjustment of the irrigation schedule according to daily weather conditions will achieve an application of higher efficiency and closer to the optimum.

Assessment of the total variation or difference between BIR and Ir provides a measure of the overall performance and potential efficiency of irrigation.

Preparing irrigation audit reports

An irrigation audit reports on the following aspects of an irrigation system:

- the condition of the system
- the operating effectiveness in terms of uniformity of application (field DU_{LQ})

- the current control program
- the recommended irrigation control program (schedule) (called the base schedule)
- the recommended actions regarding improved performance of the irrigation system
- the opportunities to improve water use efficiency of the site.

The purpose of the report is to inform the client or property manager about the current condition of the system and advise on strategies to achieve high efficiency and improve overall water management. The following list outlines the minimum information that should be included in an irrigation audit report.

1. Site description, location and contact person
2. Detailed sketch or plan of the site, irrigation system and test area(s)
3. Description of the irrigation system including design or recommended operating conditions for sprinklers, sprays and microirrigation equipment
4. Details of current irrigation schedule
5. Details of the test including prevailing weather conditions, test layout, water supply flow and pressure, and duration of test
6. Comments on any conditions or circumstances (e.g. weather, hydraulic or equipment malfunctions) that may impact on the results
7. Details of existing problems or weaknesses in the system
8. A recommended irrigation schedule based on performance tests and site analysis –vegetation, soil and climate.

Identification of potential improvements, including correction of existing deficiencies and areas of further investigation, is required. Examples of potential recommendations that could be made to assist in achieving high efficiency and sustainable use of water at the site:

- upgrade water supply – flow and pressure
- use alternative sprinkler nozzles
- implement pressure regulation
- install soil moisture sensors or rain shut off devices
- install dedicated irrigation water consumption meters
- implement maintenance program, including checklists, records and reporting
- investigate potential improvement due to soil amendment (wetting agents), dethatching, coring, and so on (suitably qualified agronomists and horticulturists may be needed to provide specialist advice in these topics).

Note that this is not a complete list: reference should be made to opportunities for increased efficiency outlined in Chapter 9. The best management practice checklist can be used as an initial reference list to identify deficiencies in the system tested. The strategies described in Chapter 10 can be used as the basis for recommendation on potential improvements.

Brief assessment of sprinkler system

Table 11.16 and the following text is an example of a condensed evaluation report of an irrigation system that can be used to inform managers of a site of the current condition of the system.

Table 11.16. Summary report on sprinkler system performance

Performance parameter	Industry best practice	Test site results	Comment
Uniformity (DU)	75%	66%	Poor application uniformity
Precipitation rate	Loam soil – 10 mm/h	14 mm/h	Risk of runoff as too high
Irrigation depth	Recommended 12 mm (from base schedule)	7 mm (average)	Inadequate application – longer run times or higher flow rates required

Report comment

Suggested summary comment:

> 'This system is not operating up to industry best practice standards in three key areas. In its current form, it is not able to irrigate efficiently'.

Detailed evaluation of each area of performance will identify specific aspects of the system that should be rectified.

Summary – key checks of an irrigation system

The following checks should be made, as an absolute minimum, on the operation of all irrigation systems:

1. Is the application and delivery of water even and uniform?
2. Does the system have the capacity to meet peak water demand?
3. Does the application depth meet the best practice requirements of the site?
4. Is irrigation carried out according to plant water demand and weather conditions?
5. Is water use recorded, analysed and reported?

Chapter 12

Water management planning

Why have a water management plan?

A water management plan (WMP) is the cornerstone to achieving efficient use of water. The WMP expresses the commitment of the organisation to sustainable landscapes and outlines the pathway and processes that will be used to achieve those goals.

A WMP identifies the works and practices that will improve all water management, including irrigation and water use efficiency, for the site or enterprise. It identifies how water can be conserved and what strategies need to be put in place to ensure sustainability of water use in the future.

The following are the reasons why a WMP will be beneficial to an organisation:

- to secure future water availability through sustainable practices
- to enhance and protect the environment
- to assess the current water management practices
- to identify strategies for water conservation and efficiency improvements
- to provide a sound base and framework for the ongoing water management of the site
- to comply with regulations
- it is an integral part of business strategy
- to provide support for investment and funding applications
- to communicate to community and industry.

Various horticultural industry sectors have identified water issues that are particularly relevant to their water future. These issues include:

- the security of water supply
- reducing dependence on potable water
- increasing use of recycled water
- sustainable water use
- improving water use and irrigation efficiency
- water quality management and protection (recycled water, borewater, storage and waterways)
- the increasing cost of water.

Most organisations involved in the management of open space are required to prepare multiple plans regarding natural resource management. These include broader environmental issues, including flora, fauna, greenhouse and soils, as well as water. Although each sector tends to have specific reporting requirements, it is advisable to centralise and standardise data collection and reporting so that there is not unnecessary duplication.

Structure of a water management plan

A WMP reviews the current water management practices and identifies opportunities for improvement in water use efficiency across the whole area. The WMP clearly and precisely states the vision and goals of the organisation in terms of achieving sustainable use of water. It establishes the targets to be achieved and actions to be implemented that will produce sustainability of water use.

Numerous WMP formats are available through government departments, water authorities and local government. The structure outlined here was prepared as part of the Victorian Golf Association (VGA) and Australian Golf Course Superintendents Association (AGCSA) projects, by the author.

The WMP has been developed to allow all of the necessary information to be readily incorporated under four main headings:

- Part A – Water policy and objectives (context)
- Part B – Information collection
- Part C – Analysis and interpretation of data including water use
- Part D – Strategies, implementation and review.

It is recommended that gaps in information be identified and steps taken to collect the relevant data and documentation. This would normally involve liaising with other organisations including water authorities, local government, government agencies and suppliers.

PART A – Water policy and objectives

Water policies

The management of water resources has become a major focus for government and industry in recent years. The reasons for change include:

- demand for water is increasing – population growth
- increasingly stressed natural environment
- extended period of drought (over 10 years in some cases)
- climate change adaptation – less rain, higher temperatures
- greenhouse gas implications.

Most governments are developing policies relating to the sustainability of water use. Examples of priority areas identified in a White Paper by a state government (Victoria) to be tackled as part of securing water for homes, farms, businesses and the environment for the next fifty years are:

- improved water allocation framework
- restoring rivers and streams
- smarter use of irrigation water
- smarter use of urban water
- pricing for sustainability
- encouraging innovation
- accountable water sector.

All Australian governments, federal and state, have adopted policies and legislation to deal with water resource management. In developing a WMP, reference should be made to the various Acts and associated legislation that establish the umbrella framework within which water resources are managed. The following legislation is expected to be relevant at a state level:

- Environment Acts
- Catchment and Land Protection Acts
- Planning Acts
- Water Acts.

The water management of the site should be put in the context of national, state, regional and local environmental and water issues. Familiarisation with the policies and the regulatory framework is a starting point in water management planning. The WMP clearly and precisely states the vision and goals of the organisation in terms of achieving sustainable use of water.

Organisation water objectives

The development of specific objectives, to achieve the organisation's water policy, provides a framework within which priorities can be determined and appropriate actions identified. The objectives will reflect the water issues confronting the organisation and the outcomes that it is seeking. The Vision of the organisation in terms of the nature and purpose of the site and the services to be provided will have a strong influence on the specific water objectives developed by the organisation

In developing water objectives, recognition of objectives contained within other documents such as an environmental management plan or a business plan, need to be taken into account. An organisation's objectives may include:

- providing high-quality sports surface while achieving optimum water use efficiency
- investigating alternative water source options (replace potable water used for irrigation)
- improving water use efficiency
- introducing recycled water for irrigation
- undertaking a detailed review of all water (irrigation and non-irrigation) used on the site
- reviewing all existing plant material in the context of water use requirements and landscape outcomes
- maximising opportunities to use stormwater to reduce current irrigation water demand
- adopting techniques and practices that protect the quality of water in the natural environment
- adopting best management practice in all irrigation operations
- providing training for all staff in water management to ensure performance standards can be achieved
- benchmarking all water use within the responsibility of the organisation, setting targets and implementing strategies to achieve those targets
- developing a Water Management Plan (WMP)
- ensuring that the organisation has adequate resources to achieve and maintain high efficiency standards.

PART B – Information collection

Information collection – site details

Property details

This section identifies all key information about the site that will provide accurate representation of the site and will allow informed decisions to be made about water management. Maps and plans relevant to the site should be recorded or referenced in the WMP. Maps should cover the entire property and include all water features and related infrastructure. Aerial maps are particularly useful.

Records of all licenses and agreements relating to water use should be referenced. This may include agreements for use of recycled water, and water extraction from bores and rivers. Key personnel representing the issuing authority should be noted. These are key documents.

Also, any reports and investigations relevant to water issues should be referenced. This may include site environmental assessments, soil reports and arboricultural reports.

General information about the site should include the:

- name and location of the site
- type of site
- site description (terrain/vegetation)
- major infrastructure – buildings, roads and carparks
- documents and plans – listing of all plans that are relevant to water management, including irrigation and drainage.

Recording the location or repository details of all plans and noting file names assists in accessing plans in a timely manner.

Responsible authority

Both the organisation and the individual responsible for the management of the land and associated water resource issues should be noted.

Climate

Climate records are important sources of information because they provide the basis on which the site can achieve high water use efficiency. Rainfall records for the site are necessary for water management planning including drainage design, water harvesting, irrigation and water management during periods of drought and flood.

Both the rainfall records kept at the site and person responsible for rainfall records should be noted. The records should include:

- rainfall – monthly and annual
- evaporation
- temperatures – average, maximum and minimum
- wind speed and direction for irrigation season
- frosts – number per year.

The Bureau of Meteorology websites for climate averages (www.bom.gov.au/climate/averages/) should be used for historical data. The location of the nearest Bureau of Meteorology automatic weather station (AWS) should be recorded. The climate average sheet for the site/locality should be included as an attachment to the WMP.

In addition to average climate data, which characterises a site generally, it is useful to have an understanding of the variability in climate because it provides an appreciation of the risks associated with water management decisions and provides a stronger basis on which planning takes place. Percentile or decile rainfall records are relevant for water harvesting, flood planning, irrigation demand estimation and drought risk assessment. Note that percentile refers to single number or integer percentages (e.g. 5% or, 26%) and decile refers to groups of 10 numbers or integers (e.g. decile 4 is 40%, decile 8 is 80%).

Example – Canberra rainfall

Mean rainfall (long term average): 633 mm
Decile 9 (90 percentile) annual rainfall: 813 mm
Decile 1 (10 percentile) annual rainfall: 441 mm

Analysis of rainfall over the recent years, for example the last 10 years, is also valuable in terms of understanding trends. Noting extreme climatic events is also valuable (e.g. the highest daily rainfall in Canberra is 74.4 mm and the driest year 302 mm in 1983).

Soil survey

Soil properties have a major influence on water behaviour within the property. The soil types in all areas of the site should be determined, because there is often considerable variability

Table 12.1. Example of turf survey for a golf course

Category	Species	Number	Area of each	Total area	Irrigated area
Tees	Couch	18	200 m^2	3600 m^2	0.36 ha
Greens	Bentgrass	18	400 m^2	7200 m^2	0.72 ha
Fairway	Kikuyu	18	8000 m^2	14 400 m^2	14.4 ha
Rough	Native grass	–	–	–	–
Landscaped areas – club house	Flowering shrubs and annuals	–	–	300 m^2	0.03 ha
Total irrigated					15.51 ha

even over short distances. A detailed soil survey is a very useful tool in assisting in the water management of any irrigated site.

Predominant soil type (topsoil) on site: ....

Soil sample details -position, depth, classification: ....

Soil composition – e.g. roots, organic matter: ....

Vegetation survey

The total area of each type of vegetation should be established. The total irrigated area should be recorded (Table 12.1).

Information collection – water resources

Site water cycle

Understanding all of the water flows into and out of the site is an essential component of the water planning process. The location and characteristics of surface water flows (rivers, streams and creeks), together with details of groundwater, is essential. Identify any areas within the property that may be of concern or may be considered risk areas. These may include waterlogged areas, unstable or erodable areas and water bodies adjacent to chemical and fuel storage areas.

The extraction of water from surface waterways and groundwater will have an impact on the water behaviour of the whole site and adjoining areas. The document should demonstrate an appreciation and understanding of the hydrology of the site including surface water, groundwater and the water movement into and out of the site (Plate 12.1). The behaviour of the property under flood conditions should be described and recorded.

The major natural water flows into and out of the site may include:

- river, streams and water courses, and groundwater, including recharge opportunities
- drainage – surface
- stormwater/runoff (rainfall).

Water sources for irrigation

The particular water sources used for irrigation and other consumption purposes should be identified. Major water infrastructure, such as water supply lines, reticulation mains and irrigation pipelines, should be identified and copies of plans (current and accurate) made available for reference. In some cases, the location of pipelines (both potable and recycled) outside the property need to be identified, because these may be required should alternative water supplies be investigated.

The location of pumping plants, control systems, communication cables and electricity supply need to be included.

Indicate current sources being used and volume available per year.

River extraction: Yes/No Permit amount: ....ML

Storages dams and lakes: Yes/No Storage volume/capacity: .... ML

Above ground (water tanks): Yes/No Storage volume/capacity: .... kL

Bore extraction: Yes/No Permit amount: .... ML

Reticulation supply – potable: Yes/No

Recycled: Yes/No Permit amount: .... ML

Major water infrastructure

Main pipelines – supply and drainage (type, size and location)

....

Pumping plants – irrigation, water transfer, drainage

Water licenses, permits and regulations

The type of license, issuing authority and relevant information, including operating dates should be recorded. Also, key requirements and conditions of operation of licenses and permits should be outlined. The water authorities that have issued water licenses to the organisation should be recorded, along with the title and date of the contract document and the main details of the agreement.

Water costs

The cost of water needs to include not only any charge for water consumption, water rights, but also pumping costs and license costs. The total cost should be determined in $/ML. In addition to the total cost of water, it is valuable to record details of any expected cost increases over time. This will influence decisions in terms of action plans, priorities and equipment investment.

Current cost: .... ($ per ML or kL)

Stormwater and environmental protection

The nature of the site, including layout, design, land and water management practices need to be assessed in terms of potential risks to the environment. Strategies should be put in place to minimise the risk of contamination of surface water and groundwater from chemicals, organic materials and solids. Potential sources of risk include:

- bulk storages of chemicals, fuels, fertilisers and materials
- handling and application of chemicals and fertilisers
- earthworks and construction works
- cleaning of machinery and equipment
- over-watering with irrigation – leachates, erosion, sediments.

Identify potential risks to natural water bodies as result of activities carried out on the site and note strategies to mitigate these risks.

Irrigation water quality

The issue of water quality is becoming increasingly important for urban landscape sites. The general trend is that turf and landscapes are being irrigated with lower quality water. Recycled water generally contains more chemicals, biological and physical matter than potable water. It is expected that increased use of recycled water will place increased management demands on irrigated urban sites. Issues relevant to the use of recycled water, such as salinity and nitrogen content, should be identified.

The quality of water extracted from groundwater supplies and water ways is also declining in many situations. Water quality is important for numerous reasons including: impact on vegetation; impact on the soil; environmental risk; and potential effect on equipment.

tion water and storage water should be undertaken on a regular
equired monthly if the quality of the water source is likely to vary.
ual testing. The testing should include the chemical, physical and
water.
ould include:

ity (EC)

gen, phosphorus, potassium)

ratio (SAR)
gical) oxygen demand (BOD).
r quality test report is presented in Appendix 7.

on techniques and practices

already taken steps to reduce water consumption and adopt water
Outline the current water conservation measures being used. It is
progress in water management and it is also valuable to communicate
ders and the community. Examples of water-conservation techniques
re:
tion system control (e.g. central control)
irrigation scheduling
gation system
rigation system
re regulation
species and landscape performance standards
ng agents.
ibility: nade to the 'Best management practice checklist' in Chapter 7 and 'Strat-
y' in Chapter 10.

tion – irrigation inventory

. The
of the
ion system is fundamental in water management planning. The types of
d the operating characteristics should to be recorded. Irrigation equip-
working order in order to achieve the required levels of performance. All
with use and deteriorates over time. Provision for ongoing maintenance
ssary. Details of dates of installation upgrades and replacement should be

t of a master irrigation system database, independent of the control system
aluable resource. This database – which may include details of products,
formance specifications, repairs, maintenance, landscape plantings, crop
se, system hydraulics, and so on – provides the organisation with water
d by nation that is broader in application and can be used and accessed by a wide
without the constraints of the irrigation control system.
on provided in this section should demonstrate a sound understanding of
plication and the key flow rates of the system. The control of the irrigation
luences the ability to achieve high efficiency in water application. Details on
trol (e.g. automatic or manual) should be included.

The irrigation system may include:

Manual equipment – e.g. portable sprays ☐
Manually operated – sprinkler and microirrigation systems ☐
Quick coupling valves ☐
Automatic –sprinklers and sprays ☐
Automatic microirrigation ☐

The age of the irrigation system should be recorded:

Installation of system – Year ....
Major upgrades completed – Year ....
Planned upgrades/replacement – Year ....

Sprinklers

Sprinkler or irrigation outlet make, model, nozzle: ....
The sprinkler system performance specifications should be recorded:
Sprinkler discharge rate:.......... L/min ....L/sec Wetted diameter: .... m
Operating pressure (design): .... m (kPa)
Sprinkler/spray system PR: .... mm per hour

Microirrigation emitter type, make, model ....
Emitter discharge rate .... L/h
Emitter operating pressure .... m (kPA)

Irrigation control system

Type of controller – stand alone, master/satellite, central control: ....
Controller make and model: ....
Number of stations: .....
Weather station: Yes ☐ No ☐
Reference ETo input: Yes ☐ No ☐
Sensors –rain, wind, soil moisture, flow, pressure: ....
Programming capability – number of programs, cycling, run time range, fle
....

Current irrigation schedules

Irrigation scheduling refers to the amount of water applied and the timing of irrigati peak irrigation schedule (mm/week) that is currently being used for each separate are site should be recorded.

Pumps, filtration and water treatment equipment

Details of the pumping plant should be included.

Number of pumps:
Pump type and make:
Motor capacity: ... kW
Design duty: ... L/s at .... kPa
Pump efficiency at Design duty: ...... %

The type of filtration equipment used for horticultural sites is principally determin the quality of the water source and the irrigation techniques employed.

Details on both the type of equipment and operating functions should be recorded.

Does the site have filtration equipment? Yes ☐ No ☐

Brief description: ....

Fertigation system

Does the site have a chemical injection system? Yes ☐ No ☐
Brief description of fertigation system: ....
Type of water conditioning: ....
Backflow prevention equipment installed. Make, size, model and location:...........
Details of servicing of backflow:..

Irrigation system maintenance

Irrigation systems comprise many operating parts that are prone to wear. Continual monitoring of the operating condition of the whole system is required to ensure efficiency. It is valuable to have a document (irrigation system maintenance plan) that outlines the maintenance requirements of the irrigation system. It should identify the equipment to be checked, testing procedure, frequency of checking and the recording details of inspections and maintenance works. Personnel responsible for system maintenance should be nominated.

Documented maintenance program for the irrigation system? Yes ☐ No ☐

Information collection – water consumption

The volume of water used for each site or landscape area needs to be accurately determined. Dedicated metering for irrigation consumption is essential. It is valuable to record water consumption over a number of years. This allows trends to be identified and further analysis of the data. However, caution is required in evaluating trends because the volume of water required in a particular year will depend to some extent on the weather conditions for that period.

Current and expected future water consumption volumes are key values in water management planning (Table 12.2).

Record volume of water used for each purpose and from each source. Sources may be potable water, river/stream, bore, stormwater/rainfall or reclaimed water.

Opportunities to replace potable water should be investigated. The roof of the sheds and office buildings provides a suitable area for rainfall harvesting. This water may be used for filling spray tanks and cleaning machinery.

Rainfall collection from roof : Yes ☐ No ☐

PART C – Analysis and interpretation of data including water use

Analysis of water consumption

The annual irrigation volume (AIV) is the expected amount of water that will be required, taking into account the total area to be irrigated, turf and landscape species, required vegetation performance, efficiency of the irrigation system and average climatic data for the period.

At the end of an irrigation season, the actual water consumption should be compared with the AIV or water budget. If the actual consumption is higher, there could be several causes. Hotter conditions and/or less rainfall than average will generally mean higher water consumption. It may also indicate that the system was poorly managed and water in excess of

Table 12.2. Annual water consumption report. Year ....

Water use	Water source 1 Potable	Water source 2 Rainwater tank	Water source 3 Dam
Irrigation	12 000 kL	–	80 000 kL
Other. Spraying/maintenance/wash down	–	5 kL	–
Clubhouse/office/residence/maintenance building	250 kL	10 kL	–

requirements was applied. When reporting annual water consumption, explanations should be provided on reasons for significant variations in values, whether higher or lower.

Evaluation of irrigation performance

Audit results

All irrigation systems should be regularly evaluated to determine the current level of efficiency. Irrigation systems can be assessed in terms of how well water is currently being applied and how well the system has been managed over the irrigation season.

The main technique used to evaluate irrigation systems is to carry out an irrigation audit. This procedure involves visually checking the system and then carrying out a can test. Other crucial information, including pressure and flow rates, should be obtained. The readings obtained during a can test allow the uniformity coefficient (DU or EU) to be calculated.

The evaluation of how well the irrigation system has been managed needs to consider the total amount of water applied over the irrigation season and the actual climate conditions, evaporation and rainfall during the period. The performance indicator used to provide a measure of the overall irrigation efficiency is called the irrigation index (Ii). Details on the procedure to evaluate irrigation performance, including auditing and determination of the Ii are outlined in Chapter 11.

The audit of fixed sprinkler irrigation system performance should include:

Has an irrigation uniformity audit been carried out? Yes ☐ No ☐

Name of auditor: ....

Date of audit: ....

Name of site/area tested: ....

Sprinkler uniformity: DU ...... %

Irrigation application rate: ......mm per hour

Monthly water budgets and annual irrigation volume (AIV) preparation

How much water is expected to be used in a typical year? .... ML

This is an estimate of the amount of water that will be required for the total landscape area, including turf, trees and landscape plantings, expected climate conditions and the efficiency of the irrigation system.

A monthly water budget should be prepared for the site. This provides reference values and target water volumes for the assessment of the performance of the irrigation each month. The method used to determine water budgets and annual irrigation water requirement is outlined in Chapter 5.

Benchmark irrigation water consumption

The amount of water used to maintain the site should be benchmarked against other similar enterprises and industry best practice. This informs the organisation about current water consumption patterns and provides a reference point for improvement in performance. Appropriate benchmarking units are ML/ha and kL/1000 m^2 (Note: 1.0 ML/ha = 100 kL per 1000 m^2). It is determined by dividing the total irrigation water volume by the total irrigated area.

The current actual water consumption for the site .... ML/ha

Optimum irrigation schedule

Efficient irrigation is founded on delivering the right amount of water efficiently to the root zone in a timely manner. The optimum irrigation schedule should be determined as part of the evaluation process. Details on the soil properties, rooting depth of the vegetation and water requirements of irrigated vegetation need to be established. Typical climate conditions for the irrigation season are also required.

The determination of the recommended irrigation depth is outlined in Chapter 6.

Month of prepared schedule (e.g. summer month): ....

Recommended irrigation depth per irrigation event : ....mm

Recommended weekly application for nominated month: ..... mm

Typical irrigation event frequency (e.g. every 3 days): .... days

PART D – Strategies, implementation and review

Strategies to address inefficiencies and wastage

To improve water use efficiency of the whole site, which may include buildings such as pavilions, hard surface areas as well as soft (turf and landscape) areas, the following best practice approach to water conservation should be followed:

1. Remedy unnecessary uses – for example, sweeping rather than hosing down paths.
2. Fix leaks – respond quickly to leaks, adopt maintenance procedures that facilitate checking of equipment to correct faulty equipment.
3. Use the minimum amount of water to accomplish a task – does not leave taps running, using trigger nozzle hoses and assess water needs of each task.
4. Recycle and reuse – investigate opportunities to harvest, use and reuse recycled water sources.
5. Replace potable water – adopt a 'fit-for-purpose' approach to water source selection.
6. Install meters to measure the amount of water used. Install flow meters in irrigation systems to monitor the correct operation of the system.

Determine what steps can be taken to prevent wastage and improve water use efficiency. List and describe each opportunity to reduce waste. Water-saving achievements and initiatives should be communicated to all stakeholders, including the community.

Potential turf and landscape water conservation strategies

There are potentially many opportunities to conserve water when irrigating turf and landscape areas, including:

- reducing demand for irrigation water – design, plant selection
- optimising rainfall – harvest rainfall and prevent runoff
- improving the application efficiency of irrigation – technology
- improving the scheduling of irrigation
- improving the soil water properties – e.g. organic matter, encourage deeper rooting.

It is valuable to initially list all of the initiatives that may be applicable to the site. The process should involve consultation with all stakeholders, including staff and the community. Each potential initiative should be assessed in terms of its potential to reduce water demand, increase efficiency, cost, ease of implementation and overall contribution to the water objectives of the organisation.

Expected savings and efficiency gains will be an estimate. It is valuable to review experiences of other organisations managing similar irrigated areas who have adopted potential strategies for the site. Assistance may be available from irrigation specialists, consultants, water authority staff and colleagues in your industry sector.

Select strategies and initiatives to be implemented and outline the expected water savings for each. Refer to Chapter 10 for details of specific water efficiency strategies.

Strategies for sustainable water use

Potable water replacement opportunities

Drinking water (potable water) is a highly valuable resource. The replacement of potable water currently being used to irrigate turf and landscapes is a priority in most states. Strategies to replace potable water used for irrigation should be investigated, evaluated and incorporated into future WMPs.

Some alternatives to potable water for irrigation include:

- reclaimed stormwater from both within and outside the property
- treated wastewater (recycled water)
- greywater (e.g. water from clubhouse showers)
- rainwater harvested from roofs
- borewater – if extraction is sustainable
- rivers and creeks.

Outline the steps being taken to secure alternative water supplies.

Capacity to achieve WMP objectives

In addition to having the policies in place to achieve sustainable water use, it is also necessary to have the capacity to implement the various strategies and achieve the desired outcomes. The key areas to consider are the quality information – (data), financial resources (both capital and operating) and skills

The area of skills training is particularly important. Courses are available through Irrigation Australia Ltd (IAL), TAFE Colleges and the irrigation companies. As a minimum requirement, personnel involved in operating irrigation systems should be able to undertake a basic audit of the system to ensure that it meets the necessary industry performance standards (Figure 12.1).

Figure 12.1. Participation in system evaluation training programs builds the organisation's capacity for high efficiency (Irrigation Efficiency Course participants, Royal Botanic Gardens Melbourne)

Drought management plan and restrictions

There will be periods when, due to climate conditions, water availability will be limited or not available at all. A plan of how the organisation proposes to manage irrigation during these periods of drought and water restrictions should be detailed.

How water is used on the site needs to be assessed and prioritised in relation to available water. The levels of turf and landscape performance that will be maintained under different levels of water supply should be determined.

Consider various water supply scenarios:

1. Time-based restrictions
2. Volume-based restrictions (e.g. 80%, 60%, 40%, 20% water volume allocations)
3. No rain (e.g. for rain-fed systems, determine what approach will be used for 10, 20, 30, 40 days without rain)
4. Restrictions on method of irrigation: review the irrigation technique and design with regard to possible limitations on the use of sprinklers and sprays.

Is there a drought management plan? Yes ☐ No ☐
What is the date of the document? ....
What are the key elements of the drought management plan?
Is there a strategy to cope with water restrictions? Yes ☐ No ☐
Brief details: ....

Planning for implementation

Key steps in the implementation process

1. Nominate actions to achieve water management goals
2. Nominate targets – water saving and efficiency
3. Who is responsible for achieving the identified outcomes?
4. When will it be done and in what timeframe: short term (1 to 3 years), medium term (5 to 10 years) or long term (20 to 50 years)?

Potential actions for implementation

There are numerous actions that can be undertaken to support the aims of the water management planning process (Table 12.3). These include:

- increasing awareness of water issues
  - newsletter to staff and stakeholders
  - reporting of water use data
  - staff training in techniques to be more water efficient
  - giving staff responsibility for specific areas of water efficiency
  - encouraging staff to participate in the process, including offering suggestions
  - rewarding staff for significant achievements in water use efficiency projects.
- measuring and reporting water use performance
  - installing meters for each separate use area
  - establishing a water use database
  - using meters to check for leaks
  - testing irrigation system performance.
- adopting technologies and practices to reduce water use
  - redesigning systems, if required
  - upgrading control systems that incorporate feedback (e.g. sensors)
  - ensuring appropriate hydrozoning of all watered areas
  - installing soil moisture sensors
  - installing pressure regulation and flow sensing

Table 12.3. Planned actions, priorities and budgets for a water management plan

No.	Planned action	Priority	Estimated water saving or efficiency gain	Date for completion	Budget estimate ($)	Who is responsible?

Table 12.4. Timetable and responsibility allocation for planned actions in a water management plan

	Key water priorities – actions	Who is responsible?	Timetable
1			
2			
3			

- matching irrigation schedule to season and site conditions
- managing surfaces for effective infiltration.

Further information on potential actions, see Chapter 10 'Strategies for high efficiency and previous sections in this chapter.

It is important to review progress and identify personnel who are responsible and accountable for delivery of the WMP objectives (Table 12.4)

Review progress

Water consumption performance data will need to be collected on a regular basis. It is important to determine the following:

- What data is to be collected?
- How is it to be collected?
- When is it to be collected?
- Who is responsible for collecting the data?

Evaluate the performance against targets on an annual basis. Provide an explanation if the water consumption performance varies from targets. The whole WMP should be reviewed at least every 3 years. Nominate the person responsible for the reviewing the WMP.

The success of the WMP is founded in the level of ownership by all involved in water management and the commitment of the organisation to its objectives.

Appendix 1

Acronyms, terms and units

Description of term	Symbol	Unit/s
Annual irrigation volume	AIV	kL, ML
Available water holding capacity	AWHC	mm/m, mm/cm, cm/cm
Available water	AW	mm
Bulk density	BD	g/cm^3, t/m^3
Christiansen uniformity coefficient	CU	Decimal, %
Coefficient of variation	C_v	Dimensionless, decimal
Crop coefficient	K_c	Dimensionless, decimal
Crop factor	CF	Dimensionless, decimal
Distribution uniformity	DU	Decimal, %
Distribution uniformity – lower quarter	DU_{LQ}	Decimal, %
Distribution uniformity – lower half	DU_{LH}	Decimal, %
Density factor (used to determine K_L)	K_d	Dimensionless, decimal
Electrical conductivity	EC	dS/m, µS/cm
Emitter interval (along pipe)	S_{ei}	mm, m
Emitter spacing (between driplines)	S_{ls}	mm, m
Evaporation (pan evaporation BoM)	E_{pan}	mm/d, mm/week, mm/year
Evapotranspiration rate (reference crop)	ET_o	mm/d, mm/week, mm/month, mm/year
Evapotranspiration rate (crop or plant)	ET_c	mm/d, mm/week, mm/month, mm/year
Field capacity (expressed volumetrically or as tension)	FC	mm/m or kPa
Flow rate – system or part	Q	L/s, L/h, L/min, m^3/h
Friction rate (pipe)	H_f	m/100 m, m/1000 m
Head (pressure head of water)	H	m
Hydraulic conductivity (saturated)	K_{sat}	mm/h, mm/d
Infiltration rate	IR	mm/h
Irrigation application efficiency	E_a	% or decimal
Irrigation water management efficiency	E_{wm}	% or decimal
Irrigation efficiency (overall)	IE	% or decimal
Irrigation depth	ID (I & Ir used in IPOS)	mm
Irrigation index	Ii	Dimensionless, decimal

Description of term	Symbol	Unit/s
Irrigation interval	Ti	Days
Irrigation water requirement	IWR	mm
Landscape coefficient	K_L	Dimensionless, decimal
Lateral (sprinklers) spacing between lateral pipes	L	m
Leaching fraction	LF	Dimensionless, decimal
Management allowable deficit	MAD	% or decimal
Mean application rate	MAR	mm/h
Microclimate factor (used to determine K_L)	K_{mc}	Dimensionless, decimal
Number of irrigation cycles	NC	Integer
Number of stations	NS	Integer
Permanent wilting point	PWP	mm, mm/m or kPa
Plant water requirement	PWR	mm
Power (pump)	Ps	kW
Precipitation rate – gross	PR_G	mm/h
Precipitation rate – net	PR_N	mm/h
Rainfall	P (R used in IEC Course)	mm
Rainfall – effective	P_{eff}	mm
Rainfall (effective) factor	RF	%
Readily available water (soil water property)	RAW	mm/m
Refill depth	RD	mm
Refill point	RP	mm or tension (– kPa)
Root zone depth	RZD	mm
Run off coefficient	ROC	Dimensionless, decimal
Run time	RT	min
Run time multiplier	RTM	Dimensionless, decimal
Saturated hydraulic conductivity	K_{sat}	mm/h, µm/s
Soil water stress factor	K_{ws} (K_s used in FAO 56)	Dimensionless, decimal
Stored water (same as total available water (TAW))	SW	mm
Sprinkler interval (spacing)	S	m
Time duration of audit test	T_t	min
Time for run off	TRO	min
Total available water (in root zone)	TAW	mm
Total volume for a control station	TVS	L, kL
Turf quality visual standard	TQVS	Dimensionless, decimal
Volume	V	mL, L, kL, ML
Working storage (same as refill depth (RD))	WS	mm

Appendix 2

Glossary of terms

Application efficiency (E_a)
A measure of the proportion of the water applied by an irrigation system that is delivered into the plant root zone and is available for use by the plant.

Available water holding capacity (AWHC)
The amount of water that is stored between field capacity (FC), the upper limit, and permanent wilting point (PWP), the lower limit. Expressed in millimetres per metre depth of soil (mm/m) and water tension or water suction.

Backflow prevention device
A safety device designed to prevent water flowing in reverse direction and potentially polluting/contaminating a potable (drinking quality) water supply.

Base irrigation requirement (BIR)
The monthly irrigation volume required to meet site performance standards using historical climate data. A scheduling term used in the Irrigated Public Open Space (IPOS) Program.

Bioretention
A shallow vegetated basin used to retain stormwater and treat the water through nutrient uptake and removal of pollutants.

Blackwater
The wastewater from a toilet or bidet, which contains human faeces and urine. It is highly polluted and infectious.

BOD (biochemical oxygen demand)
A measure of the amount of oxygen required in the oxidation of organic matter in the water. Test commonly conducted over 5 days. High BOD indicates that a significant amount of oxygen-demanding matter is present.

Bulk density (of soil) (BD)
This is the weight of soil in a given volume of soil. This term indicates the relative densities of soils. It is a term required to convert from water content by volume to water content by weight.

Capillary water
Water that is held within the pore spaces of soil following drainage and is the main source of water extracted by plants.

Cavitation
The formation and rapid collapse of air bubbles or cavities in pumped water. This results in damage (pitting) to the pump and reduced pumping performance.

Chlorination

The process of applying chlorine (gas, liquid or solid) to water to act as a disinfectant to reduce disease causing microorganisms (pathogens). This process is also used as a preventative strategy in maintaining quality of storage water and water supply.

CIMIS

California Irrigation Management Information System (CIMIS). A network of weather stations that provides real time ET data to irrigators, free of charge.

Class (pipe)

The pressure rating or maximum working pressure of a pipe. Pressure rating designated by PN number.

Coagulation

The incorporation of a compound (e.g. alum) to cause very small particles (colloids) to aggregate so that the formed larger particle can be removed through settling or filtration.

Coefficient of uniformity (CU)

The Christiansen coefficient of uniformity. A measure of sprinkler uniformity developed by J.E. Christiansen, University of California, 1942.

Coefficient of variation (C_v)

A statistical measure used to describe the uniformity of discharge of emitters due to physical variations in the dimensions of emitter flow pathways.

Colloid

The smallest of all soil particles, less than 0.001 mm in size. Clay particles include sizes up to 0.002 mm. Colloids are generally dispersed throughout the water solution.

Cross connection

The potential or actual contamination of a potable water supply from a non-potable source. Presents a public health risk to users of potable supplies.

Deficit irrigation

An irrigation scheduling strategy in which water applications are less than amounts required for full or lush growth.

Disinfection

A treatment process including heat, radiation and chemical or physical filtration (e.g. membrane) to remove, inactivate or kill pathogens in water.

Dynamic pressure

The hydraulic (water) pressure in a system when the water is flowing

E. coli

An abbreviation for the bacteria *Escherichia coli*. Used as an indicator of faecal contamination in water.

Effective rainfall (P_{eff})

The proportion or amount of rainfall that enters the root zone and is used by plants.

Effluent

The water delivered or produced from a water treatment process.

Electrical conductivity (EC)

A measure of how well the water or soil solution conducts electricity. Higher EC values indicate higher concentrations of total dissolved salts in the water.

Emitter

A microirrigation outlet device designed to deliver water at low flow rates in a controlled way. The mode of delivery may be drops, bubbler streams or sprays.

Eutrophication
The process whereby nutrients, such as nitrogen and phosphorus, contaminate water courses or water bodies and degrade the water bodies, including causing algal growth.

Evapotranspiration (ET)
The combined water transferred to the atmosphere both through transpiration from plant foliage and the evaporation of water from the ground or soil surface.

Faecal coliforms
An indicator of the amount of faecal contamination (animal or human) in water. The concentration of *Escherichia coli* (*E. coli*) present in the water is used in the test.

Fertigation
The injection of fertiliser and chemicals into the irrigation water delivery for direct application to the plant growing areas.

Field capacity (FC)
Water remaining in the soil pore spaces following saturation and drainage of gravitational water. Drainage may take 1 to 2 days.

Flocculation
The use of polymers to facilitate small suspended particles (flocs) joining together into clumps, which can be removed from the water by sedimentation and filtration. This process is similar to coagulation.

Gravimetric soil water (GSW)
The proportion of the weight of the total amount of water in the soil sample to the weight of dry soil. The dry soil measurement is achieved through oven drying (e.g. 24 hours at 105°C). GSW is an accurate measure of soil moisture content. Units of GSW are gram per gram (g/g).

Gravitational water
The water that readily drains downwards (under influence of gravity) following saturated soil conditions. Water retained in soil pores is called capillary water.

Greywater
Wastewater (untreated) from household showers, hand basins, washing machines and laundry basins. Includes kitchen waste and dishwasher waste, which are generally not considered usable unless treated.

Head (water pressure) (also known as hydraulic head)
A measure of the pressure in water expressed in terms of the height of water in metres (m) relative to a reference datum.

Hydraulics (water)
The area of engineering science involved in the study of water properties and water movement in pipe, pump and waterway systems.

Hydrology
The scientific discipline involved with the study of the properties and behaviour of water interacting with the Earth's surface. The 'water cycle' is a core component of hydrology.

Hydrophobic
A soil condition in which water is repelled, so water does not infiltrate and can lead to waste. This condition is generally associated with dry organic or sandy soils. Wetting agents (surfactants) are used to treat hydrophobic soil conditions.

Hydrozone
A section of the landscape that has its own water demand properties due to plant species, microclimate or planting density. Hydraulic constraints, such as limited capacity, also determine hydrozones.

Hygroscopic
Water held to individual soil particles by adhesion forces and not able to be extracted by plants.

Infiltration rate (IR)
The rate at which water enters the soil that corresponds to a constant steady state rate. Sometimes referred to as the final infiltration rate because water initially enters the soil at a relatively high rate and then decreases over time. The units of IR are mm per hour (mm/h).

IPOS
Irrigated Public Open Space (IPOS) is a best practice irrigation management program developed in Adelaide and managed by SAWater, Adelaide.

Irrigation efficiency (IE)
A measure of the proportion of water delivered by the irrigation system to the target area over the irrigation season, which is used beneficially by the plants.

Lateral
A pipe to which irrigation outlets (sprinklers, sprays and emitters) are directly connected. This pipe is connected downstream of a supply control valve, such as a solenoid valve.

Leaching fraction (LF)
The additional irrigation water required to be applied to the soil to move salts below the root zone through drainage.

Management allowable depletion (MAD)
The proportion of total available water (TAW) that is selected to be allowed to be removed, prior to the commencement of irrigation. It is an irrigation management decision, not a soil property.

Matched precipitation rate (MPR)
Sprays and sprinklers that apply water at the same rate regardless of the sector or arc (proportion of full circle) that is being wetted.

Membrane filtration
The use of a porous membrane, with micropores, to remove suspended matter from water. Membrane filtration classifications include microfiltration, ultrafiltration and reverse osmosis.

Microirrigation
Method of irrigation in which low flow rates and low pressures are used to deliver water directly to plants and plant root systems.

Microfiltration
The use of porous membrane, with very small pore sizes or openings – typically in the range of 0.1 to 10 microns. Used to physically remove bacteria.

NTU (nephelometric turbidity unit)
The unit of measurement of turbidity or cloudiness of water. Measured using an optical instrument. A measure of the suspended, particulate or colloid (very fine clay particles) matter.

Oven dry soil
Following extended drying in an oven, at high temperatures (e.g. 24 hours at 105°C), the soil is considered not to contain any water. The difference in soil sample weight is a measure of water in the soil.

Passive recreation
Use of open space area for recreation, purposes such as walking, resting and picnicking. Not an area used for organised active sports, such as soccer or football.

Pathogens
Microorganisms that can cause human illnesses and infections or plant disease. They are removed through the process of disinfection.

Permeability (water in soil)
The rate at which water moves through the soil. It is represented by the term saturated hydraulic conductivity (K_{sat}). Particle size, pore size and soil structure strongly influence permeability, as well as the pressure or head of the water. Examples of soil permeability units are mm/s, mm/h and m/day.

Permanent wilting point (PWP)
The amount of water remaining in the soil when plant cannot extract any more water from the soil. The plant wilts and does not recover. It defines the absolute lowest storage capacity of the soil root zone.

Pollutant
A compound or substance that degrades the quality of the environment.

Potable water
Water that is suitable for human consumption; for example, for drinking or using in cooking.

Primary treatment
The first stage in an overall water treatment process in which large suspended particles and floating debris are removed by filtration and/or sedimentation

Readily available water (RAW)
The soil water that plants can readily extract from the soil. The actual soil water level that represents readily available water depends on the plant and the soil properties. Water is more readily available at lower levels of water tension in coarser soils than in finer soils.

Reclaimed water
Water recovered and used from a sewage treatment plant. The term is sometimes used interchangeably with recycled water.

Recycled water
Water, which has been used previously, that has been collected and treated to a quality level that is suited to a specific application. Recycled water includes reclaimed water.

Reduced pressure zone (RPZ) device
A type of backflow prevention device used to protect potable water supplies. It incorporates two check valves and a chamber with pressure maintained lower than the supply pressure. The device has provision, in the form of tappings, to allow its correct functioning to be checked.

Remote control valve
A valve that can be operated, remotely or at a distance, using electricity, air or hydraulic control lines. Electrically operated remote control valves are referred to as solenoid valves.

Refill point (RP)
The value of the soil moisture level (set-point) used to initiate an irrigation event. It can be expressed as water content/volume or water tension.

Reverse osmosis
A process, using a semi-permeable membrane with very small openings (e.g. 0.001 microns or smaller), that removes dissolved salts and other solutes, on the surface of the membrane. The process requires high pressure to force the solution through the membrane.

Saturated hydraulic conductivity (K_{sat})
The rate at which water moves through the soil under saturated conditions. A measure of the permeability of the soil. The units of Ksat include mm/h and µm/s.

Saturated soil
Following a sufficient amount of precipitation, rainfall or irrigation, all of the pore spaces are filled with water. There is no air in the soil voids.

Secondary treatment
Removal of organic matter through a biological or chemical process that significantly reduces the organic content (BOD) and the suspended solids in the water.

Sewage
Waste matter, including human excrement, carried away in pipes (sewers).

Sewer mining
The extraction of wastewater solution from a sewer main for treatment to a quality level that allows reuse.

Sodicity
High levels of sodium in the soil that result in dispersal of fine soil particles and soil swelling. This condition can lead to reduced infiltration and reduced soil permeability.

Soil water potential
A property of the water within the soil and plants that can be expressed in terms of suction or energy levels. The term water potential equates to the energy level of the water. Water moves in plant soil systems as a result of water potential or energy gradients (differences).

Soil water stress factor (K_{ws})
This factor or coefficient is used to make an allowance for reduced plant water use as a result of water not being readily available and the plant is water stressed. The term is referred to as soil water stress coefficient (K_s) in FAO 56.

Static pressure
The pressure of water that is exerted when there is no water flow at that point. The water in the pipeline is stationary or static.

Station
An irrigation controller function that allows a lateral or number of laterals (supplying sprinklers, sprays and emitters, such as drippers) to be operated simultaneously.

Tension (water tension)
A measure of the force by which water is held by the soil particles. It has a negative value when referred to as tension and a positive value when it is described as suction; that is, tension of –200 kPa is the same as suction of 200 kPa.

Tertiary treatment
Further treatment of secondary treated water to remove a high percentage of suspended solids and nutrients and the reduction of pathogens to very low risk, levels, through disinfection.

Total available water (TAW) or soil water (SW)
The total amount of water that is in the root zone of the plant and is available to the plant. It is determined using the value of AWHC for each soil layer. It is usually expressed in mm.

Total dissolved solids (TDS)
A measure of the total dissolved ions (salts), such as sodium, chloride, calcium and magnesium in the solution. TDS is measured using electrical conductivity (EC). The unit of measurement is dS/m, but it is often expressed in terms of concentration (mg/L). To convert from dS/m to ppm or mg/L, multiply by 640.

Total suspended solids (TSS)
The amount of material that is in suspension in water and can be removed by filtration. The unit of measurement is mg/L.

Turbidity (*see also* NTU)
A measure of the opacity or cloudiness (lack of transparency) of water. Turbidity is caused by light scattering and absorption by colloidal matter or suspended particles. The unit of measurement is the nephelometric turbidity unit (NTU).

Vacuum breaker
A device fitted to a pipeline, to allow air to enter, to prevent excessively low or negative pressures in the pipe network. This prevents the pipe from collapsing.

Wastewater
Untreated water from industrial processes and households. The water may contain wide range of chemical, biological and physical contaminants.

Water content by weight
This is the proportion of water in the soil expressed as the weight of water in the soil divided by the dry weight of the soil. (g/g %)

Water content by volume
This is the proportion of water in the soil expressed as the volume of water in the soil divided by the volume of soil. (vol/vol %)

Water management efficiency (E_{wm})
A measure of how well the irrigation system is operated in response to meeting the landscape water needs, the climate and soil water conditions. A core component of the overall irrigation efficiency.

Water use efficiency (WUE)
A term, used very broadly, to describe the productivity or outcomes that have been achieved through the application of irrigation water. Definitions for this term vary.

Wilting point
See Permanent wilting point (PWP).

Working pressure
See Dynamic pressure.

WUCOLS
Water Use Classification Of Landscape Species (WUCOLS). A comprehensive guide, developed by L. Costello and K. Jones, University of California, in 1994, that lists the water use rates of landscape plants.

Zone (irrigation)
A group of irrigation outlets (e.g. sprinklers or emitters) operated simultaneously from an individual valve.

Appendix 3

Climate data

Appendix 3.1. Annual rainfall for capital cities (long-term)

	Annual rainfall					
Capital city and BOM station	**Average (mm)**	**10th percentile (mm)**	**50th percentile (mm)**	**90th percentile (mm)**	**Highest (mm)**	**Lowest (mm)**
Adelaide (Station # 023034)	445.7	323.6	450.6	575.6	730.8	234.6
Brisbane (Station # 040214)	1149.1	719.9	1106.7	1602.5	2242.4	411.5
Canberra (Station #070282)	632.6	440.9	604.8	812.6	876.0	301.6
Darwin (Station #014015)	1715.0	1221.4	1700.9	2147.9	2776.6	1024.7
Hobart (Station # 094008)	497.1	349.0	477.1	662.5	735.4	297.2
Melbourne (Station # 086071)	648.0	469.2	645.0	823.2	967.5	332.3
Perth(Station # 009034)	867.6	685.1	854.2	1047.2	1338.8	508.7
Sydney (Station # 066062)	1213.8	821.0	1162.2	1654.2	2194.0	583.0

Source: BOM climate averages. Website: http://www.bom.gov.au

Appendix 3.2. Annual evaporation (E_{pan}) for capital cities (long-term)

	Annual evaporation (E_{pan})					
Capital city	Average (mm)	10th percentile (mm)	50th percentile (mm)	90th percentile (mm)	Highest (mm)	Lowest (mm)
Adelaide (Airport)	1902	1789	1872	2079	2196	1541
Brisbane	1567	1390	1582	1765	1837	1357
Canberra	1372	1229	1335	1550	1217	1572
Darwin	2519	2185	2258	2819	2920	2075
Hobart	1325	1215	1337	1427	1475	1188
Melbourne (City)	1189	987	1148	1432	1592	905
Perth	1731	1583	1703	1901	2167	1473
Sydney (City)	1061	988	1053	1158	1120	969

Source: BOM climate averages Website: http://www.bom.gov.au

Appendix 3.3. Monthly evaporation (E_{pan}) and monthly rainfall values for capital cities (long-term averages)

City		Average monthly evaporation (E_{pan}) and monthly rainfall (mm)												
		Jan	Feb	Mar	Apr	May	Jun	Jul	Aug	Sep	Oct	Nov	Dec	Annual
Adelaide	Evap.	279	238	204.6	129	80.6	57	62	86.8	117	173.6	219	254.2	1898
	Rain	17.1	18.1	20.8	35.1	54.1	55.6	59.6	49.9	45.9	38.5	25.5	24.2	445.7
Brisbane	Evap.	179.9	145.6	139.5	114	93	75	75	102.3	130.2	144	173.6	189.1	1569
	Rain	159.6	158.3	140.7	92.5	73.7	67.8	56.5	45.9	45.7	75.4	97.0	133.3	1149.1
Canberra	Evap.	217	168	139.5	81	49.6	33	37.2	55.8	84	124	165	213.9	1387
	Rain	59.8	51.2	55.6	49.3	47.5	37.9	52.4	47.6	65.2	61.9	58.7	46.1	632.6
Darwin	Evap.	186	159.6	179.8	192	213.9	207	210.8	223.2	231	248	225	207.7	2482
	Rain	420.3	364.6	320.8	99.8	20.7	2.0	1.3	5.3	15.2	69.2	139.1	247.3	1715.0
Hobart	Evap.	195.3	154	136.4	84	58.9	39	43.4	62	90	127.1	150	186	1314
	Rain	41.0	36.6	36.6	43.1	34.5	30.1	43.9	46.4	40.7	47.0	43.3	53.6	497.1
Melbourne	Evap.	176.7	145.6	120.9	75	54.4	33	37.2	52.7	75	108.5	132	161.2	1168
	Rain	47.6	47.3	50.2	57.3	55.9	49.0	47.7	50.2	57.9	66.2	59.5	59.2	648.0
Perth	Evap.	257.3	218.4	195.3	120	77.5	57	55.8	71.3	99	148.8	189	235.6	1715
	Rain	8.6	13.3	19.3	45.5	122.7	182.4	172.9	134.6	79.9	54.5	21.7	13.9	867.6
Sydney	Evap.	142.6	109.2	96.1	78	58.9	36	46.5	58.9	75	102.3	129	136.4	1058
	Rain	102.0	117.9	129.4	126.4	120.7	130.6	97.3	81.7	69.4	76.9	83.5	77.9	1213.8

Notes:

1. Sourced from Bureau of Meteorology www.bom.gov.au (climate averages)
2. Annual values determined from average daily evaporation values.
3. Evaporation rates vary spatially across the geographic area of the city.
4. The data records are based on a range of years for each weather station site.

Appendix 3.4. Estimated monthly reference ET_o for capital cities (E_{pan} based)

City	BOM Station number	Jan	Feb	Mar	Apr	May	Jun	Jul	Aug	Sep	Oct	Nov	Dec	Annual
Adelaide (Airport)	023034	223	190	164	103	64	46	50	69	94	139	175	203	1518
Adelaide (Kent Town)	023090	181	150	122	74	45	34	37	52	72	107	137	161	1168
Brisbane (Regional)	040214	144	116	112	91	74	60	60	82	104	115	139	151	1255
Brisbane (Aero)	040223	181	146	144	108	79	72	79	102	132	156	173	186	1934
Canberra	070282	174	134	112	65	40	26	30	45	67	99	132	171	1110
Darwin	014015	149	128	144	154	171	166	169	179	185	198	180	166	1956
Hobart	094008	156	123	109	67	47	31	35	50	72	102	120	149	1051
Melbourne (City)	086071	141	117	97	60	44	26	30	42	60	87	106	129	934
Melbourne (Airport)	086282	198	157	141	91	60	46	50	67	98	126	142	179	1343
Perth	009034	206	175	156	96	62	46	45	57	79	119	151	189	1373
Sydney (City)	066062	114	87	77	62	47	29	37	47	60	82	103	109	846
Sydney (Richmond)	067021	146	110	99	74	55	43	47	67	91	114	118	139	1110

Notes:

1. Data were sourced from Bureau of Meteorology www.bom.gov.au (Climate averages)
2. The monthly ETo values have been determined using long term Epan readings. A pan coefficient (K_p) of 0.80 was used. This is an average value for Class A pans with light winds (FAO 56).
3. Current daily ETo values are available from BOM, however these have only been available since 2009, so there is no long term history of records based on daily data.
4. The daily average Epan value for each site was used to determine the monthly total amount.
5. The annual totals have been determined using the daily average multiplied by 365. This annual total may be at slight variance to the summation of the calculated monthly values.

Appendix 3.5. Weekly irrigation requirements in high evaporative demand periods in capital cities for selected plant water use categories

City	High demand period (January)	Plant water requirement (PWR) (mm/week)			Irrigation water requirement (IWR) (mm/week)		
	Weekly evap (ET_0) mm/week	Low (K_c 0.3)	Medium (K_c 0.5)	High (K_c 0.8)	Low water demand	Medium water demand	High water demand
Adelaide (Airport)	50	15	25	40	18.8	31.3	50.0
Brisbane (Regional)	33	9.9	16.5	26.4	12.4	20.6	33
Canberra	39	11.7	19.5	31.2	14.6	24.4	39
Hobart	35	10.5	17.5	28	13.1	21.9	35
Melbourne (Airport)	45	13.5	22.5	36	16.9	28.1	45
Perth (Regional)	47	14.1	23.5	37.6	18	29.4	47
Sydney (Richmond)	33	9.9	16.5	26.4	12.4	20.6	33

Notes:

1. Irrigation application efficiency (Ea) 80% assumed for irrigation requirement calculation.
2. Irrigation water Requirement (IWR) calculation assumes no rainfall for the week.
3. Evaporation data sourced from BOM.

Appendix 4

Plant water use

The water use ratings are provided as a general guide only. A number of sources, both Australia and international, have been used in compiling these lists. As expected, variations in water use characteristics are reported for some species. Also, the local site conditions will influence, to some extent, the actual water use rate.

Common names have been used to assist users in identifying particular plants. Due to many species having a range of common names, some common name may not be included. The associated scientific name provides a precise identification.

These comments apply to all plant lists in Appendix 4.

The following crop coefficient values and water use categories outlined by Costello and Jones (2000) have been used.

Very low (VL): K_c <0.10
Low (L): K_c 0.10–0.30
Medium (M): K_c 0.40–0.60
High (H): K_c 0.70–0.90

Appendix 4.1. Plant water use – trees

Common name	Botanical name	Water use rating
Ash (claret)	*Fraxinus oxycarpa* 'Raywood'	L–M
Ash (flowering)	*Fraxinus ornus*	M
Ash (golden)	*Fraxinus excelsior 'Aurea'*	M–H
Banksia (coast)	*Banksia integrifolia*	L
Banksia (giant candles)	*Banksia* 'Giant Candles'	L
Banksia (silver)	*Banksia marginata*	L
Birch (silver)	*Betula pendula*	H
Bay tree	*Laurus nobilis*	L
Blackwood (wattle)	*Acacia melanoxylon*	L
Bottle tree	*Brachychiton rupestris*	L
Bottlebrush (Harkness)	*Callistemon* 'Harkness'	L
Bottlebrush (weeping)	*Callistemon viminalis*	L–M
Bottlebrush (willow)	*Callistemon salignus*	L
Cabbage tree	*Cordyline australis*	M
Camphor laurel	*Cinnamomum camphora*	M
Cherry blossom (Japanese)	*Prunus serrulata*	M

Common name	Botanical name	Water use rating
Cherry tree (Tibetan)	*Prunus serrula*	M
Chinese tallow wood	*Sapium sebiferum*	L–M
Crab apple	*Malus ioensis plena* ('Gorgeous')	M
Crepe myrtle	*Lagerstroemia indica*	L
Cypress (Monterey)	*Cupressus macrocarpa*	M
Cypress (Italian)	*Cupressus sempervirens*	L
Dogwood	*Cornus florida*	M
Elm (American)	*Ulmus americana*	M
Elm (Chinese or lacebook)	*Ulmus parvifolia*	L
Elm (Dutch)	*Ulmus × hollandica*	M
Elm (English)	*Ulmus procera*	M
Elm (golden)	*Ulmus* 'Sapporo Autumn Gold'	M
Fig (weeping)	*Ficus benjamina*	M–H
Fig (Moreton Bay)	*Ficus macrophylla*	M
Flamingo tree	*Acer negundo* variegated	L–M
Flowering peach	*Prunus persica*	M
Frangipani	*Plumeria rubra*	L–M
Gum (coastal/Gippsland Manna Gum)	*Eucalyptus pryoriana*	L
Gum (lemon scented)	*Corymbia citridora*	L–M
Gum (manna gum)	*Eucalyptus viminalis*	M
Gum (mountain ash)	*Eucalyptus regnans*	M
Gum (narrow leaf peppermint)	*Eucalyptus radiata*	L
Gum (red-flowering)	*Corymbia ficifolia*	L
Gum (red spotted)	*Eucalyptus mannifera* ssp.*maculosa*	L–M
Gum (river red)	*Eucalyptus camaldulensis*	M–H
Gum (spotted)	*Corymbia maculata*	VL–L
Gum (snow)	*Eucalyptus pauciflora* 'Little Snowman'	L–M
Gum (swampy)	*Eucalyptus robusta*	L
Gum (Yellow box)	*Eucalyptus melliodora*	VL–L
Gum (Yellow, White ironbark)	*Eucalyptus leucoxylon*	VL–L
Gingko/maidenhair tree	*Gingko biloba*	M
Gleditsia (golden, honey locust)	*Gleditsia triacanthos*	L
Hakea (tree)	*Hakea erinatha*	L
Hakea (willow leaf)	*Hakea salicifolia*	M
Honeylocust	*Gleditsia triacanthos*	M
Jacaranda	*Jacaranda mimosifolia*	L–M
Japanese Maple	*Acer palmatum*	M–H
Lightwood	*Acacia implexa*	VL–L
Liquidambar	*Liquidambar styraciflua*	M–H
Lillypilly (weeping)	*Waterhousea floribunda*	M
Magnolia	*Magnolia grandiflora*	M–H
Melaleuca (snow in summer)	*Melaleuca linariifolia*	M

Common name	Botanical name	Water use rating
Manchurian pear tree	*Pyrus ussuriensis*	L
Maple (trident)	*Acer buergerianum*	M
Myrtle (crepe)	*Lagerstroemia indica*	L
Oak (cork)	*Quercus suber*	L
Oak (English)	*Quercus robur*	M
Oak (pin)	*Quercus palustris*	M–H
Palm (date)	*Phoenix dactylifera*	L
Palm (Washington)	*Washingtonia robusta*	L
Paperbark (broad-leafed)	*Melaleuca quinquenervia*	M
Pear (Manchurian)	*Pyrus ussuriensis* var. *ovoidea*	L
Pear tree (ornamental)	*Pyrus calleryana*	M
Peppercorn	*Schinus molle*	VL
Pine (Oyster Bay)	*Callitris rhomboidea*	L
Pine (pencil)	*Pinus* spp.	L
Plane (London)	*Platanus × acerfolia*	M–H
Poplar (golden)	*Prunus × canadensis*	M–H
Poplar (yellow, tulip tree)	*Liriodendron tulipifera*	M–H
Robinia (mop top)	*Robinia pseudoacacia*	L–M
Rubber plant	*Ficus elastica*	H
Sheoak (black)	*Allocasurina littoralis*	VL–L
Sheoak (grey)	*Casuarina glauca*	L–M
Sheoak (river)	*Casuarina cunninghamiana*	L
Sheoak (drooping)	*Allocasuarina verticillata*	VL
Spruce (bruce)	*Picea pungens*	M
Umbrella tree	*Schefflera actinophylla*	M
Wattle (cedar)	*Acacia elata*	M
Wattle (golden wreath)	*Acacia saligna*	L
Wattle (Queensland)	*Acacia podalyriifolia*	L
Wattle (Snowy River)	*Acacia boormanii*	L

Appendix 4.2. Plant water use – shrubs

Common name	Botanical name	Water use classification
Abelia (glossy)	*Abelia × grandiflora*	L–M
Agave	*Agave americana*	VL
Azalea	*Azalea indica*	M–H
Banksia (heath)	*Banksia ericifolia*	VL
Bay laurel, cherry laurel	*Prunus laurocerasus*	L
Bird of Paradise	*Strelitzia reginae*	VL
Blackthorn	*Bursaria spinosa*	VL
Bottlebrush (splendens)	*Callistemon citrinus 'Splendens'*	VL
Bottlebrush (Kings Park)	*Callistemon* 'Kings Park Special'	VL–L

Common name	Botanical name	Water use classification
Box (common, English)	*Buxus sempervirens*	L–M
Butterfly bush (summer lilac)	*Buddleia davidii*	M
Camellia	*Camellia japonica*	M
Camellia (sasanqua)	*Camellia sasanqua*	M
Correa (white)	*Correa alba*	VL
Crown of thorns	*Euphorbia milii*	VL
Cushion bush	*Leucophyta brownii*	VL
Daphne	*Daphne × burkwoodii*	H
Daisy bush	*Olearia axillaris*	VL–L
Dogwood	*Cassinia aculeata*	L
Dogwood (flowering)	*Cornus florida*	VH
Euphorbia	*Euphorbia* spp.	L
Frangipani	*Plumeria* spp.	M
Fuchsia	*Fuchsia* spp.	M
Fuchsia (native)	*Epacris longifolia*	L
Gardenia	*Gardenia augusta 'Radicans'*	M
Geraldton wax flower	*Chamelaucium uncinatum*	VL
Grevillea (bronze rambler)	*Grevillea* 'Bronze Rambler'	M
Grevillea (lavender)	*Grevillea lavandulacea*	L
Grevillea (royal)	*Grevillea victoriae*	M
Hakea (willow leaved)	*Hakea salicifolia*	L–M
Hawthorn (Indian)	*Rhaphiolepis indica*	VL–L
Heath (pink beard)	*Leucopogon ericoides*	L
Hebe	*Hebe* spp.	M
Hibiscus (rose of China)	*Hibiscus rosa-sinensis*	L–M
Hibiscus (rose of Sharon)	*Hibiscus syriacus*	M
Hydrangea	*Hydrangea × macrophylla*	H
Hydrangea (oakleaf)	*Hydrangea quercifolia*	M
Jasmine (star)	*Jasminum nitidum*	L
Jasmine (Chinese)	*Jasminum polyanthum*	VL
Jasmine (primrose)	*Jasminum mesnyi*	L
Kunzea (tick bush)	*Kunzea ambigua*	VL
Kunzea (crimson)	*Kunzea baxteri*	L
Lantana	*Lantana camara*	VL
Lavender (French and English)	*Lavandula* spp.	VL
Lily of the valley bush	*Pieris japonica*	M–H
Lomandra	*Lomandra longifolia*	L–M
Magnolia (Port wine)	*Michelia figo*	M
Myoporum (slender)	*Myoporum floribundum*	L
Oleander	*Nerium oleander*	VL
Photinia	*Photinia × fraseri 'Robusta'*	L
Pittosporum (golden sheen)	*Pittosporum tenuifolium*	M

Common name	Botanical name	Water use classification
Pittosporum (diamond leaf)	*Pittosporum rhombifolium*	L–M
Rhododendron	*Rhododendron* hybrid	M–H
Rose	*Rosa* spp.	M
Rosemary (native or coast)	*Westringia fruticosa*	VL
Saltbush	*Rhagodia spinescens*	VL
Sedum	*Sedum spectabile*	L
Tea tree (woolly)	*Leptospermum lanigerum*	M
Viburnum (burkwood)	*Viburnum × burkwoodii*	M
Wattle (Qld silver)	*Acacia podalyriifolia*	VL–L
Wattle (gold dust)	*Acacia acinacea*	VL
Wattle (Flinders Range)	*Acacia iteaphylla*	VL
Wattle (Sydney golden)	*Acacia longifolia*	VL
Wax Flower (pink)	*Eriostemon australasius*	L
Yucca (Spanish dagger)	*Yucca gloriosa*	VL

Appendix 4.3. Plant water use – groundcovers

Common name	Botanical name	Water use rating
Australian bluebell	*Wahlenbergia gracilis*	L
Banksia (groundcover)	*Banksia blechnifolia*	VL
Banksia (hairpin dwarf)	*Banksia spinulosa* (dwarf)	M
Clematis (evergreen)	*Clematis armandii*	M
Cockspur flower	*Plectranthus ecklonii*	M
Conifer (dwarf)	*Juniperus horizontalis 'Glauca'*	L–M
Convolvulus (morning glory)	*Convolvulus sabatius*	L
Correa (dusky bells)	*Correa* 'Dusky Bells'	L–M
Creeping boobialla	*Myoporum parvifolium*	L
Cushion bush	*Leucophyta brownii*	VL
Cast-iron plant	*Aspidistra elatior*	L
Daisy (cut leaf)	*Brachyscome multifida*	M
Daisy (seaside)	*Erigeron karvinskianus*	L
Daisy (Swan River)	*Brachyscome iberidfolia*	M
Dampiera	*Dampiera diversifolia*	L
Euphorbia (creeping spurge)	*Euphorbia myrsinites*	L
Gardenia	*Gardenia augusta* 'Radicans'	M
Grevillea (bedspread)	*Grevillea* 'Bedspread'	L
Grevillea (Poorinda royal mantle)	*Grevillea* 'Poorinda Royal Mantle'	L
Hellebore	*Helleborus × hybridus*	M–H
Herringbone cotoneaster	*Cotoneaster horizontalis*	VL
Ivy (Australian violet, leaf violet or native)	*Viola hederacea*	M
Ivy (English)	*Hedera helix*	L
Kidney weed	*Dichondra repens*	M

Common name	Botanical name	Water use rating
Lambs ears	*Stachys lantana*	L
Lavender (sea)	*Limomium perezii*	VL–L
Leptospermum 'horizontalis'	*Leptospermum 'Horizontalis'*	L
Lilly of the valley	*Convallaria majalis*	M–H
Liquorice plant	*Helichrysum petiolare*	L
Mat rush (pale)	*Lomandra glauca*	L
Mat rush (wattle)	*Lomandra filiformis*	L
Pigface	*Carpobrotus glaucescens*	L
Rosemary	*Rosmarinus officinalis*	L
Rounded noon flower	*Disphyma crassifolium*	VL
Running postman	*Kennedia prostrata*	L
Salt bush (Creeping)	*Rhagodia spinescens*	VL
Silver bush	*Convolvulus cneorum*	L
Star jasmine	*Trachelospermum jasminoides*	L–M
Sweet box	*Sarcococca ruscifolia*	M
Thyme	*Thymus* spp.	VL–L
Verbena (Peruvian)	*Verbena peruviana*	L
Violet (native)	*Viola hederacea*	M
Yellow buttons	*Chrysocephalum apiculatum*	L

Appendix 4.4. Plant water use – climbers

Common name	Botanical name	Water use rating
Banksia rose	*Rosa banksia 'lutea'*	L–M
Bougainvillea	*Bougainvillea glabra*	L
Bower climber	*Pandorea jasminoides*	M
Clematis (happy wanderer)	*Clematis microphylla*	VL
Clematis (traveller's joy, goats beard)	*Clematis aristata*	M
Clematis (mountain)	*Clematis montana*	M
Clematis (small flowered)	*Clematis pauciflora*	VL–L
Clematis (tropical)	*Clematis pickeringii*	L
Climbing guinea flower (or gold vine)	*Hibbertia scandens*	M
Climbing rose	*Rosa* cv.'Climbing'	M
Fig (creeping or climbing or creeping)	*Ficus pumila*	M
Happy wanderer	*Hardenbergia violacea*	L–M
Honeysuckle (Japanese)	*Lonicera japonica*	M
Ivy (Boston)	*Parthenocissus tricuspidata*	M
Ivy (German)	*Senecio macroglossus*	M
Jasmine (star)	*Trachelospermum jasminoides*	M
Kennedia (Cape Arid)	*Kennedia beckxiana*	L
Maidenhair creeper (mattress vine)	*Muehlenbeckia complexa*	L–M
Ornamental grape vine	*Vitis vinifera*	L–M

Common name	Botanical name	Water use rating
Passion fruit vine	*Passiflora edulisor, Passiflora* spp.	L–M
Pea (black coral)	*Kennedia nigricans*	L
Pea (dusky coral)	*Kennedia rubicunda*	L
Sarsaparilla (native)	*Hardenbergia violacea*	VL
Vine (Bengal clock, sky flower)	*Thunbergia grandiflora*	M
Vine (glory)	*Vitis*	L–M
Vine (orange trumpet)	*Pyrostegia venusta*	M
Vine (potato vine, potato creeper)	*Solanum jasminoides*	L–M
Virginia creeper	*Parthenocissus quinquefolia*	M
Wisteria (Japonica)	*Wisteria* spp.	M
Wisteria (Chinese)	*Wisteria sinensis*	L–M

Appendix 4.5.1. Plant water use – natives and ornamentals

Common name	Botanical name	Water use rating
Blood grass (Japanese)	*Imperata cylindrical 'Rubra'*	M
Blue grass (Queensland, silky)	*Dichanthium sericeum*	M
Blue or pale flax-lily	*Dianella longifolia*	L–M
Fescue (blue)	*Festuca glauca*	M
Flax (bronze baby)	*Phormium 'Bronze Baby'*	L–M
Flax-lily	*Dianella* spp.	M
Flax-lily (black-anther)	*Dianella revoluta*	L–M
Flax-lily (Paroo lily)	*Dianella caerulea*	L
Iris (wild)	*Dietes grandiflora*	L–M
Juncus	*Juncus australis*	M–H
Kangaroo grass	*Themeda triandra*	L
Kangaroo paw	*Anigozanthus manglesii*	L
Knobby club-rush	*Ficinia nodosa*	L
Lavender grass (Elvera)	*Eragrostis elongata 'Elvira'*	L
Mat rush (spiny head)	*Lomandra longifolia*	L
Mondo grass (black)	*Ophiopogon planiscapus 'Nigrescens'*	M–H
Mondo grass (mini green)	*Ophiopogon japonicus*	M
Plume (long hair)	*Dichelachne crinata*	L
Rat's tail grass	*Sporobulus creber*	L
Sedge (tall)	*Carex appressa*	L–M
Spear grass (bush tail)	*Austrostipa densiflora*	L
Spear grass (feather)	*Austrostipa elegantissima*	L
Spear grass (native princess)	*Austrostipa ramosissima*	L–M
Spear grass (prickly)	*Austrostipa stipoides*	L
Spear grass (slender)	*Austrostipa verticillata*	L
Tussock grass (blue)	*Poa poiformis*	L
Tussock grass (common)	*Poa labillardieri*	L–M
Tussock grass (velvet)	*Poa morrissii*	M

Common name	Botanical name	Water use rating
Wallaby grass	*Austrodanthonia racemosa*	L
Wallaby grass (hill)	*Austrodanthonia eriantha*	L
Wallaby grass (long leaf)	*Austrodanthonia longifolia*	L
Wallaby grass (East coast)	*Austrodanthonia tenuior*	L
Weeping grass	*Microlaena stipoides*	M
Windmill grass	*Chloris truncata*	L

Appendix 4.5.2. Plant water use – turfgrasses

Common name	Botanical name	Water use rating
Cool season grasses		
Bentgrass (creeping)	*Agrostis palustris*	M–H
Bentgrass (velvet)	*Agrostis canina*	M–H
Bluegrass (Kentucky)	*Poa pratensis*	H
Fescue (hard)	*Festuca longifolia*	M
Fescue (red)	*Festuca rubra* var. *rubra*	M
Fescue (chewings)	*Festuca rubra* var. *commutata*	M
Fescue (tall)	*Festuca arundinacea*	M–H
Ryegrass (perennial)	*Lolium perenne*	M
Winter grass	*Poa annua*	H–VH
Warm season grasses		
Bahia grass	*Paspalum notatum*	L
Buffalo grass	*Buchloe dactyloides*	L–M
Carpet grass (broadleaf)	*Axonopus compressus*	L
Carpet grass (Narrowleaf)	*Axonopus fissifolius*	L
Centipede grass	*Eremochloa ophiuroides*	L–M
Couch grass (blue) (Bermuda grass)	*Cynodon dactylon*	L
Kikuyu	*Pennisetum clandestinum*	M
Paspalum (seashore)	*Paspalum vaginatum*	L–M
St Augustine grass	*Stenotaphrum secundatum*	M
Zoysia grass	*Zoysia japonica*	L

Appendix 5

Salt tolerance

Appendix 5.1. Salt tolerance of selected ornamental plants, fruit trees and vegetables

Very sensitive (<1.8 dS/m)	Sensitive (3–4 dS/m)	Moderately tolerant (5–6 dS/m)	Tolerant (8 dS/m)	Very tolerant (13 dS/m)
Azalea Avocado Beans Begonia Blue spruce Camellia Dahlia Douglas fir Fuchsia Gardenia Lilium New Zealand flax Rhododendron Rose Star jasmine Violet	Aster Bird of paradise Blue daisy Cotton palm Deodar cedar Camphor tree Eastern red cedar Geranium Gladiolus Hydrangea Peach Pepper Pittosporum Plum Poinsettia Radish Viburnum	African lily Apple Boxwood Beach-oak Carnation Century plant Chrysanthemum Cucumber Cyclamen Hakea Lemon Orange Pear Philodendron Potato Sweet corn Walnut	Allepo pine Asparagus Beets Bougainvillea Broccoli Bottlebrush Chinese hibiscus Coolabah New Zealand Christmas tree Oleander Olive Red iron bark Rockmelon Swamp mahogany	Broad leaf cabbage tree Crimson bottlebrush Banksia Date palm Pampas grass Pigfaces Norfolk Island hibiscus Orange wattle River red gum Rosemary westringia Salt bush Salt river gum She-oak Small-leaf fig Tea tree Western tea-myrtle White correa Yucca (Spanish bayonet)

Note: Salinity values refer to measurement of salinity of soil extract and represent the upper ranges of tolerance.
Source: adapted from Handreck and Black (2005)

Appendix 5.2. Turfgrass species tolerance to various levels of soil salinity

Sensitive (<3 dS/m)	**Moderately sensitive (3 to 6 dS/m)**	**Moderately tolerant (6 to 9 dS/m)**	**Tolerant (>10 dS/m)**
Annual bluegrass Colonial bentgrass Kentucky bluegrass Rough bluegrass Centipede grass	Annual ryegrass Creeping bentgrass Bahia grass Chewings fescue Creeping red fescue Hard fescue	Perennial ryegrass Tall fescue Buffalo grass Zoysia grasses Creeping bentgrass cv. Seaside Fairway wheatgrass Slender creeping red fescue cv. Dawson	Alkali grasses Bermuda grasses Seashore paspalum St Augustine grass

Source: Harivandi MA (2000) Irrigating turfgrass and landscape plants with municipal recycled water. *Acta Horticulturae* **537**: 697–703.

Appendix 6

Turfgrass characteristics used in species selection

Turfgrass type	Water use	Drought tolerance	Wear tolerance	Shade tolerance	Salinity tolerance	Fertiliser requirements	Use
Couch	Low	Excellent	Good	Poor	Very good	High	Playing fields, parks, lawns
Kikuyu	Medium	Good	Excellent	Fair	Good	Low	Playing fields, Parks
Buffalo	Medium	Fair	Poor	Good	Good	Medium	Lawns
Seashore Paspalum	Low–medium	Good	Fair	Fair	Excellent	Medium	Foreshores, lawns
Ryegrass	Very high	Poor	Good	Fair	Fair	Medium–high	Over-seeding, playing fields, lawns
Bentgrass	Very high	Poor	Fair	Fair	Good	High	Greens
Tall Fescue	Very high	Poor	Fair	Fair	Good	Medium	Lawns

Source: Ruscoe P, Johnston, K and McKenzie G (2004) *TurfSustain: A Guide to Turf Management in Western Australia,* Sports Turf Technology, Como, WA.

Appendix 7

Water quality analysis and report (example)

Client: A-One Golf Club
Water source: Recycled water
Water use: Golf course irrigation
Water analysis by: Australian Golf Course Superintendents Association (AGCSA)

Chemical analysis	Ideal range	Recycled water sample
Water characteristics		
pH, units	5–8	7.3
Total alkalinity, as $CaCO_3$ (calc.)	<150	75
Bicarbonate, as HCO_3 (mg/L)	–	75
Carbonate, as CO_3 (mg/L)	–	<1
Calcium, as Ca (mg/L)	<100	25
Magnesium, as Mg (mg/L)	<100	22
Hardness, calculated as $CaCO_3$ (mg/L)	<150	–
Impact on plant growth (total salinity)		
Electrical conductivity (microS/cm @25°C)	<750	1226
Salinity by calculation (mg/L)	<450	817
Impact on foliage contact (ion toxicity)		
Sodium, as Na (mg/L)	<70	196
Chloride, as Cl (mg/L)	<100	272
Impact on root growth (ion toxicity)		
Sodium, as Na (meq/L)	<3	8.5
Chloride, as Cl (meq/L)	<3	7.8
Impact on soil structure (Na permeability hazard)		
Electrical conductivity (microS/cm @25C)	<750	1226
Salinity by calculation (mg/L)	<450	817
Residual sodium carbonate (calc.)	<1.25	–1.85
Sodium adsorption ratio – SAR (calc.)	see Note 1	6.9
adjSAR (calc.)	see Note 1	–
Nutrients		
Nitrate + nitrite, as N (mg/L)	see Note 2	–
Total Kjeldahl nitrogen, as N (mg/L)	see Note 2	–
Total nitrogen, as N (mg/L)	see Note 2	2.0

Chemical analysis	Ideal range	Recycled water sample
Phosphorus, reactive filtered as P (mg/L)	see Note 2	–
Phosphorus, total as P (mg/L)	see Note 2	<0.5
Potassium, as K (mg/L)	see Note 2	18
Other		
Iron, as Fe (mg/L)	<1	0.08
Manganese, as Mn (mg/L)	<0.2	0.13
Copper, as Cu (mg/L)	<0.2	–
Zinc, as Zn (mg/L)	<2.0	–
Boron, as B (mg/L)	<2.0	0.16
Sulphur, as SO_4 (mg/L)	<100	115

Source: Tests performed by ALS Water Resources Group for AGCSATech.

Notes:

1. Reference to additional information is required regarding SAR results.
2. Nutrients such as nitrogen and phosphorus need to be accounted for in the fertiliser program.

Water analysis report for A-One Golf Club (example)

Report prepared by: Australian Golf Course Superintendents Association (AGCSA)

The recycled water sample submitted for analysis has been tested and the results are attached. The following comments are made:

Recycled water sample

- The pH is slightly alkaline.
- The water has a medium salinity hazard.
- The water contains elevated levels of sodium and chloride.
- The water contains no residual sodium carbonates.
- The water has a low sodium adsorption ratio (SAR).
- The water contains 2 mg/L of total nitrogen, which is equivalent to 2 kg of total nitrogen per megalitre of water applied, which is only minor.
- The water contains very little phosphorus.

Comments

1. **Electrical conductivity (salinity):** The EC is 1226 µS/cm (1.226 dS/m), which is equivalent to 817 mg/L (ppm). As a result, the salinity is classed as medium and it would be expected that some accumulation of salts in the soil would occur by the end of the irrigation season; however, with the use of warm season grasses on sand constructions (greens and tees), the turf quality would not be affected. The use of gypsum and aeration and decompaction coring (e.g. verti-draining) would be required before and after the irrigation season on fairways to assist with leaching of any accumulated salts through the profile.
2. **Sodium:** The sodium concentration is elevated, although at levels where it is not a major concern. The use of gypsum as previously recommended will assist in displacing these ions from the soil solution.
3. **Chloride:** The chloride levels are elevated, but only at much higher levels would cause occasional leaf burn on warm season grasses. However, irrigating during the heat of the day should be avoided where possible.
4. **Nutrient levels:** The total nitrogen levels are generally insignificant and will not affect turf management.

5. **Summary:** While the water is not the quality of potable water, the supplied data indicates that the water quality would be suitable for use with all warm season grasses and on the majority of soil types. In heavy clays, the accumulation of salts and sodium may eventually cause some turfgrass decline if no supplementary soil management techniques are employed.

The quality of treatment water does vary considerably over time and regular testing is required.

AGCSATech.

Date of report.

Appendix 8

Pipe friction

Appendix 8.1. Friction loss for PVC pressure pipe

uPVC Pressure Pipe Classes 4.5, 6, 9 and 12. Flow Resistance Chart for Water at 20°C

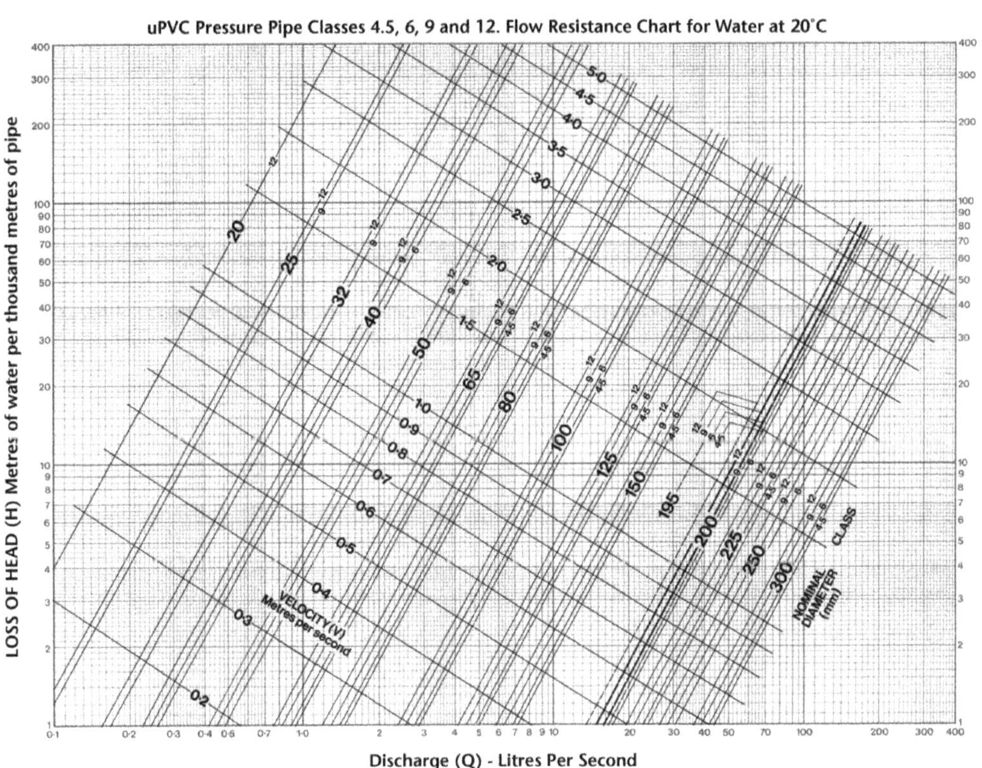

(Source: Toro Australia Pty Ltd)

Appendix 8.2. Friction loss for polythene pressure pipe

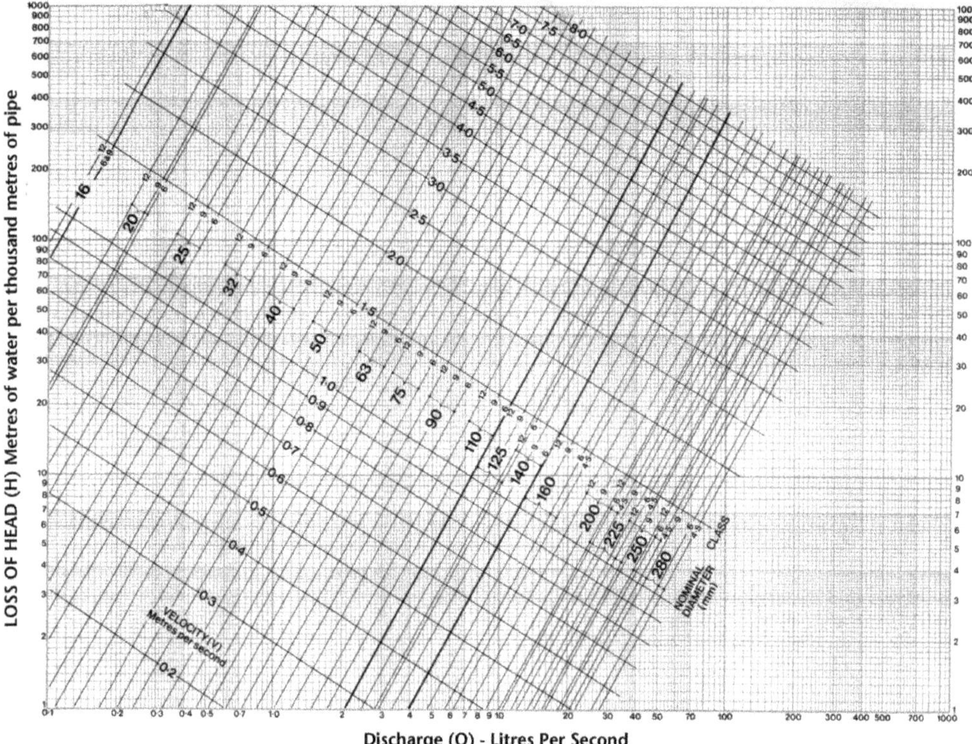

(Source: Toro Australia Pty Ltd)

Appendix 8.3. Friction loss for low density polythene pipe

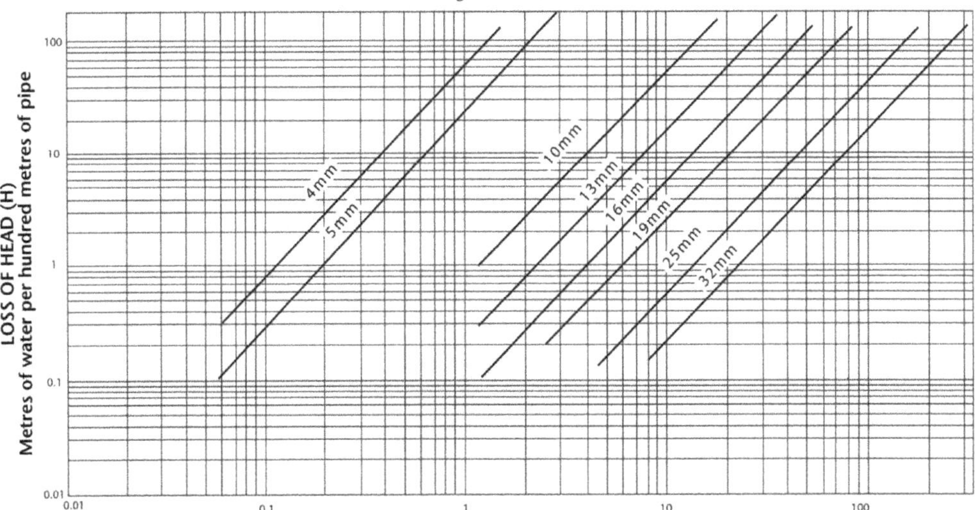

(Source: Toro Australia Pty Ltd)

Appendix 9

Pump performance curve – fixed speed, variable impeller diameter

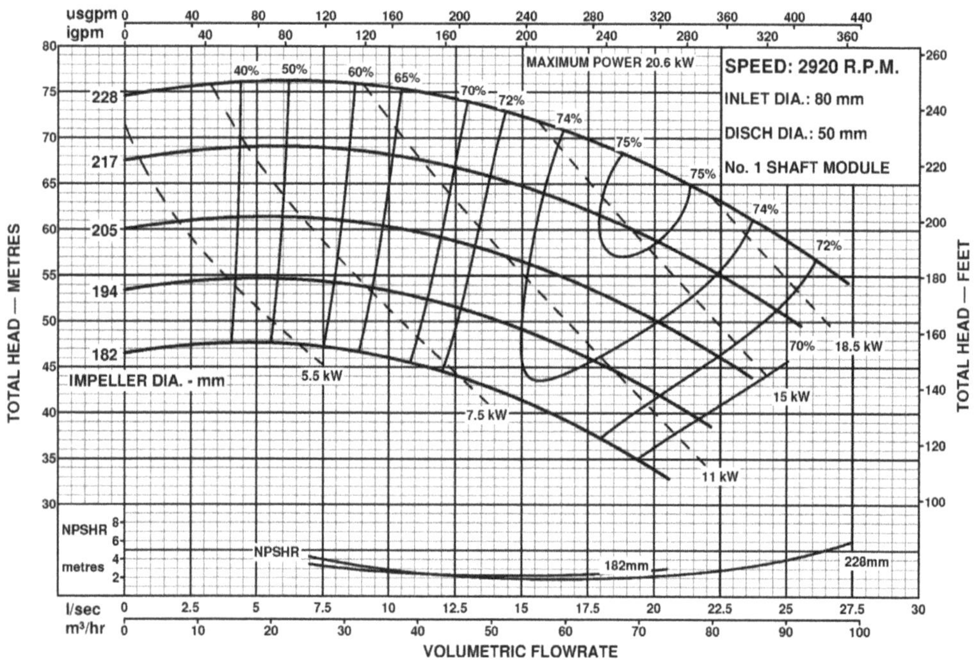

Performance to AS2417 Grade 2 for clean, cold water only. *May 2005*

(Source: Tyco Flow Control Pacific Pty Ltd)

Appendix 10

Australian standards relevant to irrigation and water supply

The following lists do not include all standards that may be relevant to each irrigation situation. There are additional standards covering product materials, fittings, valves, water treatment, plumbing, installation and testing that may be appropriate.

General plumbing

AS/NZS 3500.1:2003	*Plumbing and drainage – water services*
AS/NZS 2845.1:1998	*Water supply – mechanical backflow prevention devices*

PVC pipe

AS/NZS 1477:2006	*PVC pipes and fittings for pressure applications*
AS/NZS 2032:2006	*Installation of PVC pipe systems*
AS/NZS 2566.1:1998	*Buried flexible pipelines – Structural design*
AS/NZS 2566.2:2002	*Buried flexible pipelines – Installation*
AS/NZS 3879:1995	*Solvent cements and priming fluids for use with unplasticized PVC (uPVC) pipes and fittings*

Polythene pipes

AS/NZS 4130	*Polyethylene (PE) pipes for pressure applications*
AS/NZS 4131:2003	*Polyethylene (PE) pipe compounds for pressure pipes and fittings*
AS/NZS 2033:2008	*Code of practice for installation of polyethylene pipe systems*
AS 2698.2-2000	*Plastic pipes and fittings for irrigation and rural applications – Polyethylene rural pipe*
AS 2698.3-1990	*Plastic pipes and fittings for irrigation and rural applications – Mechanical joint fittings for use with micro-irrigation pipes*

Other pipe materials

AS 1572-1998	*Copper and copper alloys – Seamless tubes for engineering purposes*
AS 1432-2004	*Copper tubes for plumbing, gasfitting and drainage applications*
AS 1074-1989	*Steel tubulars for ordinary services*
AS 1565-1996	*Copper and copper alloys – Ignots and castings*
AS 1589-2001	*Copper and copper alloy waste fittings*

AS 3517-2007	*Capillary fittings of copper and copper alloy for non-pressure sanitary plumbing applications*
AS 1628-1999	*Water supply – Metallic gate, globe and non-return valves*

Identification and signage

AS 1319-1994:	*Safety signs for the occupational environment*
AS 1345-1995	*Identification of the contents of pipes, conduits and ducts*
AS/NZS 2648.1:1995	*Underground marking tape – Non-detectable tape*

Electrical

AS/NZS 3000; 2007/Amdt 1:2009	*Electrical Installation (Australian/New Zealand Wiring Rules)*
AS/NZS 3439.1:2002	*Low voltage switchgear and control gear assemblies*
AS/NZS 2053.1:2001	*Conduits and fittings for electrical installations – Rigid plain conduits*

Pumps

AS 2417 -2001:	*Rotodynamic pumps – Hydraulic performance tests – Grades 1 and 2*

Spray irrigation

NZS 5103:1973	*Code of practice for the design, installation and operation of sprinkler irrigation*

Soil testing

AS 1289.1.1-2001	*Method for testing soils for engineering purposes – Sampling and preparation of soils – Preparation of disturbed soils for testing*

Contracts

AS 4300-1995:	*General conditions of contract for design and construct*

International standards

There are numerous irrigation and water supply standards available from international organisations.

These include:

International Standards Organisation (ISO)
Website: http://www.iso.org/iso_catalogue.htm

Irrigation Association (USA)
Website: http://www.irrigation.org/Resources/Best_Practices_Standards.aspx

American Society of Agricultural and Biological Engineers (ASABE)
Website: http://www.asabe.org/standards.aspx

Further details on irrigation standards are available from Irrigation Australia Ltd (IAL).
Website: http://www.irrigation.org.au

Appendix 11

Metrics and conversions

Appendix 11.1. Metric units and equivalent values

Quantity	Equivalent
Length	
1.0 metre (m)	1000 mm
1.0 kilometre (km)	1000 m
1.0 cm (centimetre)	10 mm
1.0 μm (micron)	0.000 001 m (10^{-6} m)
1.0 nm (nanometre)	0.000 000 001 m (10^{-9} m)
Area	
1.0 ha (hectare)	10 000 m^2 (10^4 m^2)
1.0 cm^2	100 mm^2
Volume	
1.0 litre (L)	1000 mL (millilitre) (10^{-3} L)
1.0 m^3 (cubic metre)	1000 L
1.0 kL (kilolitre)	1.0 m^3 or 1000 L (10^3 L)
1.0 ML (megalitre)	1000 kL, 1 000 000 L or 10^6 L
1.0 GL (gigalitre)	1000 ML or 1 000 000 000 L (10^9 L)
Water application rate	
1.0 ML/ha	100 kL/1000 m^2
Flow rate	
1.0 m^3/h	1000 L/h
1.0 m^3/h	16.67 L/min
1.0 m^3/h	0.277 7 L/s
Pressure	
1.0 kPa (kilopascal)	1000 Pa
1.0 MPa (megapascal)	1000 kPa
1.0 m (head of water pressure)	9.80 kPa

Appendix 11.2. Conversions – metric and Imperial

Metric quantity	Imperial quantity	Conversion
Length		**Multiply by**
1.0 m	3.28 ft (feet)	m × 3.28 = … ft ft × 0.3048 = … m
25.4 mm	1.0 in (inch)	mm/25.4 = .. inches
1.0 km	0.621 miles	km × 0.621 = … miles miles × 1.61 = … km
Area		
1.0 ha (hectare)	2.471 acre	ha × 2.471 = … acre acre × 0.405 = … ha
Volume		
1.0 m^3 (cubic metre) (1000 L)	220.2 gallons (Imperial)	
1.0 m^3 (cubic metre) (1000 L)		
1.0 L	0.22 gallons (Imperial)	L × 0.22 = … gallons (Imp) Gallons (Imp) × 4.546 = … L
1.0 L	0.265 gallons (US)	L × 0.265 = … gallons (US) Gallons (US) × 3.773 = … L
Flow rate		
1.0 m^3/h (1000 L/h)	16.67 L/min	(L/h)/60 = … L/min (L/h)/(60 x 60) = … L/s
1.0 m^3/h (1000 L/h)	0.278 L/s	(L/min)/60 = … L/s
1.0 m^3/h (1000 L/h)	3.67 gallons/min (gpm) (Imp)	L/h × 0.0037 = … gpm (Imp)
Pressure		
1.0 kPa	0.145 psi (pounds per square inch)	kPa × 0.145 = … psi psi × 6.895 = … kPa
1.0 MPa	145 psi	
1.0 m (head of water)	9.8 kPa	m × 1.42 = …. psi psi × 2.31 = … kPa
1.0 m (head of water)	1.42 psi	
1.0 atmosphere (bar)	1.0 kg/cm^2	
1.0 kg/cm^2	1000 kPa (1.0 MPa)	
Energy		
1.0 kW	1.341 horsepower (hp)	kW × 1.341 = … hp hp × 0.746 = … kW
Salinity		
1 dS/m (deciSiemen/m)	1.0 mS/cm, 1000 µS/cm, 1.0 mmho/cm	
1.0 mg/L (TDS)	1.0 ppm (TDS)	

Appendix 11.3. Table of pressure conversions from kPa and psi

Conversions from kPa				Conversions from psi			
kPa	m (head)	bar	psi	psi	m (head)	bar	kPa
50	5.10	0.50	7.25	5	3.52	0.34	34.47
100	10.20	1.00	14.50	10	7.03	0.69	68.95
150	15.30	1.50	21.76	15	10.55	1.03	103.42
200	20.40	2.00	29.01	20	14.06	1.38	137.90
250	25.50	2.50	36.26	25	17.58	1.72	172.37
300	30.60	3.00	43.51	30	21.10	2.07	206.84
350	35.70	3.50	50.76	35	24.61	2.41	241.32
400	40.80	4.00	58.02	40	28.13	2.76	275.79
450	45.90	4.50	65.27	45	31.65	3.10	310.26
500	51.00	5.00	72.52	50	35.16	3.45	344.74
550	55.10	5.50	79.77	55	38.68	3.79	379.21
600	60.20	6.00	87.02	60	42.19	4.14	413.69
650	65.30	6.50	94.27	65	45.71	4.48	448.16
700	70.40	7.00	101.53	70	49.23	4.83	482.63
750	75.50	7.50	108.78	75	52.74	5.17	517.11
800	80.60	8.00	116.03	80	56.26	5.52	551.58
850	85.70	8.5	123.28	85	59.78	5.86	586.05
900	90.80	9.00	130.53	90	63.29	6.21	620.53
950	95.90	9.50	137.79	95	66.81	6.55	655.00
1000	102.00	10.00	145.04	100	70.32	6.89	689.48

References

Chapter 1 Sustainable water use and efficiency

Atkins D, Maheshwari B, Simmons B and Mulley P (2006) 'The sustainability challenge'. Technical Report No. 03-3/06. Cooperative Research Centre for Irrigation Futures, Darling Heights, Queensland.

Fairweather H, Austin N and Hope M (2002). *Water Use Efficiency, An information Package.* Irrigation Insights Number 5. National Program for Sustainable Irrigation, Land and Water Australia, Canberra.

Neylan JJ (Ed.) (2003) *Improving the Environmental Management of New South Wales Golf Courses.* Australian Golf Course Superintendents Association, Clayton, Victoria.

Chapter 2 The urban water scene

Aldous DE (2004) Research and Education Imperatives for Turf Management in Australia, 1st International Congress on Turfgrass, Acta Horticulturae **661**, 537–542.

AGIC (Australian Golf Industry Council) (2007) *Australian Golf Industry Council Water Survey, 2007.* AGIC, Sandhurst, Victoria.

Australian Bureau of Statistics (2006) *Water Account Australia 2004-05*, ABS Catalogue No. 4610.0, Australian Bureau of Statistics, Canberra.

Australian Bureau of Statistics (2010) *Water Use on Australian Farms 2009-10*, ABS Catalogue No. 4618.0, Australian Bureau of Statistics, Canberra.

Australian Bureau of Statistics (2011) Water Use on Australian Farms 2009–10, ABS Catalogue No. 4618.0, Australian Bureau of Statistics, Canberra.

Commonwealth of Australia (2009), *National Greenhouse Accounts (NGA) Factors*, Department of Climate Change, Canberra.

Connellan GJ (2007) Australia's water usage and efficiencies: Agriculture, golf, sports field, parks and recreation. In *Proceedings of the 23rd Australian Turfgrass Conference.* 24–26 July 2007, Cairns. Australian Golf Course Superintendents Association, Clayton, Victoria.

CSIRO (2006) *Climate Change Scenarios For Initial Assessment Of Risk In Accordance With Risk Management Guidance.* CSIRO Marine and Atmospheric Research, Aspendale, Victoria.

Dedman R (2010) *Stormwater for Health and Wellbeing. Proceedings of 2010 Stormwater Victoria Annual Seminar.* Stormwater Victoria, Altona, Victoria.

Ernst & Young (2006) *The PGA Report.* Victorian Golf Association, Melbourne.

Fam D, Mosely E, Lopez A, Mathieson L, Morison J and Connellan G. (2008) 'Irrigation of urban green space: a review of the environmental, social and economic benefits', Technical

Report No.04/08. Co-operative Research Centre for Irrigation Futures, Darling Heights, Queensland.

GHD (2007) 'Strategies for managing sports in a drier climate'. Report commissioned for Municipal Association of Victoria, Melbourne.

Hough (2005) 220 trees die in parklands since cut in water, *Adelaide Advertiser*, 18 June 2005.

Killicoat P, Puzio E and Stringer R (2002) The economic value of trees in urban areas: estimating the benefits of Adelaide's street trees. In *Proceedings of the 3rd National Street Tree Symposium*, 5 and 6 September 2002, pp. 94–106. TreeNet, University of Adelaide Waite Campus, Glen Osmond, South Australia.

Killy P, Brack C, McElhinny CG and King K (2008) *A Carbon Sequestration Audit Of Vegetation Biomass in the ACT*. ACT Government, Canberra.

McKenzie A (2009) 'Playing on – drought relief for community sport'. Presentation in Geelong, July 2009, Department of Planning and Community Development, Victorian Government, Melbourne.

McPherson EG (2005) Trees with benefits. *American Nurseryman* 7(201), 34–40.

Moir R (2004) *Strategic Development Plan – Turf Industry (Draft)*. Turf Producers Australia, Cleveland, Queensland.

Moore GM (2009) Urban trees: worth more than they cost. In *Proceeding of the 10th National Street Tree Symposium 2009.*, 3–5 September 2009, pp. 7–14. TreeNet, University of Adelaide Waite Campus, Glen Osmond, South Australia.

Neylan J (2003) *Improving the Environmental Health of New South Wales Golf Courses*. AGCSA and NSW Department of Environment, Sydney, <www.environment.nsw.gov.au>

Power (2007) 'Water security for SEQ sports fields'. Presentation by Neil Power, February 2007. Power Horticultural Services, Brisbane.

VEAC (Victorian Environmental Assessment Council) (2010) 'Metropolitan Melbourne investigation: discussion paper, October 2010'. VEAC, East Melbourne, Victoria.

VGA (Victorian Golf Association) (2005) 'Sustainable water for golf report, August 2005'. VGA, Burwood, Victoria.

Chapter 3 Water sources for irrigated turf and landscape supplies

ANZECC (1992) *Australian Water Quality Guidelines For Fresh and Marine Waters*. Australian and New Zealand Environment and Conservation Council, Canberra.

Benham B and Blake R (2002) *Filtration, Treatment, and Maintenance Considerations for Micro-Irrigation Systems*. Publication 442-757. Virginia Cooperative Extension, Biological Systems Engineering, Virginia State University, Petersburg, Virginia, USA.

Burt CM and Styles SW (1994) *Drip and Microirrigation for Trees, Vines and Row Crops*. Irrigation Training and Research Center, Cal Poly, San Louis Obispo, California.

Cross R and Spencer R (2008) *Sustainable Gardens*. CSIRO Publishing Gardening Guides, CSIRO Publishing, Melbourne.

FAWB (Facility for Advancing Water Biofiltration) (2009) *Adoption Guidelines for Stormwater Biofiltration Systems*. FAWB, Monash University, Melbourne.

Handreck and Black (2001) *Growing Media For Ornamental Plants and Turf*. Third edition. University of New South Wales Press, Kenningston, New South Wales.

Harivandi MA (2000) Irrigating turfgrass and landscape plants with municipal recycled water. *Acta Horticulturae* **537**, 697–703.

Institute of Engineers (1997) *Australian Rainfall and Runoff (ARR)*. Institute of Engineers, Canberra.

MUSIC (2010) *Model for Urban Stormwater Improvements Conceptualisation (MUSIC)*. eWater Cooperative Research Centre (CRC), Monash University, Clayton, Victoria.

National Water Commission (2008) *Rainwater Tank Design and Installation Handbook*. National Water Commission, Canberra.

Neylan JJ (2007) Turfgrass water use and salinity tolerance. *Australian Turfgrass Management* **9.2,** 28–30.

Neylan JJ and Peart AJ (2005) 'Survey of Australian golf courses using reclaimed water: supplementary report for Horticulture Australia Limited Project TU 1003'. Australian Golf Course Superintendents Association, Clayton North, Victoria.

NPSI (2009) *Using Recycled Water for Irrigation - An overview of the issues, opportunities and challenges*. National Program for Sustainable Irrigation, Land and Water, Canberra.

NRMMC, EPHC and AHMC (2006) *The Australian Guidelines for Water Recycling: Managing Health and Environmental Risks, Phase 1 (2006)*. Natural Resource Management and Ministerial Council, Environment Protection and Heritage Council and Australian Health Ministers Conference, Canberra.

NRMMC, EPHC and NHMRC (2009) *The Australian Guidelines for Water Recycling: Managing Health and Environmental Risks (Phase 2) Stormwater Harvesting and Reuse (2009)*. Natural Resource Management Ministerial Council (NRMMC), the Environment Protection and Heritage Council (EPHC) and National Health and Medical Research Council (NHRMC), Canberra.

RMIT University (2008) *Urban Greywater Design and Installation Handbook*. RMIT University and National Water Commission, Australian Government, Canberra.

Stevens DP, Smolenaars S and Kelly J (2008) *Irrigation of Amenity Horticulture with Recycled Water*. Arris Pty Ltd, Richmond, Victoria.

Yiasoumi W, Evans L and Rogers L (2005) *Farm Water Quality and Treatment, Agfact AC.2.* 9th edn, April 2005, NSW Department of Primary Industries, Orange, New South Wales.

Chapter 4 Irrigation methods

Aqualinc Research Ltd (2006) 'Irrigation efficiency gaps – review and stock take'. Report No L05264/2. Prepared for Sustainable Farming Fund and Irrigation New Zealand, Riccarton, New Zealand.

ASAE (1999) *Procedure for Sprinkler Testing and Performance Rating*. American Society of Agricultural Engineers (ASAE) standard ASAE S398, American Society of Agricultural & Biological Engineers (ASABE), St. Joseph, Michigan, USA.

Burt CM and Styles SW (1994) *Drip and Microirrigation for Trees, Vines and Row Crops*. Irrigation Training and Research Center, Cal Poly, San Louis Obispo, California.

Connellan GJ and Denman L (2009) 'Irrigation system performance and control devices'. Milestone 5 and 6 Report. Household Garden Watering Assessment Project, Smart Water Fund, Heatherton, Victoria.

Frost KR and Schwalen HC (1955) Sprinkler evaporation losses. *Agricultural Engineer* **36**(8), 526–528.

Harris G (2006) *Sub-surface Drip Irrigation (SDI) – System Maintenance*. Queensland Government Department of Primary Industries and Fisheries, Brisbane.

Irrigation Association (1997) Drip Irrigation in the Landscape, Irrigation Association, Falls Church, Virginia, USA.

Netafim (2007) *Techline AS^{TM} Design Guide*. Netafim Australia Pty Ltd, Laverton, Victoria.

Pair CH (Ed) (1969) *Sprinkler Irrigation*. 3rd Edition. Irrigation Association , Washington DC.

Smajstrla AG and Zazueta FS (1994) *Evaporation Loss During Sprinkler Irrigation.* BUL290, IFAS Extension, University of Florida.Gainsville, Florida, <http://edis.ifas.ufl.edu>

Chapter 5 Plant water use and irrigation budgets

Allen RG, Pereira LS, Raes D and Smith M (1998) 'Crop evapotranspiration. Guidelines for computing crop water requirements'. FAO Irrigation and Drainage Paper 56. FAO, Rome, Italy.

Bureau of Meterology (2005) *Automatic Weather Stations for Agricultural and Other Applications.* <http://www.bom.gov.au/inside/services_policy/pub_ag/aws/aws.shtml>

CIT (Center for Irrigation Technology) (2011) *Turf Crop Factors.* Wateright, CIT, California State University, Fresno, California, <www.wateright.org/KcTurf2.asp>.

Costello LR and Jones KS (2000) *A Guide to Estimating Irrigation Water Needs of Landscape Plantings in California – The Landscape Coefficient Method and WUCOLS III*, California Department of Water Resources, Sacramento, California.

Dodds PE, Meyer WS and Barton A (2005) 'A review of methods to estimate irrigated reference crop evapotranspiration across Australia'. CRC for Irrigation Futures Technical Report No.04/05 April 2005, CRC for Irrigation Futures/CSIRO Land and Water, Toowoomba, Qld.

Doorenbos J and Pruitt WO (1977) 'Guidelines for predicting crop water requirements'. FAO Publication Irrigation and Drainage Paper No. 24. FAO, Rome.

Handreck K (2008) *Good Gardens with Less Water.* CSIRO Publishing, Melbourne.

Handreck KA and Black ND (2005) *Growing Media for Ornamental Plants and Turf.* 3rd Edition. New South Wales University Press, Kensington, New South Wales.

Irrigation Association (2011) *Landscape Irrigation Auditor.* 2nd Edition. Irrigation Association, Falls Church, Virginia, USA.

Neylan JJ (2007) Turfgrass water use and salinity tolerance. *Australian Turfgrass Management* 9.2 28–30.

Pettinger D and Shaw D (2004) What we know about landscape water requirements. Cooperative Extension, University of California, *Co-Hort Newsletter* **6.1** 1–3.

SAWater (2008) *Code of Practice – Irrigated Public Open Space.* SAWater, Adelaide.

Short D and Colmer T (1999) Update on WA water use study: a comparison of 11 turfgrass genotypes during summer in Perth. *Australian Turfgrass Management* **1.5**, 20–22.

Chapter 6 Managing soil water and irrigation scheduling

Handreck, K (2008) *Good Gardens with Less Water.* CSIRO Publishing, Melbourne.

McIntyre K and Jakobsen B (1998) *Drainage for Sportsturf and Horticulture.* Horticultural Engineering Consultancy, Kambah, ACT.

Pair CH (Ed.) (1969) *Sprinkler Irrigation.* 3rd Edition. Irrigation Association, Washington DC.

Short D and Colmer T (1999) Update on WA water use study: a comparison of eleven turf grass genotypes during a summer in Perth. *Australian Turfgrass Management* **1.5**, 20–22.

Stevens DP, Smolenaars S and Kelly J (2008) *Irrigation of Amenity Horticulture with Recycled Water.* Arris Pty Ltd, Richmond, Victoria.

USDA (US Department of Agriculture) (1967). *National Soil Survey Handbook, title 430-VI. Part 618.87 Texture Triangle and Particle-Size Limits of AASHTO.* USDA and Unified Classification Systems, Washington DC, <http://soils.usda.gov/technical/handbook/>.

Chapter 7 Best management practices (water management and irrigation)

Irrigation Australia (2006) *Urban Irrigation – Best Management Practices*. Irrigation Australia Ltd, Hornsby, New South Wales.

Irrigation Association (2010) *Turf and Landscape Irrigation Best Management Practices (December 2010)*. Water Management Committee, Irrigation Association, Falls Church, Virginia, USA.

SAWater (2008) *Code of Practice – Irrigated Public Open Space*. SAWater, Adelaide.

Chapter 8 Designing irrigation systems

APMA (Australian Pump Manufacturers Association) (1987) *Australian Pump Technical Handbook*. 3rd Edition. APMA, Canberra.

ASAE (American Society of Agricultural Engineers) (2003) *Design and Installation of Microirrigation Systems*, ASAE EP405.1 FEB03, ASAE, St Joseph, Michigan, USA.

Commonwealth of Australia (2009) *National Greenhouse Accounts (NGA) Factors*. Department of Climate Change, Canberra.

Irrigation Australia (2006) *Urban Irrigation – Best Management Practices*. Irrigation Australia Ltd, Mascot, New South Wales.

Irrigation New Zealand (2007) *Irrigation Code of Practice and Irrigation Design Standards, March 2007*. Irrigation New Zealand Inc., Christchurch, New Zealand.

Irrigation Association (2011) *Irrigation. 6th Edition*. Irrigation Association, Falls Church, Virginia, USA.

Netafim (2007) *The Techline AS™ Design Guide*, Version 26/10/2007, Netafim Australia Pty Ltd, Laverton North, Victoria.

Pair CH (Ed) (1969) *Sprinkler Irrigation*. 3rd Edition. Irrigation Association , Washington DC.

Chapter 9 Achieving best practice – site studies

Bradbrook R (2005) Sub-surface irrigation in turf – a decade on. In *Proceedings of the Irrigation Australia 2005 Conference*. 17–19 May 2005, Townsville, Queensland. Irrigation Association Australia, Hornsby, New South Wales.

Carrow RN, Duncan R and Waltz C (2004) Overview of water conservation strategies on turf-grasses. In *Proceedings of Irrigation Australia National Conference*. May 2004, Brisbane. Irrigation Association of Australia, Hornsby, New South Wales.

Carrow RN, Duncan RR and Walz C (2007) *Best Management Practices (BMPs) Water Use Efficiency/Conservation Plan for Golf Courses*, University of Georgia, Crop and Soil Science Dept, Griffin Campus, Griffin, Georgia, USA.

Holborn S (2009) *Best Use Modelling for Sustainable Australian Sports Field Surfaces*. Project Number: TU06019, Horticulture Australia Ltd, Sydney, New South Wales.

Irrigation Association (2005) *Landscape Irrigation Scheduling and Water Management (Under Review)*. Irrigation Association, Falls Church, Virginia, USA.

Maheshwari, B (2006) 'The efficiency and audit in the Sydney Metropolitan area'. Technical report No.01/06, Cooperative Research Centre for Irrigation Futures, Darling Heights, Queensland.

Mitchell P and Goodwin I (1996) *Micro-Irrigation of Vines and Fruit Trees*. Daratech, Agmedia, Agriculture Victoria, Wellington Parade, Victoria.

Symes P and Connellan GJ (2004) Achieving high water use efficiency in public gardens – a case study RBG Melbourne. In *Proceedings Irrigation Australia National Conference*, 11–15 May, Adelaide, Irrigation Australia Ltd, Mascot, NSW.

Zoldoske DF (2003) *Improving Golf Course Irrigation Uniformity: A California Case Study.* California Agricultural Technology Institute, California State University, Fresno, California.

Chapter 10 Strategies and technologies to achieve high efficiency

Bransgrove K and Henderson C (2005) 'Best management practices for sustainable and safe playing surface of Australian Rules Football League sports fields'. Department of Primary Industries and Fisheries, Brisbane, <www2.dpi.qld.gov.au/horticulture/16834.html>

Charlesworth P (2000) *Soil Water Monitoring, Irrigation Insights Number One*. Land and Water Australia, Canberra.

Connellan GJ, Cahir B, Peasley B and Neylan J (1992) 'Progress report: turf water efficiency trial (Camberwell)'. Burnley College, Victorian College of Agriculture and Horticulture, Richmond, Victoria.

Costello LR and Jones KS (2000) *A Guide to Estimating Irrigation Water Needs of Landscape Plantings in California – The Landscape Coefficient Method and WUCOLS III*. California Department of Water Resources, Sacramento, California.

Craig I, Green A, Scobie M and Schmidt E (2005) *Controlling Evaporation Loss From Water Storages*. National Centre for Engineering in Agriculture Publication 1000580/1. University of Southern Queensland, Toowoomba, Queensland.

Gibeault VE and Cockerham ST (Eds) (1985) *Turfgrass Water Conservation*, Cooperative Extension. Publication 21405. University of California, Oakland.

Handreck K (2008) *Good Gardens with Less Water.* CSIRO Publishing, Melbourne.

Handreck and Black (2005) *Growing media for ornamental plants and turf.* 3rd Edition. University of New South Wales Press, Kenningston, New South Wales.

Holborn S (2009) *Best Use Modelling for Sustainable Australian Sports Field Surfaces.* Project Number: TU06019, Horticulture Australia Ltd, Sydney.

McIntyre K and Jakobsen B (1992) *Saving Water in Urban Irrigation*. Technical Services Unit, ACT Parks and Conservation Service, Canberra.

Pathan S, Barton L and Colmer T (2003) Soil moisture sensor-controlled irrigation for maintaining turf. *Irrigation Australia* **18**(4), 7–11.

Perry R (1992) *Landscape Plants for California Gardens*, Land Design Publishing, Claremont, California.

Solomon KH, Kissinger JA, Farrens G and Borneman J (2007) Performance and water conservation potential of Multi-Stream, Multi-Trajectory Rotating Sprinklers for landscape irrigation. *Applied Engineering in Agriculture* **23(2)**, 153–163.

Stirzaker RJ and Hutchinson PA (2005) Irrigation controlled by a wetting front detector: field evaluation under sprinkler irrigation. *Australian Journal of Soil Research*, **43**, 935–943.

Symes P (2000) Implementation of new irrigation management strategies at the Royal Botanic Gardens Melbourne. In *Proceedings of Irrigation Australia 2000 National Conference*, 23–25 May, Melbourne, pp. 684–696. Irrigation Association of Australia, Sydney.

Zoldoske DF (2003) *Improving Golf Course Irrigation Uniformity: A California Case Study.* California Agricultural Technology Institute, California State University, Fresno, California.

Chapter 11 Evaluating and benchmarking irrigation system performance

Barrett J, Vinchesi B, Dobson R, Roche P and Zoldoske D (2003) *Golf Course Irrigation: Environmental Design and Management Practices.* John Wiley & Sons, Inc., Hoboken, New Jersey.

Bransgrove K and Henderson, C (2006) Irrigation efficiency of Queensland grounds. *Australian Turfgrass Management* **8.3**,40–42.

Charlton G (2003) *Irrigation Efficiency Reporting and Monitoring.* Salisbury Asset Services, City of Salisbury, South Australia.

Christiansen JE (1942) *Irrigation by Sprinkling.* Bulletin 670, October 1942, University of California, College of Agriculture, Berkley.

Dukes MD, Haley MB and Hanks SA (2006) 'Sprinkler irrigation and soil moisture uniformity'. In *Proceeding of the 27th Annual International Irrigation Show.* 5–7 November 2006, San Antonio, Texas. Irrigation Association, Falls Church, Virginia, USA.

GreenCO and Wright Water Engineers (2008) *Green Industry Best Management Practices (BMPs) for the Conservation and Protection of Water Resources in Colorado: Moving Toward Sustainability. Third Release.* GreenCO and Wright Water Engineers Inc. Denver, Colorado, http://www.asla.org/uploadedFiles/PPN/Water%20Conservation/Documents/LISWM%20Draft.pdf

Irrigation Association (2011) *Landscape Irrigation Auditor.* 2nd Edition. Irrigation Association, Falls Church, Virginia, USA.

Irrigation Australia Ltd (2010*) Irrigation Efficiency Course: Resource Manual.* 2010 Edition. Irrigation Australia Ltd, Hornsby, New South Wales.

Irrigation New Zealand (2006) *Irrigation Evaluation Code of Practice.* February 20-06, Prepared for Sustainable Farming Fund, Irrigation New Zealand Inc., Christchurch, New Zealand.

Kah G and Willig C (1992) Irrigation efficiency – how to make it work for you. *Landscape and Irrigation* **Nov**, 44–45.

Keig S (1994) 'Water use efficiency – evaluation of selected sports grounds'. Horticultural Project Report. Victorian College of Agriculture and Horticulture, Burnley Campus, Victoria.

Mechan B (2001) Distribution uniformity results comparing catch-can tests and soil moisture sensor measurements in turfgrass irrigation. In *Proceedings of International Irrigation Show*, Tampa, Florida, pp. 133–139. Irrigation Association, Falls Church, Virginia, USA.

SA Water (2008) *Code of Practice – Irrigated Public Open Space.* SAWater, Victoria Square, Adelaide.

Skewes M and Meissner T (1997) 'Irrigation benchmarks and best management practices for winegrapes'. Technical Report No. 259. Primary Industries and Resources, Government of South Australia, Adelaide.

Symes P (2000) Implementation of new irrigation management strategies at the Royal Botanic Gardens Melbourne. In *Proceedings of Irrigation Australia 2000 National Conference*, 23–25 May, Melbourne, Irrigation Association of Australia, Sydney.

Zoldoske DF, Solomon KH and Norum EM (1994) *Uniformity Measurements For Turfgrass: What's Best?* Center for Irrigation Technology Irrigation Notes, California State University, Fresno, California.

Index